This book is dedicated to my early mentors:

David Dryke, Brian Coleman, and Romney White

Software Development on Adrenalin

Kenneth N. McKay

北京理工大学出版社

BEIJING INSTITUTE OF TECHNOLOGY PRESS

图书在版编目（CIP）数据

激情式软件开发 = Software Development on Adrenalin：英文／（加）麦凯
（McKay，K. N.）著. —北京：北京理工大学出版社，2011.3
ISBN 978 - 7 - 5640 - 4228 - 8

Ⅰ. ①激…　Ⅱ. ①麦…　Ⅲ. ①软件开发 - 英文　Ⅳ. ①TP311.52

中国版本图书馆 CIP 数据核字（2011）第 012869 号

北京市版权局著作权合同登记号　图字：01 - 2011 - 0382 号

出版发行／北京理工大学出版社
社　　　址／北京市海淀区中关村南大街 5 号
邮　　　编／100081
电　　　话／(010)68914775(办公室)　68944990(批销中心)　68911084(读者服务部)
网　　　址／http://www.bitpress.com.cn
经　　　销／全国各地新华书店
印　　　刷／保定市中画美凯印刷有限公司
开　　　本／787 毫米×1092 毫米　1/16
印　　　张／22.25
字　　　数／439 千字
版　　　次／2011 年 3 月第 1 版　　2011 年 3 月第 1 次印刷
印　　　数／1～1500 册
定　　　价／68.00 元

责任校对／周瑞红
责任印制／边心超

Acknowledgements

There are a number of people who have inspired or provided feedback on this book. I will likely have forgotten some and wish to apologize in advance.

Some of the writing in the text has its origin in other pieces I have written or co-written with co-authors such as Gary Black and Vincent Wiers. Some of the writing is extracted or inspired from research work I have done with a number of undergraduate and graduate students. I worked with Jennifer Jewer on risk management ideas for software project management. And, then there was Louise Liu, David Tse, Sylvia Ng, and Hao Xin. They worked with me as I initially formalized my value framework for a course I taught and by working with them, the framework benefited.

While I have re-interpreted and re-crafted previous writings, it is possible that some bits will bear some resemblance to scattered text I have crafted or co-crafted before. Some of the Agile Overview and Ethnographic Methods sections come to mind.

The writing of this text, at this time, can be attributed to Dr. Alan George who gave me an opportunity to do a reasonably large project that fits the methodology described in this book. This project in turn led to an opportunity to try to teach students and others associated with the project, the principles and concepts behind the project itself; hence the book. Alan also provided input on some of the content, such as the ideal characteristics of software.

Several individuals are named within the text — people who radically altered my thinking and affected my IT skill sets: Brian Coleman, Romney White, and David Pryke.

Others who have contributed to the quality and content of the text, directly or indirectly include: Will Gough (who has had an ongoing software engineering dialog with me lasting well over a decade), Patrick Matlock, Jesse Rodgers, Doug Suerich, and Trevor Grove. The first set of students exposed to the text have also contributed in a number of ways: Nick Guenther, Ivan Surya, Ivan Salgado Patarroyo, Chris Carignan, Shanti Mailvaganam, Yatin Manerkar, Scott MacLellan, Rajesh Swaminathan, and Jarek Piorkowski.

Thanks also go to the Department of Cultural and Educational Experts, the State Administration of Foreign Experts Affairs, and the International Office of Beijing Institute of Technology for giving me the financial support to publish this book.

Table of Contents

Part III   Architecture & Design

Part IV   Level VI Rapids & Mushing

Part Ⅰ

S/W Development: a Personal View

Overview

This text is about software development that is potentially risky, causing the release of adrenalin and the rush that is associated. The heart beat that drives the development is speeded up, breathing rate increases, and the blood going to the muscles has more oxygen. This means that more can be done faster and at higher levels of achievement. It is symbolic adrenalin of course. Instead of chemicals, it is processes and ideas that speed things up and improve the muscles being used to develop the software. It is also about the feeling that comes from delivering software that the user values and wants to use. Software that the user will fight to keep using! It is a great feeling when the user actually values your code. When this has happened to me, it feels like an adrenalin rush. When both adrenalin rushes occur, it is a really good feeling and I have been fortunate throughout my career to experience both rushes repeatedly; the rush while creating and the later rush that comes from usage.

I have named this process of concurrently developing high value software at breakneck speed *ZenTai Mushing*:

ZENTAI MUSHING

ZenTai Mushing refers to a holistic, unified way of viewing software functionality and usability combined with a high velocity version of Agile/Extreme. You can use the *ZenTai* design method with and without *Mushing*, and you can mush without *ZenTai*. The one is a *what* and the other is a *how*. Sometimes they are both appropriate and this book describes this situation: what they are, and how to use them together to get the dual rushes of adrenalin. When all of the variables align, the software form, function, and journey are unique and special. This is not a magical incantation though and there are many risks and possible rough spots, as not all people or processes can fit or operate in this fashion. Not all projects are suitable candidates either. It will push people's comfort zones and challenge assumptions. Good for some. Not so good for others.

Part I introduces you to the basics of the above and the philosophy behind the text.

Chapter 1

Introduction

This particular book is the result of approximately forty years of programming and development that has involved a wide variety of systems. It will not tell you everything you need to know about software development. Other software engineering books I recommend are Hunt and Thomas (2000), Glass (1997, 2003), Brooks (1995), McConnell (1996), McCarthy (1995), Jackson (1975), and Orlicky (1969). I suggest that you check out each of these and reflect upon what the authors are trying to get across. They are full of good suggestions and commonsense ideas. My own objective is to complement these other books and provide additional insights.

Who am I to write such a book in the first place? What are my credentials? I do not write witty, sarcastic blogs, issue on-line pronouncements or write about best practice, nor do I have a vast community of followers who hang onto every word I utter. I do not do self-promotion on the software engineering topic, and I try to avoid extrapolating off limited experiences. All I have done is design and code systems for close to four decades. Over the years I have quietly designed and programmed dozens of systems and software solutions ranging from operating systems and relational databases to accounting and veterinarian systems, and probably 150'ish end user applications based on custom toolkits I have created; sometimes as a team member, sometimes as a single developer. I have programmed an average of approximately 20,000 lines of code per year for more than thirty of those years, and more than once as much as 60–70,000 lines of code over a six month period. I am a geek, a code freak and have written lots of code. In over three decades I have failed once to deliver a project on time, on budget. No one is perfect. I like to build systems people fight to keep using and who find value from my designs and code. The proof is in the delivered systems and not in speculation and wild claims. I have had extended relationships with most of the systems I have been involved with, and I have been able to see how users have used the systems and to also see how the designs and code fared over time. I prefer to have demonstrations of skill, not could have, should have, would have, or will do.

I am not good at too many things. You never want to hear my attempts at music of any form. My kindergarten teacher thought my skills were so poor that she noted "difficulty in tone matching and in rhythms." How bad must you be to have this explicitly noted on your report card? In kindergarten? Nor would you like to see me dance, play sports, or attempt many other feats. And, my backyard shed went together in a scene from a comedy skit. However, I do seem to be good at software and software related activities. In my late-forties, a company executive with a firm I was interacting with nicknamed me *Code-Boy*.

During these many years I have evolved a specific style and approach. There are better programmers and there are better designers. There is always someone better. It is also good to work with better, smarter people, and I have been lucky to have worked with a number who have provided many lessons. I do not know if different is better, but it seems that I think and do things differently. I have been told this many times. Perhaps. Since I am not someone else, it is hard for me to judge others' thought processes. *ZenTai Mushing* is my attempt at describing the method behind my madness. *ZenTai Mushing* appears to be a way to consistently understand what is needed, and then craft the code that provides a unique, high value user experience that is obtained in very short order. The resulting software has demonstrated high quality, has been used for long periods of time with evolutionary changes, and has surpassed almost all costs and time expectations.

I am now reaching the end of my programming career and I finally feel ready to put words and thoughts on paper. I had actually planned to retire from coding when I turned fifty, but I have continued for various reasons. In the last two years I have done quite a bit of sustained coding, over 150k lines of code and it has allowed me to reflect once again on what I do and how I do it. The first draft of this book was completed just as I started coding again. Instead of a book written by someone with a few years of experience thinking that they are an expert and are capable of providing guidance on all matters concerning software, this is written by an old guy who has written a lot of code and who has specialized in making mistakes and learning from them.

Back in 1977 I actually had hair... really... thirty years later, I have ears... Beware, young geeks turn into old geeks...

Although I have touched a bit on some of the ideas in my academic papers, I have not directly approached software development as an author. I have always doubted my skill and ability, but in reflecting back over my career, it is hard for me to say that my repeated successes were accidental. They were not Herculean efforts with each being done via all-nighters and my face buried in the keyboard. They were done time and time again using a specific style and rhythm. I had a life most of the time. I hate egos and I hate people who go about talking about what they have done blah-blah. Especially those people who take one or two projects and extrapolate wildly to all kinds of software and projects. However, I also hate people who do not share with others any potential nuggets of wisdom that will help people following along behind. So, damned if I do, damned if I don't. I have felt uncomfortable writing most of the sections of this book.

Perhaps a few of my ideas have value and can be leveraged by others who can deliver better software products to the users with even more efficiency and effectiveness. I also do not expect anyone to pick up the whole lot and be able to do what I do. I am me and you are you. And, if you are older and have been immersed in one way for many years, it may be hard to adopt the ideas in this book. The ideas here are strictly another set of ideas to consider and you should have many in your arsenal. There is no single thing to do for achieving success, and you need many tools. I also do not know all of the causal relationships between the ideas. Sorry. I have also learned that most of the ideas in this book are not likely to be appreciated or understood by junior or inexperienced developers, or by software developers who are more technicians or assemblers than they are developers. Nothing I can do about that either.

I hope you are not the type that will read this book and say "that just can't work" or "I cannot do that." This type of attitude is self-defeating and sad. Over the years I have worked with positive, open minded individuals, and others who are closed minded and who insist in a narrow view; open to any way as long as it is their way. For effective software development in a number of areas, you need to be open and willing to adapt and change. I often give people a chance to show me their way first and see how it goes. Give them the benefit of doubt. If the results are close enough, we will both be happy and I will have learned something new along the way. If there is a large gap in results that cannot be dealt with, I will not be happy and if I am accountable for the project I will have to do an intervention and re-anchor the project and process: my way. You start off working with people, then go around them, and finally you might have to remove them from the project.

There is a risk with reading any book like this. You cannot be a perfectionist or be 100% literal in interpreting the methods and ideas you read. I have never done two projects exactly the same way. For high velocity development you need to be open and willing to experiment, lead with your chin, and develop fast responses. You cannot be literal, pedantic, rigid, or a perfectionist if you are going to apply the ideas you will read here. Over the years I have had supervisors, peers, and subordinates say that these ideas do not work, cannot work, and could not have worked. They do, can, and have. But, they have to be interpreted in the context of the project you are doing and the team you have.

If there is one key to my whole approach, it is one underlying assumption. I try to always remember that:

I never know the right or one-and-only way to do something.

I always doubt. I always question. What I am more confident about are wrong ways. I know **many** wrong ways, some of which are less wrong than others. I am an expert on *wrong*. I have learned the wrong ways by making many mistakes in my career. I seem to learn best by making mistakes, by willing to admit that I can and do make mistakes, and then by trying my hardest to learn from the mistakes and not repeat them. Luckily, most of my mistakes do not get seen or experienced by the users.

In a software project this means that I am willing to make mistakes with code and then re-write the code as necessary, when necessary; you need to know when something is at the end of its life and when it should be buried. As an undergraduate student in 1974, I wrote the worst code I have ever seen, fragile and really ugly. It was terrible.

That piece of bad code provided me one of the best lessons in my software career. The functionality was great and the users were very happy, but the code was horrendous in terms of robustness, and maintainability. It was like a plate of cooked pasta. I learned a lot from that first, major programming experience. It was my first assembler program, about 5k lines of code, and I have tried to avoid the same mistakes ever since. I had to maintain that piece of code for over two and a half years, and every day I checked with the operations group: "Did it run last night?" If not, I would skip algebra, calculus or statistics to fix the software. I was not asked by my supervisor to take such accountability. I just thought it was the professional thing to do. I built the fragile system and I was responsible. In hindsight, I should have built better software and skipped fewer classes. I buried the code in 1976 by replacing the code with a much better software program, better user functionality and better robustness. Code was designed to be robust and reliable. I learned many things because of that initial program and I used the lessons in subsequent software. How good were the lessons? I have been told that the replacement code was still being used in 2001. Several of the other programs I worked on or created in the mid 1970s also had long deployments. Some were running a decade or two later. I used the lessons again when working with a team in the early to mid 1980s. I have been told that the basic ideas and architecture developed in 1981 are still being used. I have used the same basic concepts throughout my career. I learned how to make good code the old fashioned way, by writing lots of code, making mistakes and learning from them.

If you only take three things away from this book, here are my three most important points to share. They are my humility principles:

- **Assume that you really do not know the requirements and what you think you know about the problem is partial, and possibly wrong.**
- **Assume that your design is faulty and that pieces will have to be ditched in a hurry and replaced.**
- **Assume that your code is buggy and that you are NOT a code ninja.**

Notice how these assumptions are aligned with my key underlying assumption of not knowing what is right! If you apply these three humility assumptions, I believe that you will then do requirements analysis in a certain way, that you will then design and build architectures in a certain way, and that you will then code in a certain way. The end result will be resilient, flexible, and sympathetic to the user's changing requirements, and be very robust. If you assume and act like you are an expert, your projects will likely stink and will possibly have a short shelf life. If you assume and act like you are **NOT** a hotshot, the projects are probably going to be far better than you imagined they could be.

And, be proud of your mistakes if you have learned from them and have controlled the damage the mistakes caused. Here is a phrase from a fortune cookie:

How can you have a beautiful ending without making beautiful mistakes?

I think that this is very true for software. You can indeed have beautiful mistakes and that they can contribute positively to a system. But not all mistakes are created equal. There are good mistakes that help get you to the beautiful endings, and there are mistakes with zero value. I often describe the task of management as constantly solving problems, some big, some small. This is what an analyst also does; constantly solving problems and just like a manager, he/she must develop good problem solving skills. A good manager will solve a problem once. If the manager keeps solving the same problem, being a manager might not be the best career for that in*duh*vidual. A repeated mistake is not a good mistake. A good architect and designer should also solve a problem once, or at least remember how to solve the problem when encountering it again, perhaps in a different context, going by a different name.

This book is not really about the methods and ideas for software engineering. There is probably not a new or unique idea to be found in this book. Good programming that is full of commonsense has been done since the beginning of automation and most ideas are built on other existing ideas. Some of the suggestions I will make about how to look at mission critical problems, or identify what characteristics to manage via interfaces are inspired by Babbage's 1832 masterpiece on manufacturing. Nothing is really new in terms of the individual ideas. What I am describing is how all of the ideas in the book can be used together.

At the end of the day, it is very much about what the final software provides the users! It does not matter to me how good the software is with respect to technical savvy and exotic features, or whether the programmer would have fun building the software if the user cannot or will not use it. The user's value comes first. That is the most important part of software development. This is a book about creating software that people want to use. I think it is about creating good software. But, what is good software?

Here is a brief summary of what could be called good or ideal software characteristics:

1. Software should be reliable and available when a user wants to use it.
2. Software should always focus on the user's goals and objectives.
3. Software should respect the user's time and effort, avoiding unnecessary data entry, unnecessary navigation, and unnecessary re-entry of data.
4. Software should recognize the user's knowledge and experience, and adjust the level of guidance and help accordingly.
5. Software should match the semantics of the task and problem, using the language of the user community.
6. Software should be generally self-supportive, without the need for "outside the system"

spreadsheets, documents, and databases.

7. Software should be intuitive and require minimum documentation, training, and instruction.

8. Software should naturally fit the user and not force the user to unnaturally fit the software.

1.1 ZenTai

ZENTAI (全体) basically means the essence of the unified whole and when I consider software, I include the users and interacting systems as part of the whole. I also consider the whole life cycle. The *ZenTai* way of design has four pillars and I will introduce them with Japanese words and concepts since Japanese (and Chinese) do a better job than English at describing the intent and spirit.

The first pillar is **Kachi** — 価値 — meaning value. To me, a good piece of software provides real value to the user and each step or activity must contribute to the generation of that value. No non-value added GUI, no waste, no clutter. Features and functions exist when and where you need them. Every screen and every feature on a screen needs to be considered in terms of the value chain and what it contributes.

The second pillar is **Anshin** — 安心 — comfort. This covers fear, stress, anxiety, and the general way a user feels when using your software, either because of the software, or because of the situation in which the software is used. The software should be designed understanding when and where the software will be used, and what can cause the user feelings of discomfort as these feelings can result in errors, frustration, and other user issues. A comfort analysis should be done for each bit of the interface and functionality. How does the system improve the comfort level and how does the system decrease the comfort level?

The third pillar is **Keiken** — 経験 — experience. Systems must recognize that people bring different initial experience (as in "I have experience doing that") to the system in the first instance, that they will accumulate additional experience through the use of the system, and that they will simultaneously accumulate other experience external to the system. Repeated experience with suitable feedback leads to expertise and different skill levels — which can affect how they interact with the system. What experience does the system recognize? What experience does the system develop? The notion of experience means something to me when designing the user interface. What does it mean to you?

The fourth and final pillar is **Shinka** — 進化 — evolution. Nothing remains as it is. Constants are not and variables will not. Assumptions are momentary things that should not be clung to. The software entity in its design and interface must support and accept evolution in the intended purpose, form, and function, as well as evolution in the user and the environment around the software. It is not likely that the users and the world will be standing still. Why do you think your software will not change? What are the assumptions behind each major function or software bit? What if those assumptions change?

All of these pillars will be explained in later chapters.

1.2 Mushing

The term *Mushing* is used for several reasons. I use it because it best describes what the process looks like from a few meters away. It is like ceramic art being formed. The software process I call *Mushing* looks ill-formed, chaotic and sometimes looks full of hand waving, and it seems like nothing is firm. This is not a bad interpretation. When *mushing* you are working with partial information solving partial problems arriving at partial solutions. In another sense, the high end projects I am describing in this book feel like dog sled mushing when you are in a situation going through the wilderness, careening around curves and over bumps without knowing exactly what is around the curve or over the hill. You are relying on the team and the lead dog. And, when in the wilderness, you have to be able to deal with whiteouts, wind, cold, isolation, and rely on your own knowledge and skill. You have to be able to fix the sled and make new paths when needed. Not only do you need to know how to careen, you need to know how to solve problems from first principles. For the type of software addressed in this book, you cannot rely on the web, surfing for the answer and then assembling the solution via copy and paste. You need to know how to program, really program and not rely on assembly skills. So, I like *Mushing* for a few reasons. I could have also used white water experiences for the hurtling and skills needed to survive. In fact, since there are nice rating schemes for rapids, I will use that analogy throughout the book.

The *Mushing* I describe in this text could be considered an agile and extreme version of Agile/Extreme; when you cannot find solutions online, find best practices, or just assemble solutions. These types of developments are not as common as they once were, but if you are pushing the limits, you might find yourself with one of these. I do not know of any software rating scheme that can be used for categorizing software projects with respect to agility requirements. The types of projects I typically do could be described using the international white water classification scheme (American Whitewater — www.awa.org):

- **Class VI: Extreme and Exploratory.** These runs have almost never been attempted and often exemplify the extremes of difficulty, unpredictability and danger. The consequences of errors are very severe and rescue may be impossible. For teams of experts only, at favorable water levels, after close personal inspection and taking all precautions. After a Class VI rapids has been run many times, its rating may be changed to an appropriate Class 5.*x* rating.

Although there are exceptions, most of the cases and stories I have heard about Agile/Extreme being used in read more like Class I or II:

- **Class I : Easy.** Fast moving water with riffles and

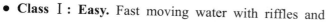

small waves. Few obstructions, all obvious and easily missed with little training. Risk to swimmers is slight; self-rescue is easy.

- **Class II: Novice.** Straightforward rapids with wide, clear channels which are evident without scouting. Occasional maneuvering may be required, but rocks and medium-sized waves are easily missed by trained paddlers. Swimmers are seldom injured and group assistance, while helpful, is seldom needed. Rapids that are at the upper end of this difficulty range are designated "Class II+."

Agile/Extreme concepts can be used at all levels, but it is my view that the processes and concepts need to be adapted to the type of rapids being run. This will be talked more about in the project management chapter. Since I will occasionally refer to the rapid analogy, here is the complete list:

- **Class III: Intermediate.** Rapids with moderate, irregular waves which may be difficult to avoid and which can swamp an open canoe. Complex maneuvers in fast current and good boat control in tight passages or around ledges are often required; large waves or strainers may be present but are easily avoided. Strong eddies and powerful current effects can be found, particularly on large-volume rivers. Scouting is advisable for inexperienced parties. Injuries while swimming are rare; self-rescue is usually easy but group assistance may be required to avoid long swims. Rapids that are at the lower or upper end of this difficulty range are designated "Class III−" or "Class III+" respectively.

- **Class IV: Advanced.** Intense, powerful but predictable rapids requiring precise boat handling in turbulent water. Depending on the character of the river, it may feature large, unavoidable waves and holes or constricted passages demanding fast maneuvers under pressure. A fast, reliable eddy turn may be needed to initiate maneuvers, scout rapids, or rest. Rapids may require "must" moves above dangerous hazards. Scouting may be necessary the first time down. Risk of injury to swimmers is moderate to high, and water conditions may make self-rescue difficult. Group assistance for rescue is often essential but requires practiced skills. A strong Eskimo roll is highly recommended. Rapids that are at the lower or upper end of this difficulty range are designated "Class IV−" or "Class IV+" respectively.

- **Class V: Expert.** Extremely long, obstructed, or very violent rapids which expose a paddler to added risk. Drops may contain large, unavoidable waves and holes or steep, congested chutes with complex, demanding routes. Rapids may continue for long distances between pools, demanding a high level of fitness. What eddies exist may be small, turbulent, or difficult to reach. At the high end of the scale, several of these factors may be combined. Scouting is recommended but may be difficult. Swims are dangerous, and rescue is often difficult even for experts. A very reliable Eskimo roll, proper equipment, extensive experience, and practiced rescue skills are essential. Because of the large range of difficulty that exists beyond Class IV, Class 5 is an open-ended, multiple-level scale designated by class 5.0, 5.1, 5.2, etc... each of these levels is an order of magnitude more difficult than the last. Example: increasing difficulty from Class 5.0 to Class 5.1 is a similar order of magnitude as increasing from Class Ⅳ to Class 5.0.

The basic forms of Agile/Extreme have certain benefits in some situations when compared to the traditional Waterfall methods and are agile, flexible, and adaptive in ways that the formal Waterfall methods are not. However, it is possible to make Agile/Extreme too formal, too reliant on artifacts, buzzwords, and prescribed how-to-do-it methodologies. When a method becomes too standard and too formal, it has the possibility of losing any agility and flexibility it once had. There are no standards or 'only way' or normative prescriptions with *ZenTai Mushing*. None. There are consistent principles and concepts, but the realization and instantiation of the process is likely to be different each and every time! If you go into a Class Ⅴ or Ⅵ rapids with a firm plan that you are fanatically attached to, you will discover the hard way what extreme and exploratory really mean.

I have worked with flexible developers and I have worked with developers who were very pedantic (i.e., rigid, black and white thinkers who are excessively rule or text book driven). People who take things literally and are pedantic will have lots of problems with this book and the way I develop software. It is like petroleum and fire. Key lesson I have learned: do not mix the two. I have had to occasionally, on some jobs you just have to accept who and what you are given, and the result has not been pretty and I have had to rely on Plan B. Plan A was just not going to work. The mismatched individuals might be fine on some types of developments, but are not well suited for the extreme variety.

To recap, I use the phrase *Mushing* to capture the feeling of going through the wilderness in the winter via dogsled, or kayaking down white water. The momentum and navigational skills. If it is a one person project, it might also be like skiing blindfolded downhill without any knowledge about what the mountain looks like. You hope that the trees will be soft and that when you feel air below you, there will be snow and not rocks where you land.

I also use *Mushing* to capture what the software looks like as it is being created. It is like watching someone work a piece of clay with their thumbs and fingers. Pressing, smoothing, extending, removing, adding... adding some water, rotating the work on a wheel to achieve a consistent finish, using an occasional tool add grooves, define shapes, remove excess clay... all the while moving towards something that cannot be initially seen. Structure and key elements take form, features appear, final tweaking is performed, glazing applied, and the work is fired. There is a flourish of movement and it looks like the person is playing with mud and clay like a child, but something different is happening. In the hands of an experienced craftsman, beauty in form and function will result. The clay does not look like anything for a while and suddenly the form can be recognized, but the path is not straight and linear, the piece is reworked, mushed back down, and re-crafted.

When someone knows what they are doing, they can mush clay accurately and consistently. An amateur can create ceramic works that can be ugly and somewhat comical. An amateur might get lucky now and then and hit the mark, but this is by accident and not by design. You can also take this fluid way with other art forms, even when working with stone. It is a state of mind and a way of going with the flow. When I was carving the piece below, the stone started feeling and

looking like a fish, then a whale, then a bird, then an eagle's head with beak, then a tiger's face, and finally…

Mushing with stone is a bit different than mushing clay, but the concept of going with natural flows and structures is the same. You have to work with the stone but not against it or forcing it. You hear the same type of analogies with white water kayaking; not fighting the river, feeling the river, understanding the river and going with the river's flow. Software can be created the same way, with the same benefits and the same risks. Sometimes the stone will crack when carving and sometimes you will hit a rock while kayaking or capsize because of the current. Sometimes the code works and sometimes it does not and you have to refactor a small bit or do major surgery. Nothing is perfect, no one is right 100% of the time. Almost everything is a compromise as well, with various trade-offs between effort, costs, value, comfort, experience, and evolution.

What is all of this discussion about kayaking and white water leading up to? Well, I have been wondering if something similar to the six white water classes could be described for Agile/Extreme software development:

- **Class Ⅰ: Easy.** Obvious requirements, many examples, and existing toolkits. Almost all ideas and features are commonly known, with simple customization being done. Few problems, all obvious and easily missed with a little training. Risk to programmers and users is slight; self-rescue is easy.

- **Class Ⅱ: Novice.** Straightforward development with occasional backtracking and recovery. Majority of functionality is well known and understood. Almost all problems can be missed by trained programmers, and individual programmer recovery is most common. Developers and users usually come out of the process unscathed.

- **Class Ⅲ: Intermediate.** Development with moderate, irregular patterns of challenges which may be difficult to avoid and which can delay or cause an over budget situation. Complex maneuvers are often required to get the functionality right and to complete the build. Increased requirements effort is required in advance for inexperienced developers. Delays and problems can be recovered, but often require a group effort involving additional developers.

- **Class Ⅳ: Advanced.** Intense, powerful development effort as the requirements and process is turbulent and unstable. Many requirements become revealed as the project proceeds. Precise reactive refactoring and redevelopment are needed during the development to keep daily builds going. There may be extended periods without daily builds as the refactoring and redeveloping may require additional requirements analysis and prototype development to test ideas. Risk is moderate to budgets and time expectations. Additional personnel might be needed at various points. Some tool work might be needed to adjust the toolkit's suitability to the tasks at hand. The work might be done in two stages; a fuller toothpick before full development is authorized.

- **Class V: Expert.** Extremely long, challenging development with many unstable requirements will not be revealed till the last minute during the development. Extended periods of tool crafting during development and re-jigging the system as major functions expose themselves. There will be periods when both developers and users doubt the wisdom of the endeavor. Developers not only need to know how to deliver the functions, use the tools, and do minor repairs on the tools, they will likely need the ability to build tools from scratch. Risk is high and project failure is a strong possibility. Initial toothpicks and fuller prototypes might be attempted to address some of the risks, but this will not be possible for all of the risks.

- **Class VI: Extreme and Exploratory.** These developments are relatively unique and few exemplars if any exist. The requirements are extremely fluid and appear to constantly change. The use of technology is also pushed to the limit. Progress is extremely unpredictable. There are many risks and the consequence of error is severe. This is for teams of experts and all precautions must be taken.

Chapter 2
High Velocity Mushing

2.1   Low Volume vs. High Volume Development

This chapter provides an overview of how to put Agile/Extreme on steroids when developing low volume software (Parts III and IV of the book will go into the details). I had to name this method, so I call it *High Velocity Mushing*. Low volume software refers to software packages that have a relatively small user base, few software packages in the same domain, and for which there are few well-known and well-documented sets of features and functions, like Class VI rapids. Many of the principles and concepts also apply to high volume software situations, but the reverse is not always true.

A high volume software situation is one in which there is an existing body of knowledge and where most of the common functions are well defined. The specifics might vary and a great deal of tweaking and customizing is necessary, but most of the basics are really known. I consider these situations to be like Class I or II rapids. For example, there are many systems for keeping track of inventory, supporting sales on the web, doing small business sales and record keeping, tracking customers, working with volunteers, managing charity drives, performing simple data entry tasks, etc. These types of systems are relatively standard and it is possible to develop formal and well structured techniques for knocking one off after another. You can have the forms and processes pretty down pat and prescribe "this is what you do" and "this is how you do it" and have the overall methodology repeatable. Many of the documented Agile/Extreme examples seem to fall into this category of meat and potato systems. The purpose of this book is not to replicate the knowledge imbedded in these practices for rapid, high quality development in high volume situations.

Low volume systems are different beasts. For example, there are few systems designed for integrated planning, scheduling, and dispatching in focused factories. There are few systems for co-operative education. There are few systems for teaching deaf children physics (i.e., there are often multiple learning disorders in addition to the obvious limitation on audio). These examples are the types of low volume systems addressed in this book. In the low volume case, there are few (if any) existing systems to base your system on. You have to develop the requirements almost from first principles and you cannot rely on "this is like ..." as much. In the high volume situation, you will likely have toolkits and libraries sitting around that can be exploited during the rapid fire builds;

but in the low volume case, you need to build your libraries first. There are other differences, but these are two key points: few reference systems, few existing tools.

It is important to remember that there is no single method, computer language, operating system, and so forth that is appropriate for all problems in all situations. You need to put together a toolbox of many methods and skills and know when to use them and when not to. There are many methods for documenting designs and although the design is properly documented and looks pretty, it does not mean that the design is appropriate or is a good design. As Emerson (1911) pointed out in the early 1900s, buying a set of law books does not make you a lawyer. To put this in context, we will paraphrase Michael Jackson's observation (1975, page 2): a good programmer does a good job with monolithic methods, with structured methods, with Object-Oriented Programming, with probably any method while an idiot will do a poor job regardless of the method used (p.s. this MJ is not the one gloved one). The actual quote is: "Programmers who had previously written good monolithic programs now write good modular programs; programmers who had previously written bad monolithic programs now write bad modular programs." In other words, the clothes do not make the person! Or, in the words of Forrest Gump: stupid is as stupid does.

At this point, I need to state one of my warnings. Do not turn methodologies into mythologies and religions. I once worked in a place where a senior software designer changed the user functionality so that the lines would not cross on his flowchart. As if computers care about what your documents look like! Be agnostic — understand the tools and use the appropriate tool at the appropriate time. Do not get so caught up with an operating system, language, or methodology that it blinds you. Keep learning systems and languages if you want to be a good architect since you need to know a variety of solutions if you wish to avoid the worse ones. ZenTai Mushing is not something to be religious about.

There are times when *Mushing* will be appropriate and many, many times it will not. There are a number of preconditions that must be satisfied, and there are certain assumptions that must be true. In an ideal situation, all of the preconditions will be met; but that is not realistic to expect. There will always be compromises and reasonable judgment must be used. You will have to weigh how close you are to the preconditions, which ones are most important in your situation, and what risks are raised because of the compromises. Sorry, no easy magic formula exists.

The basic principles behind what is now known as Agile/Extreme have been around for decades and have been practiced by a relatively small group for as long. Agile/Extreme (if you forget physical artifacts, buzzwords, and formality) describes a process in which you have a close and intense relationship with the user, almost (and sometimes) to the point of creating the software while sitting beside them. It prescribes rapid development of basic functionality and then evolutionary (and more rapid) development of the minor functions, user interface components, and other relatively small pieces of software. It requires that the coding and implementation starts before everything is documented in writing, that the development will be constantly in iteration, and assumes that things will work out in the end (the iterations assume that the highest priorities get done in order before the money runs out). You can do this and have a success; you might be very good and might have repeated successes. It is also possible to have a complete botch job, a disaster, and yet another software project characterized as over budget, behind schedule, with unhappy users. As I noted in Chapter 1, I have been involved in software development for approximately four decades and I have used many methods with the choice depending on the problem. I have also been doing projects in the Agile/Extreme fashion for over three decades. There are many subtle things to know about Agile/Extreme. You need to know when to do it, how to do it consistently, and how to do it in *low volume* situations. In these situations, it is unlikely that any two projects will be done in exactly the same way. It is more likely that the spirit and intent of fast, rapid development with intense user involvement will be followed and that formal steps and *THE way* will not be slavishly followed. There will be times when manifestos, sets of procedures, processes and methods will be largely followed and the process will look like a textbook use of Agile/Extreme, but there will be other times when you need to know how to bend, fold, and mutilate the recipe.

In the hands of a less-than-skilled developer, Agile/Extreme can be a very BAD THING! An average or poorly skilled developer will be even more so when doing Agile/Extreme.

2.2 From a Toothpick to a Decorated Living Christmas Tree

When *Mushing* in low volume situations, I have always created what can be thought of as a toothpick first, then the trunk of the system, then main branches, perhaps some decorations, more branches and more decorations. The trunk will also continue to grow to allow for more branches, and heavier ornaments. The tree has to be considered a living entity with a full, organic life cycle. A toothpick has form and function. It can bend (just enough) and you can put together a number of toothpicks to achieve greater functionality, strength, and performance. You can build technology toothpicks and functionality toothpicks. The one tests and illustrates key pieces of language or system features, the other illustrates key concepts in the application. Both toothpicks might exist in the final system or are blown away as with all code they have to be considered disposable. It is an old maxim in software development that you need to write it once, throw it out and start over. This is likely true for the Class V and VI situations where you have lots of learning to do. In these cases, be prepared to torch the toothpick. Of course, it is nice to hope that the toothpick or initial code structures will last longer than a pure prototype test, but it is best not to get emotionally attached to the code. If value has been gained from the toothpick, then be happy. Do not assume that all code you write will end up in the final product. The toothpick is similar (but a bit different) to the tracer code concept described by Hunt and Thomas (2000).

To put this toothpick analogy in the context of software, first consider size and scale. In terms of scale, the toothpick might be 3–4,000 lines of code, the initial trunk about 40–50,000 lines of code (growing to perhaps 70,000 lines of code by the time that the first release or major checkpoint is considered), with about 20,000 lines of branch code and about 50,000 lines of code in decorations and ornaments. The trunk grows and gains strength as the whole tree grows.

The trunk has to support the tree and provide life to the branches. It is the most important part of the system, the essence or core, and is where you spend the most time and the most thought. It is very hard to replace the trunk later without killing the tree! You can cut branches and graft new ones onto the trunk, but when you get disease in the trunk, c'est la vie. The trunk has to be the most generic and resilient part of the system. It has to weather storms, years of use and abuse, wet and dry seasons, hot and cold, and the occasional woodpecker.

The toothpick is very important in the low volume instance and is applicable to the first run at it and the final production push (if there is one). This is something that you want to quickly build. It has a little bit of each key concept and/or use of technology and this can quickly probe technology and potential risks. It is used to show users potential and create credibility; giving early demonstrations of feasibility and some visible progress. The toothpick can be used for training and for introducing new hires to the domain and concepts being used in the software solution. It is probably a good idea to always keep the toothpick around as a demo. You basically start with a copy of the toothpick to make the trunk of the tree. The toothpick will have different fibers and parts. There is the blunt end, the sharp end, the piece of wood between, edges, shapes, etc. and you keep adding and building on top of the toothpick.

When you look at the final tree, it is likely that if you take the shortest path from top to bottom, you will be still using the toothpick bits (if not the actual code, the spirit of the code). In a larger system, some parts may be replaced at different times and there might be some original, old wood there, as well as some recently replaced subsystems. As you go through development you might have to remove, fix, and reshape the trunk components and this is the refactoring of code. If you have an ego and get emotional over code and your effort, you might get upset and consider yourself disrespected when someone decides your code needs to be thrown out or refactored. Get over it. It will happen. Of course, if you are a true expert, you will try to anticipate and avoid refactoring (or try to minimize it) by pre-factoring and investing in building the little building blocks you *know* will be eventually needed. I do not mean just stubs either. It can be full code as you go. While I have some toothpicks break up and vaporize, I have also had toothpicks grow from a few thousand lines of code and stay largely intact to over 150,000 lines of code. And, everything in between.

2.3   Agile & Extreme — an Overview

In general, the traditional software engineering development practices have been characterized by predictable, repeatable processes, and have been described as plan driven (Beck 1999a). In these methods, there is an extensive requirements effort at the beginning and users tell the developer once and for all exactly what they want. From here, the programmers then design the system, hoping to deliver those features using conventional coding and testing processes. Sometimes it works, while sometimes it does not. On paper, this approach should work fine. However, in practice, the users did not (or could not) always tell developers exactly what they wanted and they changed their minds. To complicate matters, project management in software has always been tricky and many programmers misjudged their progress. Combined, these and other issues have resulted in many cost overruns, and schedule delays often followed (Glass 1997). Unfortunately, the pressure has remained for programmers to produce software more quickly (Aoyama 1998, Cusuman and Yoffie 1999). One outcome has been the realization that overall development processes needed to become more flexible to respond to dynamic changes.

A major goal of Agile processes is to support quick and early development of working code. As noted in previous sections, the focus is on simplicity and speed and while there is no clear agreement on where exactly the boundary lies between agile and plan-driven processes (Abrahamsson et al. 2002), there are some distinguishing characteristics (Beck et al. 2001, Cockburn 2002):

- □ Agile processes focus on individuals and interactions instead of processes and tools.
- □ Agile processes focus on working software over comprehensive documentation.
- □ Agile processes focus on customer collaboration over contract negotiation.
- □ Agile processes focus on responding to change over following a plan.

In the agile scheme of things, the emphasis is placed on the relationship and communality of software developers (i.e., the human role), as opposed to strict institutional processes and development tools. The software team is clocked or paced to continuously turn out tested and working software with new releases produced at frequent intervals with the the developers striving to keep the code simple and straightforward. While the relationship between developers and clients is given preference over strict contracts, the importance of well-drafted contracts does grow at the same pace of the size of the software project. You still need agreements, boundaries, and clear understandings about the *what*, the *when*, and the cost! Usually, the negotiation process itself is seen as a mechanism for achieving and maintaining a viable relationship. In most cases, the Agile development process results in a focus on delivering business value immediately as the project commences, and this can reduce the risk of non-fulfillment of the contract. Finally, the development group, which is comprised of programmers and client representatives, is prepared and authorized to make changes to address emerging needs during the development process (Abrahamsson et al. 2002).

Agile versus plan-driven approaches are often viewed as opposite ends of the software development methodology spectrum:

- □ Agile methods are viewed as totally adaptive.
- □ Plan-driven methods are viewed as totally controlled.

But, there is a large (very large) grey area between the two ends, and some researchers have claimed that synthesizing the two approaches can provide developers with a comprehensive spectrum of tools and options that can be preferable in some circumstances (Boehm 2005). For instance, over-responding to change has been cited as the source of many software disasters, such as the $3 billion cost overrun of the U.S. Federal Aviation Administration's Advanced Automation System for national air traffic control.

Agile development has been widely documented as working well for small (<10 developers) co-located teams that are facing unpredictable or rapidly changing requirements (Beck 1999a).

However, its applicability in the following scenarios has been questioned and criticized for use in situations such as:

1. Large-scale development efforts (> 20 developers).
2. Distributed development efforts (i.e., non co-located teams).
3. Mission and life-critical efforts.
4. Command-and-control company cultures.

Reasons for these criticisms relate to the reduced amount of documentation, the compressed software design timeframe, the trend toward globally distributed development environments, the unproven quality control mechanisms and the cultural change that is sometimes required to provide the development team with true empowerment to identify and make major changes to the software and contract. One paper has stated, "It is also clear that companies that develop long-lasting, large complex systems may not be able to use agile processes in their current form" (Turk and Rumpe 2002). However, much of this criticism has been refuted by agile practitioners as simply being misunderstandings about the agile development process (Boehm and Turner 2004).

The above has described the Agile development approach in general. However, the agile approach contains a number of different methods such as extreme programming, Scrum, the Crystal family of methodologies and adaptive software development, among others. Extreme programming (XP) is arguably the most popular agile methodology and has evolved from the problems caused by the long development cycles of traditional development methods (Beck 1999a). It started as "simply an opportunity to get the job done" (Haungs 2001) using practices that had been found effective in software development during the preceding decades. After a number of successful trials in practice, XP was theorized upon the key principles and practices used (Beck 1999b). Even though the individual pieces of XP are not new, XP has integrated them in such a way as to form a new methodology. The term extreme comes from taking these commonsense principles and practices to extreme levels (Beck 1999b).

The formal XP lifecycle consists of six phases: Exploration, Planning, Iterations to Release, Productionizing, Maintenance and Death (Beck 1999b). In the **Exploration phase**, customers write the story cards that they wish to be included in the first software release. Each story card describes a feature to be added. Concurrently, development team members familiarize themselves with the technology and tools to be used. This phase can take from a few weeks to a few months. The **Planning phase** sets the priority order for the stories (i.e., features) and an agreement of the contents of the first release. The time span until the first release normally does not exceed two months. This Planning phase itself normally takes just a few days. The **Iterations to Release phase** breaks down the schedule from the Planning phase into a number of iterations, each of which will take 1–4 weeks to implement.

The first iteration creates the architecture for the entire system. Functional tests and stories to be included are specified by the customer. At the end of the last iteration of this phase, the system is ready for production. The **Productionizing phase** requires extra testing and checking of the system before it is released to the customer. During this phase, new changes can still be identified and incorporated. Postponed ideas are documented for implementation at a later phase. The **Maintenance phase** is reached after the first release is provided to the customer. The development rate may decelerate as the customer uses the system. Customer support tasks are provided. New iterations are produced as the customer identifies new stories to be added. New people may be incorporated into the team. Finally, the **Death phase** arrives when the customer no longer has any stories to add and the system meets his/her needs in all other respects (i.e., performance, reliability). At this time, necessary documentation is finally written since no more architecture, design or code changes are required. Death may also occur if the system is not delivering the desired outcomes, or if it becomes too expensive for further development.

There are many assumptions implied by the Agile and Extreme methods. For example, there are assumptions about the user's ability to explain and understand what they want the developers to do. This is a BIGGY! There are other assumptions about the ability to interact closely with representative users. Some of these assumptions will be dealt with in the how section of the book. At this point, just remember that there are many assumptions and risks that must be explicitly dealt with and no two situations are the same!

2.4   Preconditions

In my opinion, there are a number of preconditions which must be satisfied before an agile form of Agile/Extreme approach should be considered. Some of them are:

- You must have enough knowledge about the problem and what is needed by the user. You must know enough that the final product is a dog and you must know what a dog usually implies, versus knowing that the user wants a fish and knowing what a fish usually implies. You might not need to know the type of dog or the type of fish, but you should also probably know enough if the dog is a house pet, versus a guard dog, versus a hunting dog. And, you should probably know what a house pet implies, versus the normal characteristics associated with guarding, versus what hunting implies. You need to know that there might be key and significant characteristics and know enough to identify them and to understand their implications. Too often, in general practice I have seen people think that the Agile methods are an excuse for not doing their homework, for not understanding the essence of the problem, and are an excuse for sloppy analysis. They do not know what *enough* knowledge implies and they start the actual implementation too soon and without sufficient preparation. In a low volume, relatively unique and special domain, enough knowledge can imply many months of studying and analysis! This is especially true for a strategic or mission critical system. For these types of systems, you have to do analysis and design work BEFORE you start the actual Agile process. Failing to do so will result in lots of code being thrown out and a very displeased client. It is my experience that it is impossible to build a complex, critical system without doing a substantial amount of understanding and designing (e.g., you cannot build an enterprise level system without having a good data schema and concept to start with, you cannot build it ad hoc through the daily or weekly scrum process!).

- You must know enough about the basic domain and area to provide reasonable advice and guidance to the users. You cannot simply do what the user asks you to do. They might be beating on a symptom and not on the specific problem. They might not know best practice. They might not know how to exploit information technology for even better efficiency, effectiveness, and user experiences. They might not know what others are doing or what can be done. This is your job. You are not an empty vessel. You are supposed to have a brain and use it. Agile is not an excuse for checking your brain at the door. This point and the one above harkens back to a comment made earlier. Agile is not for the inexperienced and the amateur. You need to have experience and a certain level of expertise, both in development and the domain in which you are working. You need to have enough experience and skill to work with users and to understand the differences between could be, should be, and will be.

- If this is the first bit of software in a domain without much commonality with anything else, you have to do a great deal of analysis and a lot of design work before starting anything that closely looks like Agile/Extreme (if at all). The possibility of Agile/Extreme will also depend on other factors, such as how much user interaction there is (versus imbedded), how closely related functions are (i.e., how many do you need functional to really make a prototype or toothpick useful and meaningful). If you have existing software in the domain, or some prior experience in something similar, these sources of knowledge will make Agile/Extreme more feasible.

- While noted above, it is worth repeating and expanding upon. You can have some in-experienced folks on the team, but you better have the majority of talent resident in seasoned developers who really understand how to develop applications. These key developers must have a broad and deep understanding about how to work intensely with users, how to do risk management, how to do iterative development, and how to start with a toothpick and make a tree. You should not start with all in-experienced developers and assume you can learn this stuff quickly and without mentoring. And by in-experienced, I mean in-experienced in the Agile/Extreme methods. Just because someone has been coding for decades does not mean that they will understand how to operate in this fashion. Furthermore, there is a big difference between software folks who know the technical marvels of a language and the configuration settings on a server and the folks who know how to actually develop applications. Technical skill is not the same as development skill!!!!! To be a good developer implies that you need to understand many subtleties and make decisions about interfaces, architecture, and functionality. Knowing the latest fad or library for a language does not mean that you know what development decisions need to be made or how to arrive at a suitable answer.

- You must have the right tools to build the toothpick and then be able to create (or obtain) the right tools to create the tree and decorate it in a very rapid, high quality fashion. I am talking daily or 1/2 day builds, reacting almost in real time with the user sitting beside you. You need to be able to craft non-trivial functions out of small building blocks quickly and accurately.

Most of these points will be discussed again in later chapters. To recap: you should only use Agile/Extreme when and where it makes sense; remember that not everyone is suited for development or Agile/Extreme; and try to have the team mostly made up of seasoned professionals. Above all else, remember that Agile should not be used as an excuse for not designing or analyzing.

2.5 Bite Sized Pieces

There is another analogy related to *Mushing*. From a distance (and it does not have to be too far), mush is mush and there are not too many big pieces. This is a key clue to successful high velocity *Mushing*. When you are doing short term project management, say at the week level, the week is decomposed into bite sized pieces. Remember, in Agile/Extreme, the short cycle implies **design, code, test, integrate, and deploy**. This is what you do in any single bite. The full cycle. What you can do in one hour, two hours, three or four are bite sized. If it takes longer than four hours to create a bite sized bit you should break it down. This concept will be revisited later when time management is discussed. You should think in terms of what can be done in these small bits! If you do not, you get fragmented effort, loss of focus, delays in the daily build, delays in feedback, ad nauseam.

The benefits of bite sized morsels are many. Or at least they appear many to me. I have had developers push back, refuse to try it, and generally say that they cannot think this way. In most cases they did not even try and dismissed the idea of bite sized development as the ranting of a mad man. I think that there are clear and obvious benefits associated with the bite sized approach. You have focus and concentration on a small topic and you can deal with the exceptions, edge cases, easily remember what you were doing when a bug suddenly arises, etc. You build up the system based on small blocks that are usually robust and easily replaced. If the pieces are small, it is hard to get them tangled into too many other places if you restrict the time and effort.

In addition, they help people with attention deficit disorders focus on activities with clear start and end points, totally encapsulated in one burst of effort. I know this from personal experience. As some medical folk describe it, ADD is not really about attention deficit, but attention variability and they talk about hyper-focus when some people with ADD get into a task. The hyper-focus phenomenon is often associated with doing something all at once. This might be reading a book, doing a task, and so forth. The bite sized approach to coding matches this ability for short bursts of intense productivity and works for me. It might not work for others. Before I start a programming task of one, two, three or four hours, I need to see the end. It is not a precise view of the end and I know that I will mush the code in quick iterations and refactoring as I go forward. I assume that all code is disposable, and while it should be done with the usual practice of good form and standards, it is a temporary reality.

While I might not know the journey, I have an idea about what I will get done in the assigned time window and I exploit muscle memory between the ears to pull it off — going like a bat out of hell on the keyboard. In thinking about the code I have written in the last decade or two, most of the toothpicks are about two to three days (max.) and are about 3,000 lines of code. Even those relatively short periods are broken down to two-four hour bits and I personally like to see useful things happen on the screen (or between functions) within a couple of hours of starting something. That is how I keep interest and focus. I have new values being generated to the user every hour or so. Lots of little values and functions. Lots. It makes me feel good. This also means that I do not start this foolishness until I have a warm enough feeling about what I am building and where I might end up. This is the homework. How warm is the warm feeling? This is the trick and where experience comes in. You should always know the basics and the key characteristics of what you are building; you need to know the dog'ness, the fish'ness, or the bird'ness of the system. You do not need to know the specifics of fur, feather, or scale color, but you need to know that there might be something like fur, feathers, and scales to deal with. Beyond this basic level, the warm feeling will depend on how critical the system is. If the system is a mission critical system, you need to have a warmer, a very warm, feeling that you understand the user's problem and how the user will use (or potentially use) the software you are building. You might need to do full ethnographic studies, job shadow, prepare process flows, and spend months before you can be confident enough that you can build what is required. I do not think it is possible to build a Class V or VI system without a substantial investment in analysis and architectural design. It would be like a building contractor starting off with a single family dwelling with the foundations and first floor built and then realizing that the client wanted a thirty-story building. You need the basic architecture and basic understanding before you can start building it in bite sized pieces.

This general bite sized approach also means that I create my software systems with many wrappers and sub-libraries of little tools (they have to be little if I can do them in a few hours). I think about the little tools I will need in the big picture and then I build my blocks and my tools. From a distance, the beginning of the project or software looks rather slow and painful, but when enough of the goodies exist and bigger tools are created with the little tools, the floodgates open and it is at this point I engage in daily builds and development with the users. If you are in a low volume situation (remember the theme of the book), you need to recognize if the tools exist already or if you have to build them. If you have to build them, you have to be patient and create the tools before you launch and fire. If you are in a high volume situation like a Class I or II rapids, it is likely that the tools exist and that you have used them for a variety of similar applications. That is nice, but those are not the situations this book is about. For one Class VI project, I had to build tools for about seven months before I was able to pull together eighteen subsystems into a complete flow or system test. At that time, only a few functions were exposed from the service. Over the next few months, several hundred functions were quickly exposed from the base service. Each function was constructed in minutes or hours as the end user functionality was fleshed out on the web interface.

Because you focus on one small item at a time, you can become totally engaged, become consumed, and know everything you need to know about this bit of code. If you deal with the detailed design, code and debug, test, integration and cap in one sitting, you create many small hunks of nice, clean, working code. When someone says the functionality is wrong and it has to be re-written, it is not a problem and does not hurt the ego. You have only invested a few hours, usually less than a day, so who cares! Assuming that you have been coding in nice clean ways, you can take out a few of these little bricks without affecting any other piece of software; another bonus. I will talk more about the small bits later in the *how* sections.

Chapter 3

Experience and Expertise

Chapter 2 noted the importance of experience and expertise. The term *expert* is over used in our day-to-day vocabulary. In the field of Cognitive Psychology, the term is used in a more restrictive fashion and there are a number of phases one goes through as a cognitive skill develops and before one is considered an expert. Oh yes, just because you consider yourself an expert and some people heap praise upon you and call you an expert, does not make you one.

Just because you get high marks in a course or through a program does not make you an expert either. I found some information from my high school days. The average at graduation was about 68% and only 5.9% of the students had over 80% (this was back in 1972). I found some recent information about a large school board in Canada where 45% of the students had over 80%. Hmmm. I was not aware that the brain had evolved so quickly. Imagine, in about thirty years we went from 5.9% to 45% of the students getting 80+%. Wow! The brain must have rapidly evolved. The younger generation must be smarter because the numbers prove it. The students might feel better about getting the higher marks, but they also might think of themselves as being more expert and more skilled than they really are. At university, the class averages may drop down to the lower 70's and this can be a blow to self esteem after getting those high 80's and 90's. Be careful about how skilled and expert you think you are.

See **The Cambridge Handbook on Expertise and Expert Performance** (Ericsson et al. 2006) for an extensive discussion on the subject; especially if you wish to claim to be an expert ☺. Expertise in common language is not a refined concept and is more of a relative measure. In this book, the starting point is about a decade of continuous experience with multiple success stories which have proven successful by the test of time with multiple versions, multiple users. This will give you close to the 10,000 hours of practice and reflection you will need when you factor out meetings, coffee, lunch, and socializing. I had about eight years of eighty hours a week coding and development experience before I was an official architect for a couple of product families at the age of 28. You need to put in the time to learn the trade. It takes about 20,000 hours to be a grand master! It is not just the time either. It is the type of time you are putting in. The time must be reflective and supportive of learning. Doing something in a brain dead way for 10,000 hours will not an expert make.

There is another aspect to the relative nature of expertise. If you are the only one on Earth who has done something, you might be considered the world's leading expert, but you might not understand all of the issues, relationships, or be able to repeat the something in a reliable and consistent fashion. As a relative expert, you might have spent many years investigating the question, but you might not know the answer. Always consider this before you pontificate or make prescriptive or normative statements. In my academic research area, I was the first one to do a certain style of research on a specific problem and became the world expert. After a quarter century studying this one area as my major research agenda, I still do not know the answer to the question that started me down the path, and I am still regarded as the world's expert in this niche topic. It is all relative. Part of my research topic since the mid 1980s has dealt with cognitive skill and skill acquisition and that forms the basis for this chapter.

Software development involves cognitive skills throughout the life cycle. Covering requirements analysis, through design and architecture, to implementation and deployment. How one thinks, what one thinks about, and when something is thought about is different for those starting to learn a cognitive skill and those considered true experts. The differences can be subtle (or very distinct) and will affect almost every facet of what is being developed. For example, novices often think in vertical threads and need to drive ideas down to the lowest level and the thinking patterns resemble an elevator going up and down. A true expert thinks in a lateral fashion, keeping all of the ideas at the same level of granularity, having faith in their skill and knowledge that the lower bits will work out and that their higher decisions are reasonable. The up/down thinking is not efficient or effective when compared to the expert's lateral and layered approach. The lateral and layered without any retracing is the shortest path if you are into mathematical abstractions.

Always remember that it takes time to develop a cognitive skill and to develop a deep and broad understanding. As noted above, a common estimate for developing expertise in a cognitive skill is approximately ten person-years of continual practice and learning (note the learning part) and the grand master level requires twenty person-years. For example, if you continually develop a skill eighty hours a week for a decade, and have repeated evidence of your skill, you might be qualified to be considered a Grand Master. There are three points here. The practice, the learning, and the repeated demonstrations and testing to assess skill.

Seven levels that can easily be identified in software are:

1. Novice.
2. Junior.
3. Intermediate.
4. Senior.
5. Journeyman (terminal level for most in the career).
6. Expert (Consultant).
7. Grand Master.

This model of skill development follows an apprenticeship philosophy and assumes that one must learn the tools of the trade and learn how to use them effectively and efficiently. It also assumes a reasonable progression of increasing complexity and sophistication in what is being developed. While I can only speculate, it is likely that Michelangelo worked on many tombstones before doing David. He had to learn the characteristics of stone, to read a specific piece of stone, how to use the tools, and know how to bring out the emotion and feeling in a sculpture.

Not everyone achieves the sixth and seventh phases of expertise or *enlightenment*. The normal and expected termination point for probably 90+ percent of the population is the fifth phase: the Journeyman level. In the day of the guild and trade, the Journeyman was skilled and very proficient in doing the trade, capable of instructing the novices and apprentices, but was not considered the main teacher, expert, or Grand Master. They did not possess the insights and abilities that come with deeper learning and understanding.

Each level involves greater responsibility for more important pieces of the puzzle. Each is more complex, larger, more dependencies, higher demands for performance and quality, and so forth. Each level implies increasing knowledge about the domain one is working in. Knowledge about the tools, how to use the tools, the subtleties of tool usage, the domain specifics, and the domain users. The tools are not unlike the chisels and hammers used to sculpt with, and address requirements analysis through ongoing support and software evolution.

Progression through the ranks implies repeated demonstrations at each level of skill and not just one, perhaps lucky, event. For example, Michelangelo is known for more than one painting and for more than one sculpture. Designing one protocol or one operating system does not mean that you are an expert in all forms of software, or are capable of designing a second, equally good, piece of software in the same genre. Learning does not imply that everything must be done from first principles or by a single person. For example, you can learn from maintaining a certain piece of software and reflecting upon its design, evolution, and execution. Or, you can learn about different software development concepts by working as a member of a team. The key is variety over the long term with constant reflection and thinking about what you are working with, how you are working with it, and how the software works in the intended context.

The apprentice concept can be used to think about the scale of projects as well. Code projects that are 20 lines of code, versus 200 lines, versus 2,000, versus 20,000, versus 200,000, versus — you get the idea — are all VERY different. They are different in the potential risks that could be lurking, the way you would design, the implicit amount of functionality that needs to be understood, the way you would code, the way you would test, and design for support. You do not have to be the main star on each size of project, but you would learn a great deal as you progress up the ranks. You would learn sometimes in the development mode, and sometimes maintaining or supporting a system. Time for a Zen moment to sit and reflect upon…

Eventually, many characteristics associated with master level cognitive skill acquisition resemble mystical qualities similar to Zen Masters. For example, there are Zen Buddhist writings that describe novices as seeing the water and the mountain, then gaining the skill to not see the water and the mountain, and then ultimately as masters being able to see the water and the mountain. This is similar to how expert problem solvers and designers proceed as well. When a novice looks at a problem, this is different than how someone consciously looks at a problem as they proceed, and this is different again from the recognition an expert or master immediately gets from looking at the problem. There is progression from superficial thinking to structured thinking to holistic thinking. The holistic, spontaneous thought or recognition is also somewhat similar to the no-mind described in Zen Buddhism.

In the way no-mind is described, the situation or problem is not really thought about and the mind is set free and at some level, without really thinking about the problem, the answer is realized. While Zen practice is theoretically about the greater meaning of life and becoming enlightened, the Zen principles have also been described for how excellence can be achieved in archery or swordplay. The mind does not stop, you do not think about the individual actions or what the enemy is doing, your mind takes it all in and guides your actions without fault. It is my belief and experience that many years of deliberate practice and reflection in software yields the same type of result.

There can be many characteristics which will distinguish a true expert or Grand Master in software development. Some of them are:

- *Letting Go* — an expert will not cling to any methodology or idea in the face of facts or logic that suggest other methods or ideas are better — one size does not fit all.
- *Solution Space* — an expert will not know the 'right way' to do something as they know that there is usually no single right way, but they will know many wrong ways to avoid.
- *Opportunistic Reasoning* — an expert can identify the appropriate situation and timing to exploit concepts, tools, and available resources; their expertise and level of preparedness allows them to seize the moment.
- *Problem Focus* — an expert will first form the problem space and then identify or work solutions into it — not the reverse.
- *Clarity of Focus* — an expert will focus on the appropriate issues for the moment and know what issues should be addressed later, or which ones can be dismissed in their entirety.
- *Stopping Criteria* — an expert will know when enough is enough and either cap the activity or startover.
- *Pre-Factoring* — an expert will minimize refactoring of code through anticipation and pre-factoring.
- *Humility* — an expert will know that the requirements are neither stable nor complete, that the design is not perfect and will possibly require rework, and that software bugs cannot be totally avoided — and analyze, design, and code accordingly.
- *Optimization* — an expert will know what should be optimized and what does not need to be optimized — in general: not wasting time on tasks that do not need to be done.
- *Muscle Memory* — an expert's automaticity or muscle memory will be used to identify and work with blocks of logic and solutions without consciously thinking them through — after reaching a certain point of understanding the problem, a self-realization point will occur with the solution suddenly appearing.

- *Unified Whole* — an expert will embrace the full context of the development, incorporating automatically the major elements, players, and environment.
- *Life Cycle* — an expert will appreciate the implications of time and how the environment and problem space will evolve, and know what aspects to address in the software project.
- *Risk Awareness* — an expert will understand the implications of any change in the status quo and other risk generators.
- *Risk Management* — an expert will organize work units and work sequence to avoid or mitigate risk and have contemplated back-up plans for the back-up plans, thinking several plies deep in terms of alternatives.
- *Self Awareness* — an expert will know what they do not know, while a novice will not acknowledge or recognize what they do not know.
- *Visualization* — an expert will see things others will not see — they will see patterns and relationships that define the essence and the core.
- *Essential Qualities* — an expert will intuitively recognize form, function, nature, embodiment, causes and effects when performing an analysis.
- *Model Based Reasoning* — an expert will look at the situation and use various abstract modeling concepts to capture the problem's essence and then use the model for analysis and for informing the creation of the solution.
- *Formality* — an expert will recognize when methods become the end and not the means, and when formal methods are needed or not needed.
- *Constraint Relaxation* — an expert will know when certain rules and concepts should be bent, folded, and mutilated and when they should not be — that is, not everything worth doing is worth doing right.
- *Requisite Variety* — an expert will understand that the solution space (including development methodology) must have adequate degrees of freedom to match the problem's degrees of freedom.
- *Lateral Thinking* — an expert will think in a top-down lateral or horizontal approach, dealing with similar degrees of granularity and importance and will avoid single-threading through the design bit by bit; they will selectively identify any critical aspects for early risk analysis.
- *Calmness* — an expert will understand what to worry about and what not to worry about.

Learning and developing expertise is a tough nut to crack and there is no magical answer. If you want to become a really good designer and architect with repeated successes over many years, you should learn to learn and you will need to consciously think about the learning process. You will need to use your right brain and your left brain. One side will be contributing creativity, the big picture, and identifying patterns and relationships; and the other will be providing systematic and logical thought processes about the little bits.

The best recommendation I can give you is to find a mentor. Find someone who will teach you how to fish, rather than caving into your whining and begging, and just giving you the blasted fish.

And, by teaching you how to fish I do not mean giving you a set of well defined steps. You do not want someone to tell you to use this fishing rod, use this line, tie this fly at the end of a line so long, go to this hole, lower the line exactly this far into the water at this specific point, at this particular time, and behold a fish will be caught. When we old people talk about teaching you how to fish, this is NOT what we mean. We mean that you need to understand the concept from first principles (or close enough), that you understand the science or theories below the superficial that when you do not find that specific line you can make do with another, that you can craft your own rod if you have to, that you can identify where the fishing hole might be without being told, that you can experiment and draw outside the box. Hopefully, your mentor will make your head hurt and when you ask to know something, they will encourage you to think and to test your knowledge and thought processes.

Apprenticing and working with a Master is not an easy road. Things might not always be so obvious or clear. A true expert or master has a difficult (if not impossible) time explaining what they do or how they do it. They often fall back on what they were told as an apprentice and cannot express how they actually do the task as a master. They will tell you what they were told, not what they do! Part of this is what is called tacit knowledge. Tacit knowledge cannot be expressed and this might be stuff stored as muscle memory (in the brain) and be part of automaticity. In some cases they might be too embarrassed to admit that they cannot really explain why they make one decision over another and it would sound lame or flakey. "My gut just tells me that this is the better way, trust me." Young people do not typically like this type of answer and want to know EXACTLY what to do and how to do it. So, the master wings it and fakes it.

It is hard to accept or believe, but the expert often really does not know what they do. They are flying on automatic pilot. I have studied some experts and documented their actions and when asked about the documented actions, the experts set aside about 10% of the actions and said "I did not do that." They were not aware of the decisions they made and the actions they took. 10% of their decisions. Sigh. This is another reason why field observations and ethnographic methods are important. Try learning via questions and interviews from someone who is not aware of what they actually do!

Certain Masters also use a technique where they specifically and consciously do not explain things and force the student to think about why the master might have them do something or what might lie beneath the surface. The master wants to open the mind of the student and have the student think about thinking and how they are thinking. Students often have nasty names for such masters. This lack of explanation and opening of the mind is also in the Zen way, but I doubt that Zen Buddhist students would have nasty names for their teachers, but I do not know this for sure. When you are learning, it is also hard to understand why some things are consciously done (in your opinion) in a sub-optimal way. In systems theory and practice, you have to be aware of what is called local or myopic optimization versus optimization at the system level. Unless you have the system perspective, you might think something is really bonehead when in fact it makes sense from the larger perspective.

If you are more of the typical left brain type of geek, you probably love lots of transparency and need lots of reasons while you learn. This is not a real theory, but it is based on my own teaching and interactions with students; "Just tell me what you want." "Tell me what it should look like." "What should I do?" "How should I do it?" "Why should I do it?" Here are my thoughts on how to think about learning from a harsh Master, of which I have had a few.

- Just because you do not **understand** something does not mean it is wrong, nor that it will not work.
- Just because you do not **like** something or enjoy it does not mean that it is wrong, nor that it will not work.
- Just because you have not **done it before** or used it before does not mean that it is wrong, nor that it will not work.
- Just because **others** do not do it or advocate it, does not mean that it is wrong, nor that it will not work.
- Just because others do not consider it **best practice**, does not mean that it is wrong or a worse practice, nor that it will not work, nor that it might not actually be better practice.
- Just because it is not **obvious** to you, does not mean that it is wrong, nor that it will not work.
- Just because you do not see **value** in it and think that it is a waste of time, does not mean that there is no value, nor that it is a waste of time.
- Just because you do not **agree** with it and think that there is a better way does not mean that it is wrong, nor that it will not work.
- Just because you do not have **facts** to regurgitate does not mean that you have not learned.
- Just because you do not see **immediate results** does not mean that changes in you have not been made and that results will not appear in the future — perhaps only seen in hindsight and perhaps without you being consciously aware of it.
- Just because you do not generally **believe** in it, does not mean that others do not, nor that it is wrong, nor that it should not be done.

To learn from a Master, is to have an open mind, having trust and doing lots of thinking. To develop expertise and skill, you need to be a sponge, practice a lot, and reflect. You have to wonder and have an active imagination. Why is the Master having me do this? Why am I being instructed in such a way? What skills might I be learning? Why is Master hurting my brain? The Master is never 100% right and can do stupid things too, but what if Master is not having a stupid moment and is messing with you on purpose? What is the downside if you ignore Master? What is the upside if Master is doing something that makes sense?

In any event, expertise and skill is not a single dimension and you should concentrate on developing a well rounded package. Because, every job has dull, boring, tedious, mundane bits to it, and there are often politics and people to deal with. For sure, there will be repetitive tasks which are no fun at all. In the early part of your training and learning, your instructors (and bosses) might be testing you. How will you react to dull, boring tasks? I know of some companies that consciously place university graduates into a call centre to observe how they deal with situations and deal with routine tasks. If they have the right attitude and approach, they get to move on and play more. If they moan and whine in the call centre, and have a negative outlook, they have a short career with the firm. I know of many managers who do this type of testing. This reminds me of something. If you are reading this book, you are probably really handy with a computer and technical stuff. Most employers will value this. Yes indeed. But only about 25% of the complete package. They look for attitude, work ethic, team savvy, communication, etc. as the other 75%. It is possible to get along for a while on technical facts and knowledge, but it is not a long term strategy for most people. It is high risk to assume that your pure, technical skill is going to get you through life. I will discuss this topic at least one more time in the book.

Don't get me wrong. Base skills and fundamental knowledge are necessary if you want to be a long term expert. They are necessary but not sufficient. I once heard a term describing the geeks hired in Silicon Valley just before the big boom — *brains on a stick*. These were whiz kids and hotshots who knew one (or a few) tools, really well. They had lots of facts and knew *stuff*. When the market went south and folks were laid off and terminated, the *brains on a stick* were the first to go. These specialists did not have the other skills that made them really valuable to the firm. They were disposable. If you have multiple skills and are well-rounded, you are more likely to be a flexible mechanism capable of providing value in a multitude of ways and not be considered a one-trick-pony (old guy phrase: in the circuses of old, they had acts with horses and ponies, figure out the rest of the metaphor yourself).

Part II

Understanding the Problem & Thinking Through the Conceptual Solution

Overview

Part I was a brief introduction to some of the basic ideas. What I mean by *ZenTai, Mushing*, and my thoughts on learning, how to learn, and expertise. The latter is important because there is only so much you can learn from this book. There are different ways learning can occur; you can have the sage from the stage, the drill and kill memorization of facts, Socratic Method of query and response, and so forth. For rapid learning that seems to stick best (nothing scientific about this observation), I like learning from mistakes in a controlled way. You learn by doing, engaging in the experience, creating many dimensional hooks in the brain, making mistakes, and reflecting upon them. In most cases, I do not think that there is much harm in making mistakes (again, in a controlled way, off the critical path) as long as you learn from them and do NOT repeat them. I am patient with the first mistake, less patient on the second occurrence, and not very happy if you make the same mistake more than twice.

Every chapter in this book could probably be expanded into a fuller text and a whole course dedicated to it (this is what I do with Chapter 8 actually). My goal is to prime some ideas, create some awareness, form a base from which you can remember and reflect. Was McKay right? Wrong? Close? When I teach, I like to introduce students to the concept of risk and risk management. One way is to explicitly avoid creating credibility for myself. I try to avoid explaining my background, specialties, basis for the ideas, etc. From there, the students need to think:

- Does the professor have a clue and am I going to engage?
- Or, do I think I am going to get marginal benefit from paying attention and engaging, and am I going to blow it off?
- But, what if I think the course has little or no value, blow it off, and it turns out that there was indeed something useful?
- What if I think that the course might have value, but it turns out to be useless?
- What are the costs? What are the potential benefits?

Do the math. Pick your destiny. This book is like that. I suggest you think hard about the points in Chapter 3, the *just because* points, before reading the remainder of the book.

Part II is all about the **WHAT.** What do you need to provide? Why you are going to design something? What is THE problem? Why are you going to be coding something? This is before you get the urge to start coding or start picking technology. It is for the agnostic and not for the fanatical fan of a certain language, operating system, hardware vendor, or style.

Chapter 4
Understanding THE Problem

4.1 Understanding

There is a good book to buy or look up on understanding the problem. It is called *Are your lights on?* by Gause and Weinberg (1990). I have been told by a past student (who pointed me to the book) that it reminded him of how I teach and how I try to explain problem solving. After I read it, he was right. That being said, I am not going to repeat what is in that book. If you want to improve your general problem solving skills go read it. You will have to read it several times. There are lots of subtle points in the text and it will be a good test of your reading skills.

Perhaps the first thing to understand is to understand problem solving itself. Do you understand how to understand big problems? Solving a problem given as part of a class assignment is not the same as a problem on a course project, which is not the same as a problem encountered as part of a Master's program, and a Master's type of problem is not the same as a Doctorate's. Software problems are similar. There are small problems that can be solved quickly and there are problems that are quite large and will take time to figure out. Some developers give up after a few hours, and a little bit of surfing. "I cannot find the answer," "It cannot be done." "If it could or should be done, I would have found an answer." These developers are probably quite skilled at copying and pasting solutions which do not require any inventing or creativity. They might be excellent at using tools; but not creating them. When I mentor students, I point out that doing a Master's or Doctoral thesis helps them understand how to face and approach bigger problems; a form of marathon thinking task. To do Class IV-VI problems, you need to know how to use tools and how to make them.

If all you have done is solve small Class I-III problems, the types of problems at the Class VI level can seem daunting and possibly impossible. It might be possible to find a quick solution, but you cannot guarantee this. Some of my solutions have taken many months. One took a year and a half to get sorted out with six iterations being done in the process. If I had given up after an hour or two and folded, there would have been no solution. If I had given up after the first or second try, there would have been no solution. I kept going because the problem was inherently an algorithmic one and not one based on physics or the laws of nature. For algorithmic problems, you can find approximations, reasonable alternatives or substitutes, or be lucky enough to hit upon an algorithm that does the job 100%. It just takes work and lots of thinking.

Some people have trouble doing deep thinking and cannot handle deeply nested or tangly algorithms or get their heads around recursive algorithms. It is nice to have a goal of simple code and simple algorithms, but reality might be at odds with this desire. Not everything in life is trivial or simple. Not everything can be expressed adequately as a sound bite, a one sentence headline, or a pre-digested thought. Not everything is something that you can find or generate in an hour or two. A pre-digested thought attempts to take a complicated or more detailed situation and expresses it so that others can understand it without knowing the background, assumptions, complications, or details. A pre-digested thought might be like a weather forecast. "It will be sunny today." You do not need to know the barometric pressure, detailed environmental analysis and so forth. You are happy with the one liner. Some courses and styles of problem solving also rely on pre-digested thoughts and one liner observations and recommendations. Recipe or cookbook courses are like this; you have your checklist and superficial clues and solutions. Class I -III problems are potentially full of valid pre-digested thoughts and solutions, matching the trivial nature of the problem. The simple problem solving will get you close enough.

Not so for Class IV-VI. Pre-digested thoughts do not do well here and if they exist, their value is suspect. You will have to learn how to get into the complexities and details without the aid of someone else telling you the answer. You will need to do extended problem solving without becoming too frustrated or too negative. You will have to learn how to pace yourself, how to recognize a dead end, how to dig out of a tar pit, and how to reset your thinking. You have to learn how to do this repeatedly and develop the skills and knowledge that will allow you to have more hits than misses in the long run. Over time you will need to learn what clues to look for in the problem that suggest whether it can be solved or not, or what parts of the problem should be addressed in what order. This is problem solving. It is tedious, it is work, and it is at times very boring. I have had co-workers complain about tedious, boring, non-motivating, and non-fun work and find ways to avoid it, or note that a problem cannot be solved after a brief and limited investigation, or that a junior developer could not solve it, therefore it could not be solved and therefore given up. To solve big problems will often take a big effort and it is work. It is called work for a reason. It is not called fun, relax time, or vacation.

To understand a problem as an analyst, you have to understand questions. It is through questioning and probing that you understand something. To develop the cognitive skill associated with requirements analysis and understanding the *problem*, you need to develop the dual cognitive skill of questioning. You cannot do the one without the other. The better the questions and the questioning, the better the analysis. So, this chapter is not about answers, it is about questions. My most important skill is likely in the skill of questioning. As I finished the first full draft of this text, there were about 750 question marks scattered on about 300 pages; you will find that I raise more questions in this text than I answer.

I have been informed by my mother that my favorite word as a child was WHY? Consequently, I was probably a pain to many people, as a constant nattering of Why? Why? Why? can be very annoying. I was also lucky to have great mentors in high school who encouraged thinking, questioning, problem solving, and being self reliant in terms of critical thinking. David Pryke and Brian Coleman were those mentors. The latter introduced me to computers in the late 1960s. I was in one school, he taught at another school and for several reasons, Brian took an interest in my education and started bringing me to the University of Waterloo every second Monday evening from Stratford to play on the largest computer in Canada. This was the infamous 360/75 in the now converted Red Room. This playing was done by creating a Fortran program on punched cards and then standing in line in Debug (what the area was called) and then by the time you walked around the counter, your program and hopefully its results appeared on the printer. It seemed like magic. This innovative way to do student programming was one of the things that made UW famous: the legendary WATFOR and WATFIV compilers that allowed quick and friendly turnaround.

Brian would sit in one of the study rooms and do his high school teacher thing while I played for an hour or two. How did Brian introduce me to computers? The best I can remember, it went: here is a WATFOR book, this is a punched card, those are key punches; when you have your program ready, that is where you line up. He probably included the phrase, have fun. What was a computer? What was a punched card? How did you create a program? What was a program? What was WATFOR? These were my problems, not his. This type of process really helps you focus on what your problem REALLY is. This was the start of my training in understanding problems related to information technology. Brian brought me to UW on and off for about two or three years if my memory is right. He was very giving of his time and encouragement. I owe him much. This was the left brain stuff.

David's training was of a different sort and focused on the right brain. It was thinking, design, and creative/critical examination 101. It was five years long and I had him as a teacher each and every year because of the subject he taught: wood shop. He provided what was my first, extended exposure to an intellectual mentor, who had a very sharp mind, and an equally sharp tongue. You could make anything you wanted in his shop as long as you could design it, have it challenged, and have it pass the challenge. The shop turned into the haven for many of the school's thinkers, and the classes were of more philosophy and critical thinking than anything else. All I did in the last two years of shop was design and think. Others did make things in wood shop, but I was more interested in the design and thinking angle. It was a great experience to have these verbal dialogs with David and others. I also owe him much.

Both David and Brian had the irritating habit of not answering questions directly, and they probed the process of the thought stream making my head hurt. I credit these two individuals with the majority of my thinking and problem solving skills. I also credit them with my general style of teaching and training. When I started to teach as a professor, I went and sought out David and Brian and discussed how hard to push, when to push, and how to control the pain process.

The third major influence in my career was Romney White. I worked for Romney at the University of Waterloo (basement of the Math & Computer Building) as a co-op student and also as a part-time employee on school terms. I then worked with Romney in a start-up venture for a few years building a memory-resident system that behaved like a theoretical relational database system with a context sensitive grammar. I learned much of what I know about software craftsmanship from working with Romney. I learned about creating tools to help make the development process more efficient and effective, I learned about the importance of the user, and I learned about robust systems the hard way. I learned that I could probably learn to build almost any kind of software. Romney always amazed me with what he knew about computers, software, and what he could do. It was easy to be intimidated by him and to hold him in awe. He was the best geek I have ever known, who had a sharp mind, had a sharp tongue and did not suffer fools. In terms of software, he placed high value on software that gave users value, and software that actually helped the users. He had high standards on software quality and he was never afraid to make the bits he needed. I do not remember him ever saying that something was not possible or that we could not do it. This was balanced by the wisdom of knowing that just because you could do something did not mean that you should (or would), and that not everything worth doing was worth doing right. I have liberally and shamelessly adopted and adapted Romney's modus operandi as my own.

Romney's most common response to my questions would be "it is in the book"; I do not recall him ever indicating which book. In the early to mid 1970s, there were many to choose from. Eventually, I figured out that things would go better if I looked in the book first. Which book? That was a key question! From Romney, I learned to question any barrier. I saw Romney do spectacular feats with software and while I could not do the same feats, it demonstrated to me that you could craft what you needed and that feats were possible. I learned to ask many questions of myself before presenting questions to others. He once gave me a magnetic tape containing about 780k of object code (you know, the hexadecimal stuff) representing an application running on one operating system and asked me to get it running under a different operating system (some of the first operating system was already simulated on the second system, some was not). From what I remember, that about summed up his instructions. Where do you start? How do you sort out such vague instructions? These and many other questions were keys to getting the job done!

All three of these individuals shared traits. They hardly ever answered a question with the answer you were seeking. They were also smarter than I was and I have never caught up to them. In those days the phrase 'tough love' did not exist, but that is what they practiced and they did not spoon feed. They did not change diapers. I was very lucky to have such mentors. It was at times very confusing and very frustrating. In hindsight, I now realized that they only did this because they had more faith in my abilities than I did and they knew (or hoped perhaps) that I would figure it out. They were not being bullies or rude. They were training me in their own sick way. I do not admit to always appreciating this style of training at the time. I appreciated it later. I learned a great deal about questions from them.

The fourth and fifth mentors were my graduate studies co-supervisors — Professors John Buzacott and Frank Safayeni. John was also my boss, as I was Associate Director of a research group he was in charge of. John has an amazing talent for asking questions of the most subtle type that unlock the doors of mystery, and from him I learned to hone my questions and to think more before opening my mouth. With John I worked hard to avoid looking like a dim witted soul and the old carpenter's adage about measuring twice and cutting once was appropriate. In fact, with John, it was better to measure three or four times as he really knew the research area and he was exceptional in his thinking. He helped me think several plies deeper than I had been able to before. He provided the question that has formed my research agenda since 1985: *What is scheduling?* Frank is a Master of Questions and seems to question everything. After doing a Master's and a couple years into the PhD with him and John, Frank asks "What is a decision?" The research for the previous four to five years was about decisions. ARGH!!!!!

I share this background with you because it helps explain my fetish about questions, thinking of questions, asking questions, and rarely giving direct answers. I am curious and love to work on problems which require lots of questions. Through questioning you are seeking clarity and with clarity, complex things become simple and when complex things become simple, you finally understand the problem.

4.2   Good Questions

What is a good question? Good question. Have you ever consciously thought about questions? For longer than a few seconds?

A good *question* or *questioning* session is not accidental. There are ten simple principles to follow or remember:

1. You are asking a question for which you do not have the answer or for an existing thought you want confirmed. Else, why waste someone's time?

2. You are asking a question to a more important person only because you have tried to answer it already using other resources and the cost benefit analysis indicates that it is time to impose yourself on someone else. You do not want to ask someone to do your job who should not be doing your job, especially your boss.

3. You hope that the person you are asking can or will actually answer the questions. There is no real point asking a question that cannot be answered just to show how smart you are unless you are specifically probing to see what can or cannot be answered.

4. You hope that you have the skill to understand the answers (immediately or eventually). Why seek an answer you cannot use?

5. You hope that you are asking questions that require a reasonable amount of effort from the person you are asking the questions of. If the answering requires too much effort, they might not answer, or answer appropriately. If it is too little effort, it is likely that you could have answered it yourself.

6. You hope that you ask questions that help identify, clarify, or eliminate some part of the puzzle. There should be a clear purpose to the question.

7. If you have a limited time or a limited number of questions, you hope you are asking the most important and the most relevant questions in the right order.

8. You have thought about the questions and possible answers in advance, so that you hope you can ask intelligent and useful questions once the person gives you an answer. But not too much thinking as you want to avoid bias and you want to avoid turning off your ears. You want to be open to listening and reformulating your questions and plan!

9. There are rhetorical questions which have no expected answer, but can cause the other person to think and provide other knowledge. In general, you should minimize the types of rhetorical questions where you answer yourself! Note: there are cases where rhetorical questions can be used to voice your observations and have the other person comment upon them, but this usage should not be the norm.

10. Not all questions are created equal.

The best questions have the best return-on-investment and have the most potential to decompose, disambiguate, classify, and understand. There is a large, possible solution space, and these *best* questions will help you focus on reasonable portions of the solution space and be effective and efficient in your questioning. I have never been given sufficient time to do anything in my professional career and you need to do the highest return activities without wasting your resources. This is also true for questioning. If you have one question that you can ask the President of a company, or have a limited audience with the President, you need to consciously think and design your questions. You need to determine criteria for what a good question would be in this situation. If you do not consciously think about your questions, you are not really analyzing, you are guessing and hoping that you will get lucky.

The first question in *ZenTai Mushing* is: *"What don't I know?"* In practice, this is not a rhetorical question to which there is no answer and unless you are the all knowing, all seeing sort of God or Goddess, there should be something on this list as no one knows everything.

ZenTai Mushing implies few major missteps and careful risk management. There will be minor problems encountered and mistakes made, but the patient will survive despite your best efforts. The first risk is your own knowledge and your own set of assumptions. If you assume that you are an Ace and that you are a hotshot, then you are likely to take hubris to new heights. You might get away with it for a bit, but eventually the old phrase *Pride cometh before the fall* will ring true.

The second question is: *"What does this mean?"* That is, what is the implication of your limited knowledge? This question leads you to thinking about your strengths and weaknesses with regard to the domain, preconditions for *ZenTai Mushing*, what you need to do before you can mush, and what other skills and talents are necessary to resolve the unknowns.

Thus, the first two **good** questions I can give you are:

- What don't you know?
- What does this mean?

4.3 Questioning and Understanding

When I do not have sufficient knowledge about the problem, I go into a learning mode. If the problem involves humans and I have access to the humans, I use a variety of techniques and the best ones are based on Ethnography and Anthropology. In the early phases of an ethnographic study, the emphasis is on passive involvement with the users during which time you are not studying the people with any *a priori* model or theory, but you are learning from them. This is a subtle distinction and is important. During this phase, most of the questions are silent and are self-directed. Until you know enough, you want to simply observe and learn enough that when you start asking questions of others, you might understand the answers and that you have a better chance of asking good questions.

Caveat — if I really do not know too much about the situation, I will study the area in advance, sometimes up to six months. Why? I usually work in situations where the people are busy and are under pressure. The people do not want to educate you about the basics and about the general situation; they want to educate you about their specific problems and issues. If you are studying a planning and scheduling situation in the automotive supply chain, the planners and schedulers have no time, nor patience, to explain basic manufacturing to you, general concepts of planning and scheduling, and the big picture of how the automotive supply chain works. They expect you to know about the common terms, how products are built, how inventory is physically controlled, etc. — even if what you are doing is writing software for them. Do your homework and understanding the problem is a lot easier.

When you are starting your career as a questioner, it is good to have a checklist of things you are trying to find answers to. It is likely that the questions will start off being very specific and as you learn and discover patterns and trends, the questions will become more generic and be useful in more situations. Ultimately, you will not consciously think about the questions, as you will be sensitive to relevant information and just recognize it when you see it. As I am learning about the situation, I will ask indirect questions and sometimes a direct question to clarify something, but the goal is to become immersed into the situation so that it feels natural to me. I want to understand the user's life! I need to understand the variability and variety that the user has to deal with!!!!! Variety and variability are two of the beasts that make situations hard to deal with, and they are at the top of my list in many situations.

What are some of the unspoken questions that I have found helpful? Here are some, but remember that the specific questions will depend on what the problem situation is. You can get the answers to most of the following questions without asking the question:

- What does the person do when they start the day (or task)?
- What does the person do during and after the task?
- As they go about their business, what is their current information system (formal and informal): what is the information they use, where does it come from, what form is the basic data in, do they interpret and transform the information in any way, what do they do with the information, and what do they do or what happens with the results of their actions?
- How does the person deal with ambiguity, uncertainty, missing data, inaccurate data, data that is not trusted, … ?

- When do they do the task? How often is the task done? What tasks typically precede each task? What tasks follow? Does the task change with time?
- How do they talk about the tasks? Do they use meta-terms? Do they use semantic groupings, or use specific nouns and verbs to express the situation, the issues, the task, and the outputs?
- What are the sources of variability? External variability? Internal variability? Self-inflicted variability? How large is the variability?

This process can take hours, days, or months to complete. You do not need to have all answered before starting to ask other questions, but the more comfortable you are with this knowledge base, the better your other questions will be. Some of the questions will have answers built up over time as you work in the general domain, and you can transfer the knowledge and answers quickly. Most of the questions you ask should not be rhetorical or poised strictly for your own benefit of hearing your own voice answer the question. However, there are times that a rhetorical question can be useful when doing field work. For example, you might use a rhetorical question and present an answer so that the person you are talking to can assess your knowledge and your understanding — where you present a hypothesis or thought and ask the hearer if your conclusion or observation makes sense.

Be very careful about assumptions when working with a user:

- Do not assume that you know more than the people you are working with.
- Do not assume that you know the right way or best way.
- Do not assume text books and courses have answers to real world problems, and that you can preach a solution you learned in a classroom (sometimes you can, sometimes you cannot).

Of course, do not assume the reverse either. You might indeed know more but you should not show it like a hot dog; you might indeed know the best or right way but you should not shout it from the highest roof top; there might be something useful in text books and courses but you should not preach like an academic.

You also need to be careful about making your decisions and rendering a judgement. Do not judge how people are doing their task until you know what you are talking about. In any real situation, there are many interdependencies and many subtleties. You want to understand what is actually being done, what can be done, and what should be done. You need to know this so that you can encourage the good and discourage the bad. You need to know what to strengthen, weaken, or eliminate.

To verify that I have sufficient knowledge of the domain and that I am ready to do detailed problem solving, I use two simple techniques. First, as I identify patterns of cause and effect, I will start speculating to the user about what happens next. What they might do or what the situation suggests. This is testing my knowledge and helps create credibility with the direct user. Second, I will try to give a tour or description to someone who does not know the situation with the *local expert* listening in. In this tour or explanation, I will try to demonstrate my deeper knowledge — this is blah-blah, but this is also related to blah-blah, and if blah happens here, blah-blah often happens over there.

When I do a project, I like to do the initial assessment in a reasonably short time — to scope the situation and to establish the work plan and strategy. Depending on the size of the project, the initial scoping phase could be something like six weeks. As part of the initial report, there is a problem statement that describes my interpretation of the situation, any assumptions, risks, and limitations. This description is also a test and I expect the users and clients to comment upon the fidelity of the story and the description. It is the first major checkpoint for knowing what I know (the *what*, and the quality of the *what*).

If the user says that they start work at 5:30 AM, I am there at 5:15 AM. I have found that many workers start a little early to clear the decks and to deal with the little annoyances left over from the day before or that might happen overnight. Knowing about these little annoyances is as important as other information. They also think that they are doing you a favor by cleaning things up before you arrive. In fact, you want to see all of the good, the bad, and the ugly. If the workers are in jeans and t-shirts, so am I. If I need work boots, I have them. If you do the usual *suit* thing and turn up at 9 AM in such a situation, you miss all of the good stuff and have no real clue as to what the problem really is. You think you may know, but it is an illusion and possibly a delusion.

As noted above, you have to establish credibility in the domain, in the process of doing analysis, and in understanding the user's reality. You also need to establish trust. I do not use tape recorders, and I try to avoid taking notes as someone is talking. Note taking and recording makes some people very nervous and can affect what they share with you. It can take weeks and sometimes months to establish the trust credential. Why is this so important? Another good question. In my research on planning and scheduling, I have documented at least four schedules that a planner or scheduler uses. If you do not have the trust, you only see one of them and you will think that you are playing with a full deck of cards when in fact you are not. Your observations, recommendations, and solutions will be accordingly flawed.

For example, in every factory situation I can think of that is not 100% automated, multiple schedules are at play. There is the published, what I call the *political* schedule or plan. This is what is posted on a wall, reported to and by management, and broadcast wide and far. This schedule does not always represent what the scheduler knows or what the scheduler is really doing. It might simply reflect what the scheduler thinks the management will want to hear and accept. The scheduler is likely talking to key parts of the factory floor asking them to do this, or do that, or try to do this a little bit faster or better. This manual influencing and manipulation actually results in a second plan or schedule that is not too easy to see or to capture. This is the *communicated* plan. However, this is not what the scheduler really thinks is going to happen. They are asking for 80, expecting 70, asking for the work to be done in 2 hours, expecting 2.5 hours. The expected versus communicated schedule creates a third schedule in the scheduler's head. What they actually think will happen, the *expected* plan. The fourth schedule (and not perhaps the last) is very interesting. If you ask the scheduler to get the job done, without compromising cost and quality, but ignoring seemingly arbitrary policies and procedures, what would the schedule look like? This one actually demonstrates the scheduler's ability to sequence and get the job done. This fourth schedule appears at magical moments when the bosses remove constraints and tell the scheduler to just get the parts out the door and to the customer — just get it done. This is the *unconstrained* plan. It takes time and effort to understand enough about the domain to realize that there might be different plans at play. It takes trust and credibility with the users to get access to all of these plans, especially the unconstrained one. This is an example of using the ethnographic process. I will re-visit the idea of four plans later in the project management sections. The idea also applies to software development.

Often the scheduler has to follow many written and unwritten rules about the work flow and these possibly hidden rules constrain the decision making. If you looked simply at the political schedule, you would not easily see these constraints and guiding policies. It is only through establishing trust and observing the detailed decision making that the factors become visible. These rules and issues are rarely (if ever) supported by software or captured during normal requirements analysis and it is easy for researchers and developers to criticize the scheduler's skill based on the political schedule. The scheduler in turn, criticizes the system generated schedule as not being feasible or making sense.

Using the ethnographic methods, you are trying to build up a process map that captures the tasks. This process map or task understanding is crucial to the *ZenTai* design process. *ZenTai* is extremely task focused and it is important to know:

- How tasks relate to each other.
- Which ones are repetitive.
- What aspects of the task vary.
- What a task relies upon.
- What are possible sources of error.
- What skill or knowledge is needed to do the task.
- What information or functions are needed to do each part of the task.

It is also important to recognize and understand the comfort, experience, and evolution dimensions of the task as it might relate to information technology. You need to be able to separate out arbitrary artifacts that might exist because of historical precedents from necessary and real requirements. There is another old saying that comes to mind: if you cannot do it right manually, do not try to automate it. You also need to know where you are and what the current situation is before picking a future target and introducing change. If you do not know where you are, how do you know where to go? How will you know if you are making progress? How will you know if you have actually met any objectives? How will you know when you get there? Doing a good problem analysis helps you with these questions.

Doing an ethnographic study is an excellent way to sort this out. Unfortunately, such methods are not often used. I took approximately a dozen courses in Psychology and Sociology as part of my undergrad and I also studied cognitive skill acquisition at the graduate level. To put it bluntly, it is hard to encounter serious social sciences training in the world of geeks. You have to go to where the experts are: the sociology and psychology departments. An ethnographic study is more than simply sitting with the user. It is much more. This style of analysis also requires patience and support from the people paying the bills. It costs more up front and there is an initial time period during which nothing appears to happen. The clients and users want the system yesterday and too often the assumption is made that everything is already known about the problem and the solution is to just start coding. It is my own experience that this initial effort is well worth the cost and the results turn up later, ultimately delivering systems earlier than expected, at a lower cost, and with happy users. On average, I like to spend about 1/3 of the total expected time thinking about the problem and sorting out the key requirements. This time can be reduced with experience in similar domains, but time is still needed to understand the subtle issues. The other 2/3 of the time can be relatively structured, or can be *Mushed*.

The questioning and understanding can be assisted by what is called abstract modeling. You need to know what to ask questions about. This leads to the topic of what is a good analysis. Good analyses have certain characteristics. They look at the problem in a reasonably balanced way. They do not forget or downplay significant portions of the problem space while obsessed with some other portions. They look at the problem in terms of complementary units of analysis. For example, not worrying about the knobs on the stove being chrome or black while the question of having a stove or not is also being discussed and not resolved yet. A good analysis is the best possible analysis given the constraints mandated by the situation. For example, if you only have an hour, you do the best possible thinking in an hour. If you have several days, it is the best possible analysis that can be done in several days. As a consultant or problem solver, you will always have time and resource constraints. This is where abstract models help.

4.4   Listening Is Reading

"Go talk to the user, get the requirements!" What is so hard about that? You go ask some questions and record the answers. You gather some physical artifacts. Voila — the requirements. No problem. Not only that, your first requirements will be the final ones too. Consider what was written, way back in 1956 by Kozmetsky and Kircher:

"After an application has been placed on a computer, it usually becomes evident that the machine can do more than that particular function." P. 116

"Whenever anyone prepares a program for a computer, there is always the danger that some important data are being omitted." P. 117

If you ask me, fifty years of research and thinking about user requirements have not changed these two basic observations. While you cannot be perfect, you can at least take some precautions and do a better job through the skill of listening.

Earlier in this chapter I stressed the importance of good questions, questioning sessions, and understanding the answers. It turns out that listening is very key to understanding the answers you are given. If you cannot listen (I mean **L**isten with a capital **L**), then there is no point asking the question in the first place.

When observing and soliciting requirements, you are essentially reading the situation, and the people and situation are telling you a story, perhaps a saga, hopefully non-fiction, but perhaps containing bits of science fiction. To read a situation implies a process whereby a reader reads something. A process of input recognition and analysis. As Roger Schank (1991) in *Tell Me A Story* notes, a listener places what is heard in the context of existing facts, stories, and it is not possible to understand something that is totally new. This by implication relates to the notion of expertise again. An expert analyst will have a very broad and deep storehouse of experiences, stories, facts, analogies, and relationships that will help quickly understand what the client or user is talking about. A good analyst will ensure that they have done appropriate homework before engaging the client, else many key input bits will simply go in one ear and out the other.

The input processed by an analyst can be words uttered in voice or in printed text, or presented in other aural or visual forms. Accidental processing of input may lead to the occasional success story, but it will not stand you well over time. This section is about the conscious processing of information and the conscious process of reading.

If you were to ask me how to improve your analytical skills, one of my answers would be for you to improve your reading skills. Easy to say, but what does this mean? For many years I have described to others how I listen and understand a situation. "I pay attention to this," "I look for ..." etc. and I never had a nice framework or structure that explained what I actually did; how I read the situation. Recently, I had a rare moment of enlightenment and I think that the framework that revealed itself to me makes a lot of sense. It is after all, all about 'sense'. The enlightenment moment happened during a workshop I was attending at Stratford's Shakespearean Theatre, conducted by Ian Watson who is a text coach with the theatre company. The workshop was on how to read and interpret scripts (especially Shakespeare) ultimately leading to a voice rendering that does justice to the script and allows the audience to understand the sense of the prose or verse. What does learning to read Shakespeare have to do with software requirements analysis? The answer to this question is simple and brief — **LOTS**.

An author of verse or prose is telling you a story, trying to get your attention, and engage you. They are bringing you into their world and hoping for the suspension of belief. The words on a page are passive till the moment of interpretation and reading. Pitch, tone, pause, emphasis, urgency, meter, and other tools are used to tell the story. There are transitions, qualifiers, modifiers, antitheses, rhetoric, metaphors, and all kinds of grammatical contrivances used to create emotions and visual images for the reader (or listener). When you are working with a user in situ, all of these are in play. They are what will convey the sense of the piece, the intent and meaning. Not all of the bits are relevant or pertinent to the current situation and over time, you can develop a skill that allows you to focus on pertinent parts of the story and ignore others.

For example, I once had to design a simulation for an electronics firm. The simulation was of their surface mounted technology line area. They were interested in flows, flow times, and where queues would build up. In talking with the engineers, they would go on and on about the brand of device, types of technology, and many low level details about glue, heat, placement tolerance, and many more aspects. To me, most of this information was noise. I was listening for anything that played with flow. Do the machines process one board at a time, two or more boards in an linked-indexed fashion, multiple boards in a conveyor fashion with fixed spacing, conveyor like without spacing, how boards were placed into the line, how they were removed, was it possible for any board to leave from a point within a line, could a board enter a line in the middle of the line, and other related aspects of flow control. I did not stop the engineers from telling me the details I did not want to hear, but I was certainly not remembering them or cluttering my brain with them. My ears perked up whenever a flow issue was mentioned. I needed the engineers to tell me their story, in their way. You can do this when you have enough time and you are actually doing a longer term study. If you have to do an analysis in an hour or two, that is another problem. I was listening to them, but I was listening to them in a very conscious and very focused way. In a process situation, I was interpreting and listening for issues related to quality, quantity, speed and variety. These four factors have been noted as being very important for choosing technology and designing a manufacturing capability for a long time (Babbage 1832).

Think about it, when you listen to someone or are observing someone, are you just listening to them, or are you consciously parsing, collating, categorizing, interpreting, and thinking about what they are saying, and how they are saying it? This type of conscious awareness is the key for a good analysis.

You look, but do you see? You listen, but do you hear? You read, but do you understand? This goes back to the Zen again. First the water and the mountain, then consciously not the water and mountain, and then ultimately you see the water and mountain again (but with a different, holistic understanding). The cognitive skill called *analysis* takes the same path. There are at least two levels of seeing, hearing, and thinking. You hear people talking about someone developing an eye for something, or they have an ear for something. These statements recognize the existence and importance of the higher levels.

When you see a play or watch a performance, you can get caught up in the moment and experience the event without conscious analysis. This is after all, what the actors and director hope for. As an analyst, you cannot do this. You need to listen and watch at two levels, having the two channels open. The person is communicating with you and while the conversation may contain facts, the form may be literally storytelling and not figuratively (see Roger Schank's 1991 work for a good discussion of how people communicate through stories and narratives). You get the sense and feeling of the story as it is being told and you need to analyze the components of the storytelling itself. The focus has to be on the *what*, *how*, and *why* of the storytelling. If possible you want to consciously pick up and trap the adjectives, adverbs, the transitions, the contradictions, any qualifiers or modifiers, things emphasized and things prioritized, and you focus on the principle thesis or thread (if you can find it). The principle thesis is like the *ZenTai* — the core essence of the structure. There will be little journeys as the story unfolds — journeys that will add color, qualifiers and modifiers to the thesis. The qualifiers and modifiers might have differing scope, priority, and granularity. Related qualifiers and modifiers might be in a list, might be sorted by priority, or might appear randomly along the journey. You need to capture them and weigh them. You need to capture qualifiers that restrict via inclusion or exclusion. For example, inclusion implies that you choose one from the following; and exclusion is to choose anything but the following.

There will be noise. Lots of noise and possible misdirection and it is sometimes deliberate (like the fools in Shakespeare's plays). People will try to bias your analysis and point you to solutions that are really addressing symptoms, or to solutions that will be myopically optimal for the person but not so great for the overall situation. People who have not learned to read (or listen) at two levels have problems with noise. They do not like noise and are often solving the wrong problem or answering the wrong question. When I write word problems for tests and exams, I play with the words. I want to see how well the students can read and consciously understand what is being asked. Too many text book questions ask you the direct question and do not contain noise or all of the static you will see in the real world. People do not present you problems in a nice, case study fashion, or in a cute little paragraph. In a real situation, you have to consciously find the problem and the bits of the problem which are relevant. On tests, I often had nice answers to questions I did not ask. I once had a mature student in a course. When he handed in his first midterm, I noticed that he had written RTFQ and ATFQ on every page. When I asked him what it meant, he said that he had learned from previous studies that he often jumped to an answer before really understanding the question and this was his mechanism to slow down and think first. **R** and **A** stood for Read and Answer. **Q** for Question. **T** for The. You can guess at the **F**. Yes, this student was not politically correct, but he was honest and blunt. He also did quite well in the course.

There are a lot of abstract and philosophical *thoughts* in the previous paragraphs. Sorry. There are many things to attend to in a concurrent fashion when you are doing analysis and if you miss the pieces, you do not get the full sense of the story. The ability to be sensitive to the various bits is developed over time. You can have a checklist, but it might take explicit practice to better your skill. Every day, you can pick up an article, publication, script, book, anything. Go through the words identifying the sense or key message. Identify the adverbs and adjectives. Note the priorities or how the words narrow the scope or enlarge it. Look for words like should, would, could, and must. May, can, and will are also important words. If you can tape or repeatedly play something, listen for emphasis, trigger words, sequences of words. All of these are key clues to understanding the problem. You can also pick up a text or book written for actors and reverse engineer the instructions and concepts. For a good source look for Patsy Rodenburg's books (e.g., Rodenburg (2002)).

Clarity may elude you. Not all stories are clear. The teller might use analogies, metaphors, or extreme points. The story teller might be beating on a relatively recent example which is not representative. Humans are also notoriously bad at estimating probabilities and frequencies. You should pay attention and be careful when asking for frequencies and probabilities, and when possible, seek secondary sources and supporting evidence. The stories might include gossip, fiction, and what I have sometimes called amplification and attenuation. Amplification is when the storyteller adds information that was not originally part of the information flow. Attenuation is when they leave something out.

When the qualifiers and modifiers are not relevant, perhaps you can guide the storyteller back to the thesis. If the tangents are not lengthy, just let the storyteller ramble. The rambling itself is information and can provide insights. You can develop the ability to pick up the scent of the main thesis and quickly separate the relevant from the irrelevant.

Believe it or not, there are also books on how to read critically. Check them out. One of the famous books is *How to Read a Book* by Adler (1940). Below is my view of reading and as part of that, I will use the ideas of active reading and comprehension. There are several levels of reading and analysis and each has its role to play.

— First, the most superficial reading or thinking is like reading math equations, many text book descriptions, phone book pages, stock tickers, most weather reports, etc. You read it, you might memorize it, you might use the information, but you really do not think too much about the information and it does not create any emotional or personal angst or issue. Neither does it generate any intellectual a-haa! or intellectual reaction.

— Second, you might read something (or hear or see something) and it creates a response. You like it, you hate it, you find it amusing, you find it interesting, your heart rate increases, etc. But, you read on. You really do not think too much about what has just happened. It is a gut reaction. "I do not like it." "I like it." "I think it is stupid." This is typical of reading pulp fiction or light novels for easy reading while on vacation. It might even be the way that you read text books and course material (not recommended though).

— Third, you read or experience something that creates an emotional or intellectual response, but you start reflecting and enquire as to why it registered such a response. Why did you find it funny? Why did you find it annoying? This is the start of a higher level of comprehension, thinking, and cognitive skill. You are aware of the response and start to think about causal relationships. That made me laugh, why? This level can be superficial or a bit deeper. For example, I know I laughed because the author used a funny example. I know it is funny and that is why I laughed. That is a superficial level of analysis. Why do you find it funny? Is it a cultural bias? Is it a slapstick example of humor with a face plant into a wall? If you laugh at something, why?

— Fourth, you register the response, are aware of what might have made you respond, and you also think about the intent of the author. Was the response planned? Was it a conscious and planned structure that created the response? The author used clever puns or clever wording or certain circumstances, or the author accidentally created a situation that is funny when other information is known.

— Fifth, if it was a conscious effort by the author, why? Why did the author want us to feel or react this way at this point in time, and the reaction caused by this type of mechanism? Why did the author choose this particular method to illicit this specific response? Why this response? Why not a different method? Why not a different response? Why any response? Why? The response might be emotional or it might be an intellectual point of enlightenment. This is the deeper a-haa! point.

Thinking back on how you have been reading this book, have you been reading at level one or two? Three or four? Five? Level five is what I would call critical reading.

If you are reading for a purpose, analysis, or learning and you wish to be an efficient and effective learner or analyst you will need to develop the skills to quickly assess what you read and be aware of what you read. If a reaction, intellectual or emotional, occurs, what triggered it? What was the a-haa!, eureka!, or wow! trigger? Why did such a trigger result in such a reaction? Was the trigger conscious and intentional, or was it accidental? If it was conscious and intentional, why? How are you reading?

How many times do you read something when trying to understand it? Do you make one pass through the material? Two passes? If more than one pass, do you change your reading process for each pass? To read something critically and really understand it takes several passes, focusing on different aspects with each read.

Chapter 5
Modeling

5.1 Abstract Modeling — the Art of Seeing

Abstract modeling is where you use a simplification of the problem to help you think about the problem. There are many types of abstract models and surprise, there is no single abstract model or abstract modeling method that does everything. A simple model helps you create transferable tools or techniques from one similar situation to another. A simple model helps you describe and explain the core, essence, or Zen of the problem to others and keep focused on the Zen. A simple model frees your mind of unnecessary and irrelevant information so that you can do a balanced analysis in the time allowed. How do you know when a model is simple enough? What does a simple model look like?

A simple model is like the classic elevator speech. What you can talk about in a very short ride between several floors. It is also like a description you would give to a four-or-five-year-old. Another analogy is what I call the blunt stick and sand diagram. Imagine yourself on a beach trying to create an image for this four-or-five-year-old and all you have is a blunt stick to draw in the sand with. This forces you to leave out fine details and go for the big elements that matter. This is the level to strive for. Unfortunately, simple models are actually hard to create and the consistent creation of good, simple models is itself a cognitive skill. Note, a simple model is not always visual. It can be a list of attributes and characteristics that help to classify and organize the problem space. These models are often called frameworks or taxonomies.

The term analyze means to "tear apart" or "to loosen." When you are analyzing, you are loosening or decomposing a big messy thing into understandable bits. You are creating little groupings and giving shape and form to the amoeba. Classifications, taxonomies, and semantic groupings are human creations and as such, there is no pure or right classification or modeling scheme. We make these things up for convenience. Do not cling onto any schema, or develop a delusion that your taxonomy or model is the one and only possible.

When people start a cognitive skill, it is common for them to think in what is called concrete terms. Beginners also think in specifics. For example, if you are thinking about a problem about a whale eating people, beginners often think about the specific details of the image or picture before them. They will discuss the whale's teeth (if toothed), size of the stomach, if they like a specific type of human or not, if they want to eat humans, and so forth. If the problem is to figure out how many students a whale could eat, the terms and examples would be very whale specific. The first question of course is "How do you go about figuring out how to estimate the number of students a whale could eat?" That is, the first question (remember questioning?) is to address how to question or derive the answer. It cannot be accidental and it must be a conscious step in your thought process.

In this sentence about *how to go about figuring out*, there is the domain of whales eating students. The first qualifying tidbit is the "how to estimate," but this is further qualified in the question by "figuring it out," which in turn is specifically qualified by the "how do you go about." Think back, did you consciously read and parse that sentence? Or did you just read it? I never recorded the percentages, but my memory suggests that many students did not read the sentence when they got it (or a similar one) on a test. They quickly got what they thought the intent was, perhaps from the last phrase, without catching the qualifying phrases and gave a delightful answer to a different question. They never really analyzed the sentence and understood the problem. On a test and in the real world, you do not get too much credit for supplying a good answer to a different question.

The point of this type of analysis is the following. For the general whale problem, you cannot apriori state how many students a whale can eat, but you should be able to describe how you would go about estimating the number of students any given whale could eat. Can you list the factors or issues that would assist you to effectively and efficiently think this through?

When you look at a whale, what do you see? Do you see a whale? Remember the Zen Buddhist concept of seeing? When you look at a whale and can consciously think through that what you are really seeing are things like a funnel leading into a holding tank with a drainage pipe (e.g., the open part of the funnel is the mouth, the narrow end of the funnel is the throat, the tank is the stomach, and the drainage pipe is the … you know what that is), you are at the second level of seeing. When you can look at the whale and immediately see the whale and the model at the same time without effort, you are at the expert or master level. It takes time and effort to learn how to create and craft good abstract models. People who can do this reliably do the big picture stuff. Not everyone can craft simple models, just like everyone cannot play music, bridge, or do complex mathematics.

Even when looking at the same problem, different models will come to mind. The model will depend on what you are focusing on! There might be a core model capturing the essence, but even this could be different. Once you have a core or Zen model of the problem, you can quickly **work** the model. I use the term *work* to describe the process of going around the model, looking at each element in the model, and asking the same question (or theme) of each element or filling in the same info for each element. A good model will have the same granularity throughout. For example, all strategic elements, or all tactical, or all operational. The level must be consistent and if you are thinking about having a stove or not, the model does not include the number of elements or the color of the stove. The question is about having a stove or not having a stove.

Here is a brief test. Do it before going to the next page. Think about doing an analysis that includes a tree (a real tree, not a software tree). Now, draw a picture, as accurate as you can, of the tree as it might be part of a study or analysis. If you do not want to actually draw it at least think about what you would draw. Then turn the page.

Hopefully, you actually tried the exercise, but as an author, I am not holding my breath. Did your image look like the one on the left, or the one on the right?

Did you forget the root structure? A tree is not just what is above ground, it is the whole tree — the Zen. A tree is not a tree without its root structure. I think that this is a good example of explaining what you have to see when you do an analysis. You need to understand that a tree has a root structure, and that often the root structure is as large as (if not larger than) what you see above ground. I would like to take credit for thinking of using a tree for this illustration, but I recently heard an interview on CBC radio with an environmentally oriented professor (I did not catch his name or affiliation) and he mentioned that he did this exercise with his students. Not for the same purpose of course, but to illustrate that most people forget to think about what they cannot see, especially with trees. He noted that most of his environmental studies students would start the tree at the bottom of the page like the left image. What did you do?

Let's try another seeing and modeling exercise. Consider a Zen model for restaurants that prepare pizza. What do you see when you **SEE** a pizza making process? It does not really matter what type of restaurant you are talking about. There are some basics involved with pizza making. The Zen model captures ***all that makes it what it is*** — no more, no less. Take something away and it is no longer a pizza making process. If you can take it away and it is still a pizza making process, it is not part of the Zen. A process model for pizza making would probably have:

- Some form of trigger mechanism for knowing that a pizza needs to be made,
- A spur or linking process for the dough since it must be made in advance and let rise,
- A process for prepping and readying ingredients for the top of the pizza,
- The process of putting the toppings on the pizza,
- Some form of heat application,
- Some processing after the heating is done,
- Delivery to the target of the trigger.

These are the seven basic steps:

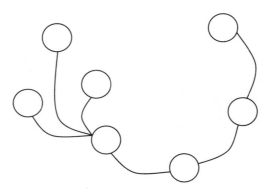

With such a model, you can ask questions such as: what are the quality objectives for each step, what is the volume going through each step, how fast must each step be, and how many types of things must be considered at each step. With the answers to these questions at each step, you can start identifying your actual processing requirements. You can also look for alternatives. Do you make or buy at each level and thus outsource one of the key parts of the pizza making? What are the inventory or space requirements for each part of the process? How can each part or step be made consistent or standardized (if that is an objective)? What are the possible mistakes or errors at each stage? With such a model and a question to be answered and a given time, you can ensure that you spend some time on each key part of the problem. For example, if I have ten minutes to think about quality in the pizza situation, I will probably spend one minute on each of the seven key pieces, then think about what are the most important parts of the process and then spend the remaining time on those. I would be able to give a holistic answer to the question without forgetting the biggies.

How long does it take to create a simple model? That depends. It depends on how much you know about the general topic area and how much experience you have. If you are a specialist in factories or process flows, the pizza model comes to you in literally seconds or less than a minute.

You have learned to recognize the key parts. If you do not know the domain and you must educate yourself, it can take months. In at least two cases I can think of, it took me about six months of research and analysis before I was able to create a simple model of the problem. Other models have been almost instantaneous to see.

Someone once asked me at a conference about what my area of specialization was — my answer was "professional simpleton." I cannot handle complexity and it gives me headaches, so I need to work with simplicity. As an analyst, this liability can be an asset as I often work with people in brainstorming or rent-a-brain sessions who can deal with complexity, but who have difficulty making complex things simple. This is also captured in a very useful Japanese sentence I have used to describe myself:

私は馬鹿です。(watashi wa baka desu.)

Or, when translated *I am a simple minded individual.*

In the next three subsections, I will briefly explain finite state automata diagrams, process maps, and Ishikawa's fishbone diagrams. These are three techniques that help create abstract representations. After I briefly explain each one, I will explain ethnographic methods a bit more.

5.2 Finite State Automata

Finite state automata (FSA), Turing Machines, etc. are interesting ways to think about processes. A Turing Machine is a FSA. So is a computer at the instruction level. You can use FSA to help organize the problem domain as part of the analysis process itself, and as part of the communication vehicle.

In a FSA, you have a set of states ($S_i \ldots S_j$) which can represent active processing states that indicate start-stop processes (i.e., things with durations and journeys), and you have a set of triggers ($t_i \ldots t_j$) that form transitions from state to state. Triggers can be external where someone sends a message, or internal where some internal process is completed and the process or software can now be considered as entering another state. There are two ways to view states and triggers actually: is it an action or a state of being? Think of nouns and verbs. You can use a FSA diagram with nouns as the nodes (or states) or nouns on the arcs between the states (the triggers) or with verbs as arcs or nodes accordingly. The general rule of thumb is to be consistent as you do not want to mix apples and oranges in a FSA diagram or model. Either all verbs and actions are nodes or they are all arcs. If you get them mixed up, and mixed together, you will need much help to fix the resulting mess. No doubt that there are purists that will mandate verbs on the nodes etc., but the key point is to separate the concepts of states and transitions.

A good FSA model will have the same granularity throughout and will not bother with minor states. If you have to, a node can always be drilled into and another FSA created for the next layer, and then again for the next layer, and again and again. A process flow diagram can also be considered a form of a FSA. You can take a rigorous or closed process flow and write a software program that will play with it like a FSA. A program designed as a FSA has a little engine or supervisory bit of code that knows about states and knows about triggers and knows the transition rules.

A FSA can take a bowl of cooked noodles and straighten it out for discussion purposes and visualization. A number of years ago, I had to analyze the Canadian search and rescue process and design a simulation program for it. If you perform the analysis without any modeling aids, it just gets uglier and uglier. By using the FSA approach, it was possible to capture and describe the search and rescue process in a systematic and clear fashion. States in the search and rescue became things like "resource in transit to search zone," triggers became "need fuel," "night time," "reached zone." The "need fuel," "night time," and "reached zone" forced the resource into a different state and the resource was no longer transiting to the search zone. The FSA diagram fitted on a page and it was easy to walk through the process with the stakeholders and validate my understanding. It is important for the clients to believe that you might have a clue about what you have been hired to do, or what you are about to do. It took about six months to understand search and rescue to the level where I could get a feel for the problem and when the model started to reveal itself to me. It was a busy and intense six months (lots of reading and research), but the final diagram was very useful. It clarified the problem and then I designed the simulation model to match.

Several years ago I had a Master student (David Tse) looking at the use (i.e., the value) of an IT system which was part of a hospital situation. He created a nice finite state model of the medical process (independent of the IT), used the model to describe the information model (i.e., he did an information audit using the model), and then matched the existing IT solution to the model/audit identifying where the IT system could be extended, where information flows were linked (manual or computerized), and how the quality or value of the system could be measured. The hospital was very pleased with the research, but the staff seemed to be more pleased to have the medical process laid down in a nice, clear fashion as per the finite state model. The hospital staff did not know it was a finite state model, but they could easily follow and understand the process and for them, it was the first time they had seen their process diagrammed. The systematic process was anchored by the process model and the dialogues surrounding the diagram ensured that we were reasonably close to understanding the problem. I talk a little bit more about David's work in section 8.8 and you can find his map there.

Exercise: take something simple and try to describe it as a finite state machine or automata. For example, what is the process by which you try to arrive at your first class of the day on time, or try to get to work on time or a meeting (notice I say try, not will or do).

5.3 Process Mapping

Process Mapping is not to be confused with Process Models. Mapping may lead to a model, but there is not a one-to-one situation here.

Mapping out processes has been around for a long time. At least a few hundred years in terms of systematic and purposeful forms. Organized factories started appearing in the 1700s and the early industrialists (such as Boulton and Watt — the famous Birmingham Soho Factory) were quite aware of flowing work through a series of processes, as was the Venetian Armory in the 1400s for making war galleys. A process map can be used to capture an existing situation, or it can be used as a design tool for helping describe and understand a new situation. Whenever I can convince a client, I like to take the time to create a process map of the existing situation. It is very useful to know what you are doing before thinking about where you should be going. A process map can capture the current world and help you understand your issues or challenges when you need to separate symptoms from core causes.

A good map can also be used to show in stages how a situation will evolve (in a controlled fashion) from A to B. Too often, people set a target without knowing what they are doing now and have no clue as to how they can get from here to there (or what is involved). I have sat in many meetings where senior management promise to improve something by 25% within so many months. In reality, they do not have a good metric or handle for the current situation, and they have no support for the 25% claim, nor would they know if they were actually static, improving or degrading during the effort. A process map can help you determine what can be counted or measured (and where), and it can highlight critical processes which are in need of analysis. I personally hate the phrase re-engineering because it implies that the situation was actually engineered or thought through in the first place and usually it was not.

So, what does a process map look like? It can take various forms and there is not a single, ideal, or golden form (or process) for process mapping. You will have to adapt and play with the form and notation to fit the situation. In the following subsections I will explain my own approach to process mapping and we will slowly build up a map to this final representation. The toy example captures the following simple process:

- Someone needs a brochure or pamphlet or manual.
- If not on the desk, get it from the author.
- If the author does not have one handy, ask for a copy to be sent from stores.
- If stores do not have it, retrieve the document from the computer and print it.

In the simplest instance, this process can be captured in the following diagram:

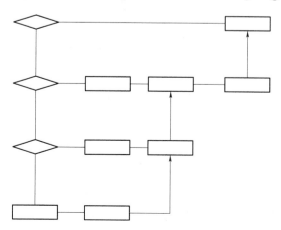

The top layer being the requester, second being the author (or owner), third is the storeroom and the bottom layer is the computer or system that would print the document.

This is a very simple process map. It has the major decision points and it could describe the key aspects: tracking what happens, to whom, for whom, by whom, how often, when, and under what conditions. The simple process map is something that should allow you to explain the process to a five-year-old! It is the reference point and it is the starting foundation for creating other value-added bits.

Consider the real situation of the document flow. There can be quality checks, opportunity for error, queues of work piling up, delays encountered, forms to be used to request documents from stores, and so forth. Some tasks might require certain skills or a deeper knowledge and some sub tasks might be needed, but not actually helping with the end request. All of these could be captured on the diagram:

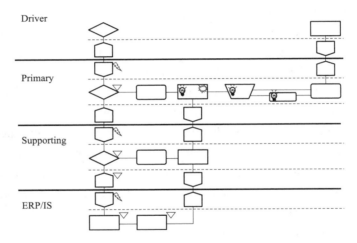

This is a fully populated map with a lot of the stuff included. It has visual clues for what is visible or invisible between layers (the dotted segments), the interfaces, and the queuing of work, etc. A diagram can have some or all of the visual clues.

In this diagram, you will see levels (driver, primary, supporting, and ERP/IS), a bunch of shapes, additional horizontal lines, and some icons. The person wanting the document is on the top level. If the document is not available, they will seek one from the author at the primary level. If the author has one available, proceed along the horizontal path, to prepare the package to send to the requestor. If the author does not have a copy, they have to get one from the supporting layer in the general stores. If there is not a copy in general stores, the IS system (perhaps an Enterprise Resource Planning — ERP — system) is needed to create the document. There are clear contact points and interaction points between the requestor(s) and the supplier(s). Some tasks require more skill and knowledge than others (the light bulbs). There can be requests or work queued up in a larger situation where there can be more than one requestor (the little triangles). Requests and interfaces have two bits: what one layer does and what the other layer does. This can be captured via different visuals or representations (interfaces bits are the boxes with up or down pointed ends).

Some of the boxes indicate value added tasks versus non-value added tasks (e.g., direct or indirect activities needed to get the larger request satisfied). Other boxes can indicate quality checks or where possible errors can be made. As this simple example illustrates, a process map can be used to capture many aspects of a process.

All parts of the map have a specific purpose. The specific levels will depend on the process being mapped. This example corresponds to a business situation. The idea of levels is generic. The number and meaning is not. The concept that there are interactions between levels is generic. What is on the interaction and what the interaction means is not. And, we will eventually get to what the shapes and icons mean. They are also generic in concept. Their possible presence and significance is not generic. In a real situation, the process map will have multiple layers, some aggregate, some detailed and can be quite extensive. For example, a process map was recently created by one of my team for the employment process involved in the University of Waterloo co-operative education process. This was an interactive document of over a hundred pages created with Visio and supplemented by Word documents; it took about four months to create. Doing a serious and quality job on a process map is a lot of work, but worth it for critical or important situations.

This leads us to the first question about process mapping. Why are you doing it in the first place? There are quite a few reasons that you, as a possible IT analyst, might be asked (or want) to do a process mapping exercise. For example, you might be looking at:

- Quality issues, related to error sources, rework, and corrective action.
  - Where, when, why, how many, what kinds.
  - Where are errors detected, how.
  - What is done about the error.
  - Cumulative, progressive effects.
  - % right, first time through, errors per.
  - Out of spec — time, data, physical attributes.
  - Value chain losses.
- Efficiency issues; looking for and understanding sources of delay, resource usage.
  - Unnecessary delays (queues, # in queue).
  - Unnecessary material usage.
  - Unnecessary resource usage.
  - Unnecessary waste of all forms.
  - Unnecessary steps, loops, entry, lookups.
  - Flow time, system cost per transaction.
- Wanting to mistake proof the process; avoiding and detecting mistakes.
  - Two ways to look at mistakes and errors.
  - "We pay them enough, should be right" vs. "Mistakes will occur."
  - Poka Yoke concept — take head out of sand.
  - Identify sources, potential for errors.
  - What can be done to avoid the problem.

- o What can be done to detect the problem.
- Information auditing; understanding sources, usage, flows, and outputs.
  - o Who touches it — read, write.
  - o What is it used in — reports, synthesis.
  - o Relationships between data.
  - o Necessary, unused, duplicate data.
  - o Ambiguous, erroneous, redundant data.
  - o Rogue systems, workarounds, pen & paper.
- Studying bottlenecks; identifying gating processes, how to avoid starvation or blocking.
  - o Stationary or moving — the gating task.
  - o Minute gained/loss — system wide impact.
  - o Time saved elsewhere — no system impact.
  - o Work always in front — not starving.
  - o Nothing downstream blocking work flow.
  - o Everything ready for it to do work.
- Reviewing boundary points between systems or organizations; focusing on the interfaces and interactions between system components (human or computer).
  - o Interactions and dependencies between organizational entities, systems.
  - o Potential sources of conflicts, delays, errors.
  - o Solicited vs. unsolicited interactions.
  - o Multiple "in" points.
  - o Impact of schedules, workloads, special interests.
- Separating myth from reality; symptoms from core causes, what can be quantified, where metrics can be gathered.
  - o Basic and fundamental — black holes.
  - o Succession planning — key tasks rich in knowledge and creative problem solving.
  - o Contrary views of tasks, flows.
  - o May take as a theme another purpose — to provide focus.
- Re-Engineering (ugh); documenting the starting and end points, migration path.
  - o Dewey noting an old and familiar saying: a problem well put is a problem half solved (Dewey 1938).
  - o Re-engineering requires known starting point and a target situation.
  - o Progressive, evolutionary process descriptions provide guidance.
  - o Why re-engineering — should tie to agenda.

Unfortunately, one or more of these purposes might be at play and they might be conflicting or ambiguous. There might be hidden or political agendas driving the effort, and in some cases there is no one at the appropriate level or position who will own the effort or will drive it. You might be lucky and have sufficient time, resources, and funds to do the process mapping adequately, or you might be told to do the mapping on a shoe string budget.

There are three questions you should ask before starting a process mapping exercise. First, what is the purpose and what are the expectations. Why you are doing the exercise will color the tasks and process you will use. Second, you need to know the scope of the activity and understand the breadth and depth. Third, you need to define or understand the unit of analysis, the granularity of detail and what will actually be represented on the map.

All three are linked together. The purpose is linked to the scope and is also linked to the unit of analysis. This will affect who the key sources will be; the questions and data gathered; the types of things you will find documented, discussed; and the resources, effort and expected elapsed time. The scope will affect the number of people and systems involved, and the effort and style used. It will also affect the skills and knowledge needed to do the study. The unit of analysis will affect the amount and type of data; type of questions that can be answered; what can be compared, extrapolated; and what conclusions can be drawn.

It is best to start with the driver level. For the middle of the sandwich, the driver or top level represents customer or driven requirements. This is what the customer or client sees and interacts with. The *why* question is very good to use at this level. The lower levels must think of what they can do to help the customer or client level with inputs or process. The driver level can also be reviewed as to how it can be better done or structured to make the lower levels better if that can result in a better experience for the top level.

The primary level is what you are really analyzing. This should be your basic value chain or the critical path as perceived by the client. On this level, things like task or process ownership are important and there are many functions to consider. This is also the front-line role for the department or area and uses the lower supporting areas as required.

The supporting layers are not seen by the user, client, customer, or requesting processes. What they are and how they operate should be invisible and totally transparent. However, they are important to map and understand as they can drive lags and problems upward through the layers.

Within each layer, you can further decompose the processes (if necessary). You have two interaction layers which are the visible contact points for the layers above and below. Between those two layers, you have the invisible tasks and processes required to deliver the product or task objective. The interactions can be bi-or uni-directional.

The simple example above uses a number of conventions for the activity shapes. The shape conventions are:

Direct	▭
Indirect	▭
Decision Point	◇
Audit/Quality	▽
Input	▭
Output	▭

The symbols or icons used to annotate the map are:

Critical	
Unsolicited	
Synthesis (deep knowledge)	
Queue Possible	

When doing a process map, you need to think about processes and the types of processes. You clue into processes that transform inputs into outputs, take one thing and produce multiple outputs, take multiple inputs from various sources and create one output thingy. You pay attention to accumulation processes (active or passive) that seek and obtain information (and if the pieces stay together or flow freely again later). You also pay attention to dissemination processes (again active or passive) that distribute information. You look for processes that must be triggered by an active solicitation process (I am ready for work; give me something) or are fired in an unsolicited way (here is something to do). There are processes that synthesize or create information from an internal algorithm or process, and there are auditing or checking processes. The point is that a process has a purpose and does something, and that this something matters when you are doing a process map.

Activities also imply resources (e.g., people or machines), materials (e.g., the paper), secondary materials or processes such as oiling the machine, and tertiary or third level activities such as cleaning the area after work or regular maintenance. Errors and mistakes associated with activities are also fun. They can pop up immediately or later, and can have wild side effects. Bottom line: what you focus on will depend on the main purpose of the process mapping exercise. Always keep that purpose in front of you.

There are some simple checklists to use when creating a process map. The first set of things to ask, observe, or probe are:

- How do you do it?
- How are you supposed to do it?
- How would you like to do it?
- How would you decide on what path to do or choose next?
- How are things initiated, terminated?
- Are tasks done fully? Partially? Handed off for partial work and then returned?
- What tasks are done before? After? — by the person and for/on the function!
- How can things break, stop, or fail?
- Are there checks, feedback loops, queries? Work re-entry?

Then, you can consider:

- Is that the only way to do it?
- What is rarely done?
- What happens when things are bad, short, long, late, early, missing?

- How is it done if you have to rush it or expedite it?

These questions will help you identify the flows, the norms and exceptions. However, there are usually various ways to do most things and you need to watch out for these too. You watch for the formal, somewhat formal, quick and dirty, and the get out of my way processes. There might also be different ways for each and multiple alternatives. Messy? Yep. And, you might find that different people are doing the same things differently.

You can also look for the smoking gun or possible indicators that not all is quite right. For example, do people (or systems) keep going back and forth getting data or having data checked? Are multiple or different sources accessed for the same, common tasks? How often are things reversed or forced to re-enter? These are all problems associated with process flow.

As you are analyzing an activity or process, you should create descriptive notes that capture the activity process. You can collect and organize various bits of data, some of which will be generic and some will be specific to the process. Some of the data will be qualitative or quantitative. For example, you can collect:

- What can be measured, when, how.
- Expected values (i.e., standards), realized averages, std. dev.
- Elapsed times, flow times.
- Skill levels required.
- Frequency, volume patterns.
- Special knowledge needed.
- Risk factors — causes and detection.
- Impact of risks, costs, recovery, avoidance.
- Boundary transitions — links, references to other tasks and processes.
- Policies, constraints acting upon activity.
- Who, how are constraints relaxed and problems solved.
- How are policy and constraint relaxations tracked and monitored.

Finally, what can you expect in a world class process? Perhaps it is more of what you do not expect to find. You do not expect to find multiple contact points and split accountability for a single request or function. You expect a single point of contact for a particular type of service and that this person is responsible and accountable for its completion. You do not expect to find multiple copies of data entry or re-entry. You expect one-and-done. You do not expect to find assumptions like "I pay you enough, so do it right." You expect to find mistake proofing for avoiding or detection. You do not expect to have to do things twice as the process should reliably process requests or functions once and handled error free. You do not expect to see partial, incomplete, or ambiguous results or data because you expect to see everything where and when it needs to be without hunting and error correction. You do not expect to see work piled up or resources starved or blocked. You expect work to smoothly flow and for bottlenecks to be fed. You do not expect to find everyone doing their own thing for the same task. You expect consistency and people to follow one best-practice for each function.

World class processes are a thing of beauty when they are both efficient and effective as viewed by the customer. World class operators basically ask one question:

How do our policies, practices, procedures, and systems affect the ability of our customers or drivers to achieve their own goals?

5.4   Ishikawa's Fishbone Diagrams

Ishikawa created a number of methods and concepts for improving processes and concepts. He is best known in the Western World for the fishbone diagram. When it is described, some people might think that it is simply a decision tree turned on its side. But it is different. Here is one I have used in a manufacturing setting:

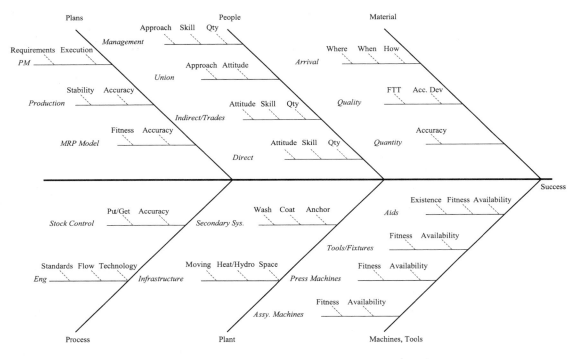

The target or focus of the fishbone in this example is success in a factory. That is the main driving force and is the main spine (this is one of the things making it different than a decision tree). It is a diagram of relationships. The next sized bones are the next largest or most critical factors (categories or groups) that contribute to success. These are the necessary, but perhaps not sufficient, factors. The factors in this example include the who, what, where, and how. Depending on the problem, the items in the fishbone diagram will be different, as will the number of minor bones and their focus.

Building the fishbone can be done as a group decision process, and this is a very useful exercise as the people think through what contributes to what. Such a tool can be used for a number of decision making tasks. For example, in one software situation, I have used the fishbone to help a software organization think through the creation of a project plan when they were in a crisis situation. They could not afford wasted time and motion or a mistake and the fishbone helped identify what should be priorities, what needed to be done in advance, etc.

I used this specific factory fishbone as part of a senior management decision process involving a continuous improvement activity. The terms on the chart were explained to make sure that everyone was using the same vocabulary. For example:

Acc. Dev

- Acceptable deviations or specs that need updating to match product.

Accuracy

- Closeness to desired or expected number, ability to have executed match plan.

Approach

- Management methods or techniques to personnel direction, development, problem resolution.

Attitude

- Work ethic, team or shared fate philosophy.

The diagram also was explained:

The diagram is a hierarchical approach to organize topics and ideas as they relate to the factory's success. The six highest topics or categories are: plans, people, material, process, plant and machines/tools. For example, if you stand back and ignore most of the detail, you need material to work on, people to do the working; the people use machines and tools; the plant provides the environment to work in; the process describes how the material proceeds through the plant and the whole thing is orchestrated with plans. The next level of the hierarchy breaks down each of the six into their sub-topics (e.g., people into management, union, indirect/direct, direct). The lowest level on the fishbone identifies characteristics or facets of the sub-topic that link up to success. Each of these smaller bones can be expanded into a relationship diagram of their own.

The management team was then asked to use the chart as follows:

a) Pass I — ignoring your functional role and thinking about raw to customer, identify what you consider to be the top five areas for improvement — place numbers on the chart 1 for highest, 5 for lowest — do not put name on chart.

b) Pass II — again ignoring your functional role, identify what you consider to be the top five areas to address in the immediate future that can indeed be addressed and resolved — place numbers on the second version of the chart — 1 for easiest and quickest thing to change, 5 for the more difficult or longer to fix — do not put name on chart.

I also asked them to identify what they were currently doing and what they were thinking of doing. The analysis was very revealing and demonstrated bias, selective execution, and a very unbalanced approach to the problem. They had not used a holistic view of the problem. They had not used a holistic view of a team solution or a shared fate approach. The organization was filled with little empires and silos. The analysis demonstrated the result of such a process. This illustrates one of the benefits of any visual model or diagram. You can hang stuff on it and discuss the stuff. You can put numbers, durations, priorities, people's names; all kinds of stuff. Visual models are great tools for group decision making.

I could have given a software example as well, but in this instance, I want you to think about how this diagram could be used in software craftsmanship: when understanding the problem, when organizing requirements, when creating plans.

Chapter 6
Field Analysis

6.1 Ethnographic Methods

What is meant by ethnography or ethnographic methods? A simple primer can be found at http://www.aiga.org/. This is the professional association of designers. The primer does not go into what social science training is needed or what a trained eye means, but the primer is a good 50,000 meter overview. Ethnographic methods are part of the larger Anthropology field when you get down to it. It is a style and set of methods for scientifically understanding situations, specifically cultures. Why is this appropriate for *ZenTai Mushing*?

Software is usually deployed in a cultural setting — the workplace, the home, and so forth. It is deployed in the cultural setting for a purpose. Understanding the setting and what the software is supposed to satisfy is important to ensure that the software will actually satisfy its intent. The ethnographic approach looks at the whole and learns about the context of the situation. For software this means many things:

- The formal and informal information flows.
- The vocabulary of the workers.
- The semantic bundles or items used by the workers.
- The timing, sequence, frequency, and interdependency of tasks.
- The sources of errors, uncertainty, ambiguity, and incompleteness.
- The formal and informal rules, policies, or constraints that govern the tasks.

And so forth. All of these things influence the design of the software; internal structure and external user interface. This is the icky stuff. The irrational, the unpredictable, the warm and fuzzies. Many geeks get into software because they do not do ickiness well and like the preciseness of mathematics and software. Too bad, if you want to build software for humans and have the software valued by humans, you will have the ickiness factor to deal with.

You learn many things by using ethnographic methods and using this approach has probably been the most influential and important tool in my career in terms of creating software people value and want to use.

A key to using the ethnographic method in s/w development is to view the users as having cognitive skills and that they are likely using the s/w to solve a problem (or to avoid one). There are of course simple computer applications which are simple data entry tasks, but many are far richer than that and require substantial knowledge and expertise on the part of the user. This is also what you are trying to capture if you want to create a kick-butt system. You are not only doing user requirements, you are understanding the user. You need to recognize that the user might use the software constantly, periodically, or in stressful situations. This is what the trained eye sees. You do not just see the user typing or entering data, you see the context and situation as well. Rarely do software systems and user tasks exist in isolation and you need to understand all of the possible contact points. Draw a circle around the subject area and identify all of the people and systems that might touch the system during its life cycle, from cradle to grave. The cultural situation is not static and it is not deterministic. It is organic and works on oxygen, blood, and nourishment. View the system as a whole and you will start to view software systems as organic entities.

Software developers are usually quite skilled in programming and have been taught how to make entity-relationship diagrams, use UML, etc., but how many have studied (more than just a lecture or two) survey design, field methods (e.g., Gummesson 1988, Easterby-Smith et al. 1991), interview methods, skill and expertise, field data analysis, and ethnographic techniques? Depending on the project, you might have to do population studies and involve these methods, including statistical analysis of surveys and collected field data. Big problems usually require big analyses.

Three topics often need to be addressed when trying to figure out the user's problem. You need to use the appropriate field methods, including the possible use of social science statistics, and account for the implications of skill and expertise. For more general problems, the first two topics will possibly suffice. The real world is not a perfect situation though, and when humans are involved, you might also need training in skill and expertise.

As noted above, there are two courses or sets of subject material which are recommended for any field based analysis. The first is field study methods including ethnographic methods (e.g., Schensul et al. 1999, Grills 1998, Spradley 1979). This area of research provides a solid foundation for how to observe and how to understand a field situation. In addition to such basic knowledge, additional material on interviewing and survey methods is recommended (e.g., Lavrakas 1993, Stewart and Cash 2003). Courses and material can often be found in sociology departments. While undergraduate material or courses might be offered, a graduate level course is recommended for the serious professional because of the emphasis on research and rigor. The focus on ethnographic methods is also very strong in many sociology and anthropology departments. In the ethnographic approach, the concentration is on learning from the subjects and understanding their culture and situation, not on studying them with a previously generated theory per se. Observer bias is dealt with in the methods and that is important in studying people in the field. Even though my graduate students are in the Faculty of Engineering, I send most of them to a graduate course in Sociology on ethnographic methods.

A researcher in an ethnographic setting will likely be a participant with a role to play. Either the role will be in the Action Science sense of introducing a causal effect or will be passive. For example, a researcher learning what independent spreadsheets are used and creating an integrated

when teaching. Every student is capable of getting an A in my course until they prove otherwise. Same rule as a consultant and analyst. If you do this, you will avoid making the maximum number of enemies and minimize the damage.

- Give everyone a chance to demonstrate their knowledge and their abilities. That also works to keep the flak down.
- Never confuse education with actual ability and knowledge. Just because someone spends time at an institution of higher learning does not mean that the person has a clue about solving a problem in the real world, or has developed a deep and broad understanding about the problem and solution space. You do learn things at school, but it is possible to learn things on the job as well. Someone doing a job for a couple of decades might just have something to share with you that you do not know. If they do not like you, they will not share!
- Make sure that you go in prepared and that you understand the basics. Part of the hostility may come from the person's frustration at having to teach you the fundamentals while they are under pressure to do the real job. For example, if you are going into a factory for a planning and scheduling gig, understand the products, customers, and basic processes before you open the door.
- Keep quiet for as long as you can — no one likes a one-hour expert who knows the solution before knowing the problem.
- Try to focus on how you can make each person's task or job better for them — more effective and efficient — thus giving them some value from your exercise; they will want to know what is in it for themselves. If you can give something to each person or part of the system, the hostility will be less than otherwise.

There are other rules of course. Many. But, you will never, ever make everyone happy to see you. And while you should operate under these rules you cannot be naïve about the possible limits of the person's expertise and credibility (see previous section). You need to be observing and learning about the users on several levels.

Because you can assume a hostile environment, you need to cross-check data, and have multiple sources for key data. You want to get the three sides to every story — his, hers, and the middle truth. The truth might require finding people who are at arm's length to the situation and who do not have any alliance or bias to help you interpret the stories. People will not trust you and it will take time to turn a hostile person into a friendly one (if at all). People will only give you the real story and the real data if they trust you. Otherwise, they might only share the official stuff or the stuff presented on a chart. As an analyst, you want the good stuff, the data that is really raw.

While there are not easy ways to deal with all hostility, there are ways to deal with issues caused by credibility gaps. A credibility gap can result in people not wanting to waste their time listening to you or answering your questions. You can do your home work and read the financial reports, study the industry, understand the common mechanics, and know their basic business before you start. You can pay particular attention to the ya-but's. This is when you see something or comment on something and your client or user says "yes but…" Understand the second and third level relationships and occasionally when the time is right, talk about them. Prepare and present a summary of the situation to the local experts so that you can show them that you really do understand their problem. This can be done formally in a meeting, or by giving someone a tour of the situation with the expert tagging along, listening. But, never start off a meeting early in an engagement (especially the first meeting), within minutes of starting the meeting, with "this is what you should do" and "what you have been doing is wrong and implicitly stupid." You ask questions, put together the problems. Eventually, describe to them what you think you understand, but do it in a non-threatening fashion without claiming that you know everything, they know nothing, and that you are now the expert and are right with divine wisdom. For example, present a summary at the end of a meeting, and ask them to rate your understanding. Ask them if you understand enough, do not TELL them that you understand enough.

If people you have to get questions and answers from start work at 5:30 AM, you arrive at 5:15 (or 5:00) AM and be cheerful about it. If they have to work the occasional weekend, you also turn up on the weekend to observe and talk to them. If during the analysis process you can do little things to help them, do it. Remove your ego and give everyone else credit, not yourself. Remove the "I," "my," "mine," etc. from your vocabulary and make it/them and we/us. Allow your contacts the chance to demonstrate their knowledge and their skills.

Another technique is the visit with coffee cup in hand. Early in my career I spent a few years in Software Quality Assurance, and there was occasional hostility between QA and the engineering group. No problem if you did simple testing and gave a pass or fail. But lots of hostility if you dared question or comment upon design, development process, or actual code quality. You could not go to a developer and say this is bad, or this is wrong, or why are you doing that? Well, you could have, but the result would not have been very friendly. My approach was to rely on questions again (remember them?) while visiting the targeted individual over a coffee. You develop the skill to ask a bunch of dumb questions that gently guide the target towards self realization and awareness both in terms of problem and solution. You lead them to the door. You let them take credit for the discovery. You let them think you are as smart as a tree stump and you are successful when this happens. To do this really requires you to bury your ego and your need to get praise and recognition. You have to remember that this is not about you, but it is about the project and what your job is. You have to remove the personal at-a-boy/girl from the equation. The people who need to know will know and the others do not matter. If you can do this during the design and solution discovery phase, you have high impact. Be gracious and generous with true praise and let others get the halo!

Meetings can also be hostile events or at least hostile if you are trying to have an organized meeting with clear outcomes and directions. How many meetings do you go to that are well organized and people come prepared? They are rare. You can control this situation somewhat. In this case, you can prepare a small note or working paper on the concept in advance. At the meeting, you let the meeting develop while you keep your mouth closed (for the most part). Eventually, after things have deteriorated to the suitable level, you casually note that you did a little bit of work on the idea, that the work might not be relevant or right, that the idea might be hare-brained, but here is the idea. You have one or more copies with you and you distribute and then give up ownership. The idea will likely be beat up a bit, but usually, about 80% of what you started with will survive. The people who did not come prepared will be quite happy since they do not have to think from scratch on something. It is always easier to work from an idea than to create it. The key is giving up ownership, not grandstanding, and giving the idea to the group. Let others beat it up (do not overly defend or take it personally). You can orchestrate the play, but keep it low key. If you end up with the idea at the end as a task, fine. If someone else takes it and runs with it, fine. Remember, it is not about you. The right people in the room will be taking note of the fact that you come prepared, that you have ideas, that you know how to contribute without ego, etc. Again, it does not matter what the other people know, the right people will know without you telling them!

Another technique is similar to circling the wagons in an old Hollywood cowboy movie. Before you do something that might upset others, you can try to bring those people (or their bosses) on-side first. You can identify the key decision makers and approach them privately about the ideas or concepts you are pursuing before you raise the point in a meeting. In most cases, you probably do not know all of the facts, interdependencies, and subtleties associated with the situation and you can honestly and sincerely ask for their help and advice in understanding the problem and in crafting a feasible and reasonable solution. If you do this with honesty and sincerity, while giving credit where credit is due (sometimes in a subtle or indirect way), there will be less hostility. It is likely that the people you have asked for help will recognize that you did indeed listen to them, valued what they said, that some of their advice is in the proposal or concept, and that they can have pride and ownership in your proposal. They are part of the solution and will possibly even actively support you in the meeting. If you get enough of the wagons circled in advance, it will be easy to deal with any threats and issues.

There are many ways to create hostility. It is really easy and can be done without much practice, skill, or thought. For example, one way to create hostility is to surprise people. Never surprise your boss (especially your boss's boss). Aside: *you should never create a situation where your boss (or boss's boss) has to apologize on your behalf for something you did or did not do. This will not be a good thing and might create hostility of another sort.* You should also always let people know that you will be arriving, what the agenda is, what and who will see the data, etc. If the surprise or issue might be associated with a policy or guideline that exists within the organization you are working with, work with the boss and the corresponding boss to first discuss the policies and issues before raising them in a public forum. If someone is surprised in a public forum, they will get defensive. If they have been approached privately, are on-side, and are part of the solution, they are not likely to be defensive. Some people who like open field warfare do not like this divide and conquer, or backroom style. They would prefer to just have ineffective meetings where the quick witted and sharp tongued debaters dominate — superficial discussion and resolutions without depth or analysis.

Being defensive and taking it personal is another way to create hostility. It is better to turn the other cheek, let the other person have the last word, and do not get into a "my boss is bigger than your boss" discussion or anything similar. Not good. If you bump into a policy or issue, you grin and bear it. You note it, but you do not try to argue the point if it is not in your scope of authority. You quietly deal with it via the proper protocol. Similar point: understand the proper protocols and try to deal with them in the proper order. Always assume that the first problem might be a simple communication problem between two people. NEVER ever fire off an email to someone's boss or boss's boss complaining or ripping about something that started at a lower level without first dealing with it one-on-one in a professional fashion. Assume that the communication fault is yours and that you did something to cause the other person's misinterpretation and misunderstanding. It might not turn out this way, but make this the first assumption and review your email, your body language, your tone, your points, and think about how someone else might have interpreted them. You have to take responsibility for what you supervise and if someone beneath you made a mistake, you have to accept the blame without sidestepping and pointing fingers. There is nothing to gain from pointing fingers and lots to lose. Your maturity will be questioned and your reputation will be damaged. It is also easy to create hostility if you quickly and widely point out faults in others or make accusations only to be very quiet when it turns out to be your mistake and you apologize (if you do) in a private email. If you are going to spread blame widely, also accept your own failures in the same fashion and let everyone know that you made a mistake, made a wrong accusation, and were an active contributor to the problem.

Good architects and designers do not get emotionally involved with their creations and they need to step back and deal with them rationally and in a detached way. If not, you get into the NIH (not invented here) syndrome. As they say, hold your own counsel and keep quiet and think about how you are reacting. How you act and react will create an equal or larger reaction! Telling secrets and gossiping is yet another way to create hostilities. No one likes a blabber mouth. They might like you to your face, but they will not trust you and they will not think highly of you. They will wonder if you are talking about them when you talk to others.

Boxing people into a corner is another bad thing to do. Always allow people to say no without losing face and possibly to say no without really saying no. I once heard that the Japanese have at least thirteen ways to say no and based on my travels to Japan, I would have to agree. If you box someone in during an analysis, they might feel threatened, forced to participate, forced to contribute, and generally will not be your friend. In Japanese business discussions, you always give outs and options, and a way for them to say no without saying no.

If you create a hostile relationship through disruptive behavior, lack of trust, lack of respect, gossiping, talking out of turn, or boxing people into situations they do not want to be in, you have a problem. They might turn into passive aggressive players, or even maliciously obedient. The latter is when they will, on purpose, do exactly what they are told to do. So far so good except that they know that what they are doing will not work and that a bigger mess will result and if they are lucky, the bigger mess will end up stuck on you. They did exactly what you asked them to do. They know that a bigger problem will result, but it is not their problem and you got exactly what you wanted.

Now, back to that credibility gap and ideas for how to close them. At the Italian car maker, there was a credibility gap and a small hostility issue. The plant manager who supposedly could speak English quite well, just talked in Italian to my host and ignored me. I thought I was there to help him and he did not seem to care at all. It turned out that he was annoyed with the vendor bringing in tourists (so to speak) and he was not fond of this particular engineer (her being a female and this being Italy in 1990). After leaving us on a catwalk overlooking the machining line and returning a short time later, I just started talking to him in English (I had verified with my host that the manager did indeed speak excellent English). At this point, he started giving me what I would call the political pre-recorded tape of how the line worked and how everything was so good. The parts went here, this happened, then the part moved, and then this happened, and there was not a blemish or issue to be spoken about. Everything was perfect!

I had enough general background in the industry and technology to know better, and I also had some inside knowledge from the vendor. I figured if he wanted to play, so could I. So, I started pointing out how amazing his story was and how this had to be written up in journals and publicized because blah was not happening, blah-blah was not happening, blah-blah was not happening, etc. As I kept exposing deeper and deeper understanding and possible value, his eyes started to brighten up (like the buttons on an elevator) and he started talking about his challenges. I had partially crossed the credibility gap by going head to head with him. Playing with him as he played with me. This can be dangerous though and you better know what you are talking about. You have to use the right tone of voice and be very respectful while the listener reads between the lines and realizes that you realized what they were doing and that you also expected them to realize this. It is a game with many non-verbal messages. After I pulled his chain, after he pulled mine, we had what I would call professional respect for each other — to a point. About an hour later as he and the vendor engineer were having a nice, heated discussion in a meeting room in Italian (which I do not speak), he turned to me and asked me a question in English. He said that out on the line, he had said he had four challenges, numero uno was blah, due was blah, tre was blah, what was quattro? The plant manager was testing me. I was able to tell him the fourth issue and then gave him a short analysis of the issue before giving him back the floor. After that, we were buddies. I had passed the second test for credibility and worthiness. This story would have turned out differently if I had not been listening and listening with a purpose! I would have looked like a fool which I would have been for not listening to him when he was telling me his stories!

6.3   The Quick "Drive-by" Analysis

In theory, you should always plan well and have the adequate time for planning, understanding the problem, and getting your act together before you open your mouth with recommendations and observations. In reality, you will never have enough time or enough resources or enough access to the people you need to talk to. This is where the various methods of abstract modeling and so forth help. In the extreme case, you might be asked to pull the proverbial rabbit out of the hat and do an analysis in an hour or two, or find the magic sword overnight to save the damsel at dawn.

If you are asked to do the rabbit trick, try to first say No! Provide suitable reasons for why an analysis might take longer and point out the risks associated with a quick study. I learned to say no early in my career and while it is true that some managers hated to hear the "I do not know." or "No, we just cannot do that.", or "I will have to check into that.", I am still alive and I am happier with the result. If people (the big people) insist on you proceeding, you will have to — of course. In this case, document the situation, the risks, issues, and weaknesses associated with the rush job, and then do your best job.

In some cases, it is indeed possible to do a quick analysis — something like a two hour visit (from start to finish). There are some pre-conditions. First, you better have true expert qualifications in the general field being sought (say flow or process analysis, s/w quality assurance, etc.). Second, you must have more specific knowledge and expertise in the situation's basic business, methods, processes. For example, issues facing any precision machine maker, or any compiler writer. Third, you better get a little bit of specific, context knowledge about the agenda, issues, purpose of analysis, and some history. Fourth, you might need more specifics about the business of the particular situation and how the situation evolves each day. If you have these skills and this knowledge, you can consider (with credibility), the quick drive-by analysis. Otherwise, do not try it. If you do, it is probably that you will be recognized for the fool you most likely are.

I will give you two examples of quick hit-and-run analyses. One from a system's viewpoint and one from a pure s/w situation.

First, the software example. At one point in my career, I was at a company that was hooking up its equipment to a large financial institution in San Francisco. The large institution was IBM based (big iron) and the company I was with, was another big player, but not IBM. This was a special, multi-year project with this large institution and it finally got down to brass tacks of getting the two systems integrated and talking. There were few IBM-wise developers at the company I was with, me being one, and I got sent to San Francisco at short notice (i.e., tomorrow) to sort out the problem of connection. The large institution's patience was exhausted and my employer was told to either hook up or shutdown the multi-million dollar project. I was not directly associated with this special project, but I knew the basic idea that they were trying to pull off. Our company had the hardware hook up figured out, while the trick was the software. The software package on the IBM mainframe was about one million lines of code if my memory is correct. Not tiny. Some Cobol, lots of assembly language. We met during the day and I had to provide a solution by morning. Where do you look? Where do you even start? What kind of solution do you look for? How do you quickly find a needle in a hay-stack? Not even being sure that a needle exists?

By this time, I had close to a decade of intense low-level IBM experience and awareness. I had seen enough IBM system code to know that usually the IBM developers put in a certain type of code deep in the bowels for testing and for re-direction. This would allow the main application to get input from various sources, not just the obvious, main source. I did not know if this existed in this application or not. Before leaving the partner's site, I borrowed a half dozen or so manuals and program listings, and headed for the hotel. Long story short, I found the expected stub after an hour of two of rooting about the logic, looking in specific types of modules, specific bits of logic, specific places in the documentation. This stub or bit of code I was looking for allowed us to hook our equipment into the main system via a side-door and pretend that we were another piece of IBM equipment handling the same type of data. I was lucky, but I was also equipped with many years of knowledge and understanding about how IBM systems were crafted and what I might expect to find. This type of knowledge is mandatory for such quick analyses. You need to have depth and breadth if you are going to do hotshot problem solving with tight timeframes. Luck was there, but so was a solid base that could be exploited in an opportunistic fashion. Without the base I would have needed a lot more luck and I am just not that lucky.

For the second example, I will share an experience I had about a decade ago. A senior mechanical engineering professor I knew at a USA university asked me to fly down, visit a factory with him and his graduate student, take a look at a situation and give an assessment. We would have about two hours, start to finish, at the factory. This would cover the first how-are-you's, to the nice-visiting-you goodbyes. A lunch might follow. Before I said yes, I needed to know if I could actually do it. It was a nice thing for the ego to be asked, but if I said yes, my reputation and the other professor's would be at risk. The reputation of two institutions would also be at risk. After asking a bunch of questions, I figured that I could indeed help out.

It was in a basic industry I had some knowledge about (discrete part manufacturing), and methods and processes I knew well (precision metal working). The problem was in my theoretical and practical specialty, planning and scheduling. The professor and the graduate student had been working with the factory for about a year analysing and re-designing a flexible manufacturing cell (with lots of automation and workstations) and they were able to give me more info about the cell and products while driving to the factory. They had come to the realization that issues beyond the straight cell design might be important if they were to achieve the efficiency and effectiveness targets, and they also realized that they did not have the expertise. I knew the professor from previous contacts at his institution and he remembered that I worked in such areas, on such topics. He asked for a favor.

Upon arrival at the factory, we had a short, perhaps 30–45 minutes' meeting with the senior management team. I then wanted to meet with the production planner, privately for about an hour, and then I would report my findings back to the management team. The first part of the visit with senior management was critical and I had a list of questions I wanted answers to. There was certain information that would tell me a lot about the operational environment in the factory. Things like how old was the factory, when was the last major upgrade done, how many engineering changes occurred each year, how many different types of parts flowed through the factory each year, and so on. About twenty questions. The questions were designed to quickly identify sources of uncertainty and sources of complexity. These are two topics that create lots of planning and scheduling challenges and directly impact efficiency and effectiveness. With a short time span the questioning session had to be tightly managed with no tangents. In fact, I had to be slightly rude and interrupt and shutdown folks when they did not answer my question (people often like to answer the question that is in their head, and not what you asked) or wandered. I had to say things like, "That is interesting, but I am really interested in ...". At the end of my questions, I indicated that I wanted to meet alone with the planner, without other factory people. Why without any other factory personnel?

I had to bond with the planner and get down to the good, bad, and ugly — my way. I had developed a theory about the situation and needed to test the idea out on the one person who would know if I was right or wrong: the planner. My gut feeling was that the factory and the other researchers had got so focused on this fancy machining cell, they forgot about the upstream processes, specifically the ones that fed the cell. If you do not have enough inventory of the right type to create the appropriate mix going into a multi-machine cell (where not everything is as it seems), you will have big-time performance challenges. My belief was also that there were sources of uncertainty and constraints in existence that most of the senior management (and researchers) were not aware of or dismissed.

Understandably, the planner was a tad (more than a tad) suspicious and really did not know why I was being ushered into his office. Who was this guy in a suit? Why is he being left alone with me? He really did not want me there. On my side, I had limited time, so after taking a deep breath, I had to first establish credibility and trust — fast. From what I can remember, the conversation went along the following lines. I commented on how interesting the place was and that probably on the way to his office every day, he looked at this and checked that out. He looked at me a bit strange and said yep. I then commented on the problems all of the engineering changes probably caused, especially on those machines that are supposed to be the same, but really weren't. That got his attention. I commented on how machines change over time, develop personalities and some parts like some machines and not others. I noted that probably some parts needed certain workers as well since the parts develop personalities and can be a bit cranky or quirky. On each of these points, he smiled more and more, and was nodding away. I asked him how the feeding processes were controlled and how the inventory bank got built up ahead of the cell. He expressed displeasure with that process and talked about all of the problems upstream with getting a good bank. We then talked about how hard it would be to ever have a good mix of parts for the cell unless that mix was driving or pulling upstream production instead of upstream production making what it wanted, when it wanted and pushing it ad hoc into the bank. There were lots of parts in the bank, and the problem was that the parts were not the right parts to make up a good mix. Bulls-eye.

We had a grand chat and talked about possible solutions and how he was under valued, not listened to, and likely under paid. His bosses did not understand his problems and did not know what he actually did each day. At the end of the discussion, I told him that I had to go back and tell his bosses my thoughts and I gave him veto power. If he thought I did not know enough to represent his situation, he could tell me and I would not say a thing. I was serious. I was not bluffing and he realized that. His response was music to my ears. He laughed and said that I knew his problem before I even entered the plant and gave me his blessing. I then proceeded to the management meeting where I gave my assessment (including the fact that I thought the planner was undervalued and under paid).

For this drive-by to work, I had to have the general and specific knowledge. I had to know enough to create my questions that would help identify problem characteristics. I had to manage and control the management meeting to make sure that the talks remained focused and that I would be prepared to meet the planner. I controlled the meeting with the planner and ensured that no bias was in the room (i.e., his boss was not invited to the party). I had to establish, quickly, that I understood his domain, his life, and the challenges he faced daily. This included subtle characteristics about machines and processes that would not be obvious on a tour or walk about. I had to demonstrate knowledge and understanding that would not be expected in a normal, suit-wearing, consultant. Once this was done, he trusted me with other information and we were able to have a good discussion as peers. I was able to test my theory and to reach closure. I then gave the main participant a veto power to stop my mouth from doing further damage to his life or situation. I could have really upset his world with loose lips and sloppy thinking. I would not have agreed to help this other professor unless I had some clue about how to pull off this type of borrow-a-brain analysis. It was high pressure and was high stress. In addition to the business success of the company, its stakeholders, customers, and employees, there were a lot of reputations at stake, as well as the working relationship between this firm and this professor.

There was some luck, but as the old saying goes, you make your luck. You think it through. You think about what you do not know. You think about what you need to know. Can you know what you need to know before you have to do the analysis? Can you enter the situation and control the situation to the point where you can get the other bits you need to make an assessment and test it? If you can answer these questions with a positive, then you can do a quick assessment. Else, once again you will be a fool and recognized as a fool if you try it.

Chapter 7

User Engagement

7.1   Stating the Obvious

It is somewhat embarrassing to have to write this chapter. Since the 1950s people have been writing about what is created with information technology and the failure to make systems and solutions that users value and want to use. Yes, yes, yes. There is the occasional great system, but on the whole, replication of successes has eluded IT professionals. Why does yet another chapter have to be written? Hasn't there been enough written about this topic? Apparently not.

There are of course many theories and ideas about the consistent failure of the IT profession to deliver the goods, and there is more than one valid reason (e.g., see Brooks 1995, Glass 1997, 2003). When talking to developers, one reason that keeps cropping up is *the user*. I have often heard developers blame the user. The user does not know what they want; the user has not read the manual or does not use the online help; the user does not pay attention to what he/she is doing; the user does not use the functions the developer provided; the user makes input errors; the user is computer illiterate; the user does not remember the sequence nor the training; the user does not remember where the function or data is; the user did not provide the right requirements; the users changed their minds; the user did not explain themselves adequately; the user does not ask for help; the user is lazy; the user is stupid, the user is old-school; the user does not understand how neat something is; the user, the user… Guess what. The user is not the problem in most cases. It is the developer.

It is the developer's job to interact with the user and understand the user. It is the developer's job to understand how requirements are gathered, and to ensure that the requirements are understood. It is not the user's job. This was the focus of the previous chapters in this book — how to work with the user to understand the problem. It is the developer's job to engage the user and to work with the user but not the other way around. And, it is more than just "going and talking to the user" and "asking a few questions." As the previous chapters highlighted, you need a method (or multiple thereof) and you need to do the requirements analysis very consciously and not in a sloppy or ad hoc fashion.

Almost all of the whining noted above comes from the developers not doing their job right. The developers did not do their homework adequately before meeting the user, did not engage the user sufficiently and adequately enough to get a handle on the requirements, and failed to develop a system that fitted the user and most importantly, failed to deliver a system that provided sufficient value to the user. In many cases the user is expected to fit the system, the system that was comfortable and easy for the programmer to develop. There is something wrong with this picture. Programmers and developers are essentially a service industry and while we should not ask if the user wants a larger order of fries, we should be asking how we can help them and how we can make their job easier, not ours. Developers should not be smug, condescending, elitist, and should not imply to the user that the developer thinks that the user is stupid, out of touch with the latest and greatest (and likely short lived) fads, and buzzwords.

Developers should never impose what "I like," or "I would do" decisions or ideas on the user. I have seen developers and so-called user interface or user experience experts dismiss and insult the user, talk down to them in a condescending way, and to impose their own personal likes and dislikes on the user. Worse still, I have seen people try to impose the latest fads and ideas on the user or system because the developer would be embarrassed if any of their inner-circle geek friends saw what was being delivered. It has to be cool and be like the latest, coolest interface or web-widget. Imagine, designing and making decisions based on what you think your friends will think instead of what the user wants or needs? I have seen this on more than one occasion and I am always surprised by such ego-driven thinking. It is about the user and the functionality the user needs and is not about what you think your friends will think. This type of thinking can appear with internal or external aspects of the system too. I have seen developers advocate using certain features and functions of a software library or tool because they are cool or show extreme geekdom, and then they can be added to the developer's resume. More notches on the sword or badges on the sleeve. Hopefully, a solid strategy for user engagement helps defend against these types of developers, but not always. Remember, it is about the user and not about your friends or your claims to technology superstardom.

As the development continues past the initial understanding, the user is a very big part of the process. As you think about solutions, designs, and perhaps the crafting of a prototype, the user has to be involved and that is the focus of this chapter. The ideas and suggestions in this chapter are applicable when figuring out the problem, sorting out the initial set of functionality requirements, and thinking through the technical or functional architecture that will satisfy the requirements. Many people use similar ideas during the actual coding and scrums, but you can use the same ideas throughout the system development!

7.2   Styles of Engagement

Remember: the engagement with the user does not stop at the requirement phase. The user must be engaged throughout the development and this is one of the many benefits associated with Agile style developments. Quick prototyping and examples of functionality can clarify ambiguity and identify overlooked subtleties. There are all of the warm and fuzzy benefits as well. The user becomes one of the team members and the quick iteration model helps with maintaining focus and interest. Since you are hustling and working hard to understand and respond, this can create synergy and a positive force within the user environment. Users want their expertise to be respected and it is important that they see some of their ideas in the final deliverable. It is not likely that all of their ideas will be present, but it is also not likely that none should be either. I have yet to meet a user that did not provide me some "Ah…" moments that then got translated into the system. However, I have met a few managers and others who work at arm's length from the actual operation and have failed to provide a single bit of useful input. There have also been managers who really contributed lots of good stuff. So, you never know who will give you the goodies, and you should initially assume that everyone is capable of providing insights.

The Agile philosophy is not restricted to the code phase and rapid delivery of functionality. You can also do the understanding and designing bits in an agile way. The next set of chapters probe the *ZenTai* concepts and it is just as important to engage the user during the *ZenTai* design phase as it is in the code crafting phase. There should always be some form of thinking before doing, else you are really wasting people's time and money. But, many people seem to bail early on the thinking side and use the Agile concept as an excuse. At least that is my own personal observation. Thinking is hard and it hurts. It is not fun. It is a lot more fun to start wailing on some poor code and showing ill-thought through functionality under the premise of this is how to figure out the functionality, claiming that this is how the Agile manifesto says how to do it (not). You need to think enough to get to a reasonable starting point and user engagement during the design helps you do that. There is nothing in the Agile texts and guidelines that says you have to be an idiot, that you should not think, and that you should use it as an excuse for sloppy development. When you take the time to engage the user and to understand the problem, it helps you understand the user's value chain, the most important comfort issues, areas where evolution will be important, and the potential and impact of experience. It helps you understand how the computer technology can help the user.

When you are building relatively small systems for a limited audience or specialized functions, it is usually easy to identify and engage the key or representative user. For example, when I build planning and scheduling systems for a factory, there are usually several users involved and that is it. You know their names, where they sit and you can dialog with them. Management might want certain fields on reports and certain features to be included, and there might be a number of these folk putting in their two cents worth. However, when it comes down to banging on keys, there are few users to really engage. This is the same in small businesses, or focused applications such as customer database development, charity donor systems, etc.

Unfortunately, not all systems are so localized or limited in use. Your deliverable might be used in many countries around the world and there is no single user that represents the rest. For example, years ago I worked on a system that had about 60% of the world market in several classes of machines. At that time (late 1970s, early 1980s) countries and regions of the world used the technology very differently and we received conflicting and contradictory sets of requirements from the various area representatives. Everyone wanted to do things differently on the machine. The domain and type of user was not that different, but the details were really different. To understand the problem, we did meet and talk with the various regions, but we also studied the previous generation system. We looked at how the different regions customized and extended the software. In some countries, they made over two hundred modifications!

Side point — a very important one: It was and still is very important for the design team (note, this does not mean every programmer) to talk to the users or someone very close to the user. If the user requirements are filtered by the field representative, then the corporate representative, then the plant representative, and then finally presented to the design team as requirements, there is a clear danger of having each layer filter and alter the real user voice. They might escalate, reduce, add, eliminate, or otherwise bend, fold, and mutilate the requirements. Not on purpose, but by accident. In the project now being used as an example, we had good client contact at the beginning. We (the functional architect and myself as technical architect) also had experience and good contacts with the field as the functional architect had been involved in both the design and support of several prior products and I had experience in supporting the previous generation through my Quality Assurance role. Things were going along quite well. Then, at one point it was decided by the big people in the organization that the developer groups should NOT go see key customers or have direct contact with the users. The user's voice would come through the appropriate corporate channels and be presented. We were being presented by information that we suspected had been severely interpreted, or even invented along the way and that we did not really trust — specifically one key telecommunications requirement.

We could not talk to the users, but we did have contacts in the field and we did our own phoning to a number of the key areas and accounts. What we heard about this one requirement was 100% different than what we were being told by the corporate representatives. We were forced to accept the official requirement as mandated by the product manager but I did write a memo to my immediate manager noting our own conclusions and disagreement with the requirement. Several years later after I had left the company, I was having lunch with this former manager. He mentioned the memo and noted that we were right and the corporate views had been wrong. This was not rocket science. All you had to do was to do your homework and think about what is happening in the field and what the field is telling you. You cannot sit back and wish requirements into being. The corporate people had not actually checked with the field on this one requirement; they thought they knew and they imposed their own beliefs and desires on the requirement. It was what they wanted to see happen in the field. You might like a certain telecommunication technology and think that this will be adopted by everyone, but you better check to see what technologies are actually being worked on and adopted in the field. As a designer or architect, you NEED contact with the users!!!!!

Back to the story. Based on this analysis we created a specialized language for the technology and then re-engaged the user representatives. We discussed with them their unique requirements and showed them how the language and system could be used to achieve their goals without actual programming and changing the system. Simple tables were also used as much as possible. There was a great deal of up-front analysis and design in this project and then the coding and development was done in rapid iterations with lots of testing of user scenarios and options. Since we had tested the design of the language against existing functionality and user requirements, the development team itself acted as the user in the iterative cycles. It was not possible to have hundreds of customers from around the world co-located at the factory, nor was it possible for the developers to be co-located with the customers. When the product was finally installed at a customer, the special language was used on-site or with close user involvement — in an agile way. A regional starting point can be used or the system can be tailored from scratch. This type of system development illustrates that you can do hybrid systems. Doing enough analysis to create the toolkit, doing iterative development on the toolkit using the previous analysis as a proxy, and then using the toolkit to do daily crafting with the user. The key is the toolkit approach and not trying to craft a final, fixed product based on the initial analysis. If the users understand this approach, the engagement will be more likely to be positive as they will realize that you intend to be flexible from the beginning and that the toolkit itself can probably be extended and played with. User engagement will not be positive if they feel that you are painting them into a corner that cannot be changed in the future.

> *Another side note: this project also illustrated the importance of doing your homework. We had 60% of the market with the existing products, but we knew the real effort and costs involved. It took a long time to install and it was a costly approach to support. Our two main competitors were also doing new platforms at the same time, trying to increase their market share, etc. As we were developing the new approach, the competition was copying the way we were and their new products looked like our old ones in terms of software options and tailorability. Needless to say, we were somewhat happy with that, knowing their future costs and difficulties.* ☺

Tight user engagement can also be difficult when the software system involves many types of stakeholders. In the above example, there was really one type of stakeholder in the beginning until the people in marketing figured out that the machines could really be used in many situations, but there was high variability within the one stakeholder group. This required a specific way to deal with the users and how to engage them.

It is also not always possible, but if you can discover who your stakeholders really are, outside of the actual IT product, this can help you understand the users' context: What they will find important, what their expectations might be, and other juicy items. You need to view the IT situation as holistic and this includes the users, where they have been, why they are users now, where they expect to go, and other personal drivers. You do not pry or be a nuisance, but if you listen and communicate with the user, you will discover who they are! You need to bond with your stakeholders (when possible).

Is every user a good user to bond with or to engage with? A politically correct answer is yes.

Of course. A user is a user. The real answer is no. Not all users are alike. It would be nice to pick your users according to your own criteria, but most real situations are mixed. You can have Joe, but Jane is also going to be on the user team, and we want Zoey on the team too although you only wanted four. We want eight users to be part of the team and you must have one user from each area, if not two. Except for these specific groups and you cannot have anyone from those areas. And, a number of managers also insist on being on the user team. Too many, too few, not the ones you specifically want, and ones you specifically do not want.

A good user for the Agile build process is someone who is open minded, not pedantic, not literal, not close minded, who has an imagination, who is not a control freak, who has patience, and is not ambitious (empire wise), and who is not driving their own agenda. They should have good reasoning and analysis skills, and be able to remove personalities and emotions from the discussion. They should know the problem space, understand the holistic issues, understand how other people do the same task, know where things come from, where things go, and have a sincere interest in doing things at the desired level of quality, efficiency, and effectiveness. Notice how the last part was phrased — the desired level — not ideal or optimal or perfect. The desired level must match the client's agenda, not the individual's. The paying party needs to set the objectives and goals and you need users who will fit that profile.

If you are lucky enough to have the appropriate user engagement, it can be a real joy. The number of users you need to engage depends on the type of software you are creating. It might be as few as one, or as many as six, or more. Be careful once you get above six though; the dynamics are hard to manage. Let us assume that you can pick the ideal users and get the perfect crew. How could you approach a larger team of users?

I would first make sure that an initial study was done to understand the requirements and that there were focus groups, structured interviews, job shadowing, process maps, surveys, etc. Almost all of the stuff in the first six chapters. I would make sure that there were semantic models of the user groups built and lists of assumptions created about each of the groups. That is, when we talk about an average user in one stakeholder group, what does this user look like and how do we expect that user to behave. In some cases, you might find a 80–20 split with 80% of the users being similar. Or, you can be unlucky and discover that a major user group is composed of six or so groups of 10%–15% representation each implying that you will not be able to do anything simple or common that will satisfy 80% of the users 80% of the time. Let us assume that you can get a reasonable split and that you have a good 70%–80% feeling about the requirements and that you are ready to start creating end user functionality. Note, if the system is very specialized, you might need a functional architect and a separate technical architect working together, and it cannot be assumed that a technical architect (more of a structural engineer) understands how to do functional design or analysis.

But, we are not ready to start engaging the user until we have the right tools at hand that allow us to engage the users in the appropriate fashion. You want tools that allow rapid code and function creation via rapid iterations of about two weeks per major bit, if not faster. If you have a good set of tools and a solid infrastructure, the development could then be done with daily builds engaging the stakeholders. In our ideal case for a substantial software project involving many stakeholder groups, I would try the following:

- There will be a dedicated task force in the main client group who services the other stakeholders with respect to the development project. The task force members will spend about four hours per week working with the team. The task force members could be asked with reasonable warning to look at something at any time and give feedback. The task force members would have sufficient expertise and respect from other task force members (and the stakeholder group at large) that an individual task force member's feedback would be accepted as being good enough and close enough for the first attempt.
- The task force would work in a loose fashion and not as a committee where almost all of the work is done together, with members of the task force trying to influence others or trying to get 100% agreement on everything. The task force is not a decision making entity per se; they can comment, provide feedback, and make suggestions, but the software cannot be 100% designed by the task force (unless you are very lucky about who is on the task force).
- There will be additional user teams from the other major stakeholder groups. They will be expected to spend one to two hours per week on the system development — testing scenarios, suggesting terminology, flows, etc.
- A number of staff from the main stakeholder or client would be full or half time members of the development team — adding further stakeholder feedback. They are also responsible for documenting the internal business rules and for helping plan the deployment of the system.
- A demo room would exist where the various stakeholders could be engaged — a common traffic spot. Any user walking by would be invited to the room to see the latest version and to discuss the interface and functionality with the developers. If the stakeholder groups are geographically dispersed, you might need a web-portal for the demo room and the user feedback. On-line discussion groups will also be useful for all levels of user engagement.

For a larger project, you will need multiple forms of user engagement. As the types of users increase and the geographic dispersion increases and the variability within a type of users increases, the forms of engagement become more and more challenging. Even with ideal users. One of the problems relates to the group decision making process. Some people think that the term consensus means 100% agreement. Check a dictionary. It means that most of the people involved or concerned agree, not 100%. When designing a large, complicated software system, you cannot make one person 100% happy all of the time, let alone 100% of the people. It will just not happen. As a designer or architect, I do not assume that I will personally like 100% of "my system." It is not my system to start with and I am not the user. We try to make things balanced, consistent, and work in most cases and it is not possible to be 100%, or cover 100% of the cases, or make 100% of the people happy. However, you will sit in many meetings where management or the team insists that 100% of the people in the room must agree on something for something to move ahead or be done. They will likely use the word consensus for this case; they are using the word in its most extreme interpretation. This is a classic recipe for paralysis, deadlock, gridlock, and inaction. Management will want everyone on side or having buy in. With a very small group of the right kind of users, this might be possible. It will not be possible when you get a committee composed of the less-than-ideal users. You will be trying to swim upstream and there is not even any water in the river. You are laying on the bottom of the dry riverbed with arms and legs moving but nothing is happening.

Many of these concerns also exist when you are building a simple system, perhaps with only one user. You have to think beyond the single user and this may imply that even the single user is not 100% happy 100% of the time with your decisions. You have to consider the bigger picture. What happens if the single user quits, retires, is sick, or is replaced? What happens if someone else is suddenly involved and also has to use the software? What happens if the task splits? The single user may also not know the management plans, and certain wishes and desires of the individual do not fit into the agenda. As a designer you are faced with all of these facets and it is possible that the single user is not aware, or might not care about any of them. They want their little tool, their way. If you give in or make the user happy, the user will feel engaged, will love you, and will give you hugs and kisses, but that does not mean that the client got value or that the system will deliver on all expectations. Just because the immediate user is happy does not mean that it is a good system. There is something called myopic optimization.

7.3   Other Sources of Insight — Support and QA

Believe it or not, there are other parts of the organization equally as important to delivering awesome functionality to the user as the geek squad hammering out the code. These are the other stakeholder connections to the organization. They deserve their own little section.

While most of the writing in this book is from the developer's viewpoint, there are other parts of organizations who might have useful insights into the user's mind and requirements. These are the direct user support folks, the people who install the solution, the people who do the training, and the people in Quality Assurance who might also do some form of support. In some companies, these folks may represent 50% or more of the company staff. This is a lot of eyes and ears!

Depending on how these groups are organized, the individual people will hear from many users about all aspects of the IT they are using. They will hear the good, the bad, and the very ugly. They will get a feeling for who likes what, who uses what, what is used, what is not used, and many other things. They will hear many of the *whys*, and the whines. They will hear real and imaginary issues and problems. Some responses will be from nice people, and some will be from people not so nice, rude even. In some cases the information will be in the form of complaints, but it might be also in the form of serious suggestions and advice. Some users, in fact many, have a clue and know what is bugging them and what is not working well for them. Who gets to hear all of this? The support personnel and in some cases, the Quality Assurance people. These people will hear it way before the developers will, unless the project is so small that the developer is doing all of the tasks. As soon as the project gets to a decent size and multiple projects are being done over a period of time, undoubtedly some form of support team must be deployed. It is unlikely that the original programmers and designers will do support for all projects they have worked on.

Hopefully in most organizations there will be mechanisms to gather and retain this information. It is invaluable stuff. Even if you have an issue with it, you can gain knowledge and insights from it. Always assume that the initial cause for any user angst is caused by you. That is the first step towards realization. It might be accidental angst, but it is still angst.

When the support team gives you input, do not dismiss it or blow it off. It got raised for a reason. You might agree with the input and welcome it with open arms. Or not. But all input is useful input. You can learn something from every piece of input.

Note, this information might not help a Class V or VI venture where there is no prior system. However, the existing support information may be very useful in the Class I through IV developments. During the deployment and training for a major roll-out of a Class V or VI system, the support team can provide many invaluable insights.

7.4 Stakeholder Checklist

There is probably no easy, single way to set up task forces and get stakeholder input. There are many decisions to make: what task forces or groups are needed, who will participate, when will they participate, how will they communicate and interact with the development team, and how will their expectations and involvement be managed? Instead of long prose, here are a bunch of bullets to consider:

- What are the stakeholder groups?
  - Can you clearly see the differences by title or by task? You might need teams from each group, or you might need a team composed of representatives of the different groups.
- How much variability exists within each group?
  - Is the variability relatively small but important? Lots but minor? Small amounts of insignificant variability will not matter.
- What are the relationships or interdependencies between the groups?
  - Will one group's ideas affect another group? There might be points when both groups might need to be engaged at the same time.
- Will the users be seconded (full or half time) into the development team?
  - Who **will** do performance evaluations? Where will they sit? What about their other duties? You will need to think about this stuff in advance and also identify when the people will resume their normal work.
- Will the users have part of their job task and work week dedicated to the development?
  - This is when you ask for a few hours a week from someone. This has to be factored into the work load and some technique needs to be used to ensure that the hours actually happen. The timing might vary during the week and create challenges for interaction and feedback.

- How many users are needed to gain a representative sample for each stakeholder group?
  - You might be able to get away with one or two lead users from each group who can then convey developments back to the larger group, consolidate their input, and interact with the developers. Or, you might need many to find ways to have many voices heard (e.g., web demo site with feedback hooks).
- When will the users interact with the system and developers?
  - If the users are asked to do specific deliverables (versus simply answering questions, reviewing screens) it might be best to have a regularly scheduled time unless the users are imbedded in the system. It is too easy to be busy one week and then be busy for another week and so forth, if one relies on ad hoc feedback mechanisms.
- What will be the frequency of contact?
  - If there are one or two users and the users are actually sitting with the developer in a dedicated setting, there is constant contact. This is obviously the best if the developer has the right tools and knowledge to ensure that the users' time is not wasted. In this situation, functional development should be measured in 1/2 hours of effort, or a few hours at most. If the user has to watch you craft the solution, you have to be as quick as a master pickpocket. However, the frequency can be once a week or once every two weeks as well. It all depends. I would not recommend longer than a two week cycle. The longest I personally try to use is two days.
- How will feedback and dialog be accomplished?
  - Show and tell is always a good way if co-location is being used. The feedback and suggestions should be recorded and noted. If remote feedback is being given, video-conferencing, remote desktops, etc. can also be used for the show and tell. The most disconnected form is where the user sees or plays independent of the developer and then sends in suggestions. This is not the best way since the developer can learn many things by watching the user.
- Who will manage, collect, digest, and interpret the user feedback?
  - This will depend again on the type of groups and numbers involved. There should be someone acting as the user's advocate or lawyer. I used to call myself the user's lawyer when interacting as the middle person between the user and IS creators. If the project is large, you can have a dedicated advocate. If the project is very large, you might have an advocate who will then work with the domain expert who will do the final interpretation.
- How to deal with conflicting requirements or suggestions between members of the same group?
  - As noted in the previous section, everyone has an opinion and not every user will be an ideal task force member. If you are unlucky, everyone is a designer; everyone knows the best and right way, and everyone will know the only way, their way. On some projects I have been lucky, and on some projects I have been very unlucky when it comes to getting a good user group. We will return to this problem later in the book when software

is actually written. The advocate has to listen to all parties, do homework, and sort out truth from fantasy. In some extreme cases, it is necessary to build prototypes of all ideas and then review them. It is sometimes possible to reason with individuals and to discuss things in an abstract way, and in other cases it is usually faster and easier to just build the ideas and then illustrate why one idea is better than another. This assumes that you have the right tools for rapid prototyping, etc.

- How to deal with conflicting requirements or suggestions between groups?
  - This is more tricky. Within a group, there might be a party line or a group think. There might be politics and inter-and intra-group fighting going on in the organization and the new system will act as the lightening rod for the conflict and arguments. You need to clearly identify contact points and the expectations and responsibilities of each group. You can discuss how one decision impacts the other group and again, you might have to build part of the system to really make the point if there is some silly stuff going on. If the politics are serious, not much you can do except for putting your head down and doing your best job. You will be hated by all parties in this case. You hope that it will all blow over and that the system will be close enough in the end. You try to build in sufficient flexibility to change various aspects later. Not much else you can do with politics and personal agendas.
- How to deal with suggestions that deal with symptoms and not the problems?
  - This is easier to deal with as the issue does not pertain to personality traits. You can usually discuss and probe the situation in a rational process, and point out to the people involved what the real problem is. It might take repeated discussions and it might be necessary to unveil the real problem in slow, painfully slow, layers. You might have to explain how A is related to B which then causes C, and if you do not fix A, you are not going to get anywhere. There is something also called system driven behavior that must be thought about when observing and looking at a situation. A system design can drive what people do and what the output is regardless of who is sitting in the driver's seat. On more than one occasion I have been asked to sort out someone's area because "someone" is considered incompetent. After the studies were done, the conclusion was that it would not have mattered who was in the situation; the output would have been the same. The people were basically OK, while the system was sick. Always consider system driven behavior when thinking about symptoms and problems!
- How to keep users engaged when not ALL of their ideas are being used?
  - Egos are involved of course. But users are often using the best of intentions and really think that their ideas are **all** very useful and everyone else should think they are also useful and that their ideas should be in the final product. Try to emphasize the ones being used and explain how the other ideas might be good ideas but that with these blah-blah other ideas, they might not be needed now, but will be kept on record and might be needed later. And, it is best to do this, because the ideas might actually end up being

useful in the end! If the users see that you do keep and possibly use some of the initially discarded ideas, it can be easier to keep the users involved and in a positive state of mind. However, there are those users who will get annoyed and frustrated if you do not immediately respond, if they cannot see the changes immediately in the software, and if they do not see all of their ideas. Nothing you can do, or will do, will make them happy.

- How to keep users engaged when sometimes NONE of their ideas are being used?
    - This is very difficult and more tricky than when you use some of the person's ideas. In the latter case, you can always, eventually, point out their contributions. This is not possible if none of their self-perceived valuable, necessary, and mandatory contributions are in the final product. If the users have had **previous** ideas used during prototypes or piloting, you can emphasize those points while downplaying the current ones. But, if the person has not had any ideas accepted yet for any purpose, or if someone else has already made the suggestion, feelings will be hurt and the user might walk away. They might shut down on you, and even be a negative force within the organization. In a large project, you will likely have a few of these no matter what you do. If reasonable, try to use something, demo something, and at least keep the ideas on file. You can try to minimize this problem by picking better people for the team (the perfect users talked about in 7.2).

Of course, there are many other issues. Every project and every team will be different.

7.5 Feedback and Suggestions — Caveat Emptor

As a developer, you will get many suggestions and lots of feedback. Caveat emptor, let the buyer beware. Section 6.1 and other parts of section 7 have highlighted possible problems associated with user engagement. Obviously, you have to interpret and think about the feedback and suggestions that might flood in. This short section adds a few more points to the discussion about feedback and suggestions. It could be elsewhere, but it seems to lead naturally from the other topics in this chapter.

Feedback can tell you many things. It can give you direct and indirect information and knowledge. The direct knowledge will come from the obvious bits. "This is confusing; this is awkward; these two things should not be on the same screen." The direct knowledge needs to be put in the context of the mandate given the people giving the feedback. It is feedback and it is not always a decision on the part of the user. A number of the Agile/Extreme books talk about doing what the user wants and suggests this in a religious, or sacred tone. That is nice, but is not always the reality you are developing in. There are three levels of decision making to consider while getting user requirements and figuring out what you should do as a developer. I talk about these three levels a few times throughout the book because they help you sort out levels of decisions, scope and responsibility. They are useful things to frame things: the strategic, tactical, and operational concepts. Within an organization, the people will typically reside in one of these levels and it represents their scope of knowledge, responsibility, and decision making. Being in one layer does not prevent people from commenting on the other layers or making suggestions beyond their proper sphere of influence.

The user of the system may or may not be the person making strategic decisions. A number of the features and functions might relate to a strategic decision that the lower user may not be aware of, does not like, does not agree with, and does not think should be in the system. The lower level can comment on it, but the goal is to make it work. The user can have a great deal of input on how it will look perhaps and how it will function, but the presence is not an option. This also applies to the tactical level. The middle managers might have made decisions as well about certain factors and these are not open for discussion either. It is a tricky business. You ask for feedback, but what is the point of the feedback. In some cases, the feedback will actually look like a decision. The user makes a point and you implement their suggestion. As the architect or designer you have a responsibility to listen and consider the inputs, but who decides *what* is always a tangly bit of work. You need to understand the strategic, tactical, and operational dynamics and situation. You need to understand the objectives and agenda at each level and interpret an individual's feedback in this context.

It is best, in my opinion, to ask for feedback on everything — even on the stuff that the user will not have authority over in terms of being in the system or not. Or, if you do not explicitly ask, allow the user to volunteer and make the comments as they choose. As an architect or designer, you will need to know what might be strategic, tactical, or operational and what various people will have influence over. You will have to filter and sort the feedback and understand the weight of each piece of data. Not all users provide the same weight, and not all comments from the same user carry the same weight. Users may or may not understand or appreciate this situation. Some users believe in equality on all dimensions and feel that their comments should carry as much impact as anyone else's. This is not a realistic expectation and it is possible that these users will become frustrated and feel ignored over time.

As the architect or designer, you combine the original requirements and business analysis, with specific knowledge gained as the system is built, and all of the feedback from the users and analyze it. All of the feedback is information. You need to know what users are uncomfortable with, what is confusing, what causes stress, what they accept, and what they like. This will help you understand where additional explanation is needed or training. It can also tell you what to phase in and when to phase it in. For example, you should avoid too many issues at once. Part of this is understanding what will be new to the user and where the challenges will be. There is much to learn from the feedback.

At one extreme, you might get lots of good suggestions and respond to all feedback. This will make everyone feel good. The users will feel good; the people interacting with the users will feel good; the developer will feel good, and the client will feel good as they will hear that the developers are paying attention to and responding to the users. The client will always hear about your interactions, and if you are lucky the users will be of the desired type. At the other extreme, you will have a user team composed of dysfunctional individuals who will complain and whine about how they are being ignored, etc. and how their valuable suggestions are not being implemented, or are not being implemented quick enough. It all comes down to perceptions. There is nothing you can do to correct a person's perceptions about their suggestions. They will think that they are right and you are wrong and they will be happy to tell everyone about the great injustice that is occurring. Hopefully, you will have enough good users to work with that the final product will be satisfactory and appropriate. Remember, you are not assuming that you are all-knowing and that you can exhibit divine wisdom about what the system should do and how it should behave. You should be making your decisions based on an informed analysis, a careful study, and with the assistance of some "good" users. You should not be making your decisions based solely on your opinions. If you do, you are not any better than the users you are dismissing.

In some cases, the poor quality of user feedback is not 100% the user's fault. They may be asked to look at something for a few minutes, or a few seconds and provide feedback or a suggestion. The thoughts come out of a short decision process that is not based on any quantitative analysis or solid base. They have a busy job and it is like a helicopter landing, dumping out a bag of supplies, and flying off again. The architect or designer has to factor in the feedback process when considering the feedback. That is, was the feedback obtained through a quality process? What are the assumptions underlying the feedback? These factors can affect the quality of the feedback and how you should interpret it.

Then again, sometimes the quality of the feedback can be linked to the individual. Not everyone is a good thinker, a good decision maker, or a good analyst. Unfortunately, the more people are given authority, or are asked to be on a committee, the more they think that they are good thinkers, good decision makers, and good analysts; even though they might not have had any training in thinking, decisions, or analysis, and even though they might have been promoted or asked to be on the team for reasons having nothing to do with their thinking skills.

I have seen situations where some software was crafted and a number of inappropriate users asked to provide feedback. If the users are not suitable users, fitting close enough the user mode that is driving the development, it is possible that you might get 100% of the users not liking something or disagreeing with something. In this case, you have to ignore 100% of the feedback. It is like the gambler's fallacy in statistics where you see a pattern of events that suggest something is predictable, but in reality it is a random sequence that gives a false confidence. Ten users may like something or hate something, but that does not mean that they speak for all users just because they are users. User profiling and modeling is necessary when interpreting feedback. If the people gathering feedback and interacting with the users do not understand this phenomenon, or if they are dogmatic and have the 100% listen-to-the-user mantra, there will be unhappy people in development as well.

Interpreting feedback is a balancing act and the best counter-measure or defense I have come up with is doing a decent problem analysis to start with. If you do a good up front study, you will have a better set of tools for understanding the types of users and for understanding what feedback to listen to, or to discount. It will not help you with the frustrated or annoyed users on the task force, but it is likely that your decisions and compromises will be better in the long run. The next four chapters document how I do the front up analysis work that helps prepare the foundation, and helps me when I am unlucky with the users assigned to provide feedback persevering through the rough periods when all of the feedback is not going my way. In these unlucky situations, you hope that your homework and understanding provides the clues. For example, you might have spent months, days, or weeks thinking about some aspects before reaching a design decision or concept that is immediately, without thought or analysis dismissed by a user or group of users based on an initial emotional response. If you have faith in your analysis, you have to weigh it more than the user feedback, especially if you have assessed the situation correctly with respect to the quality of users. You have to be careful though not to erroneously assume that you are right and that the user is wrong. It is best to assume initially that the users are of the appropriate and suitable type and will give excellent feedback in a balanced and thoughtful fashion, until they prove otherwise. Once they have demonstrated their feedback quality, it is important to factor this into your feedback interpretation process.

Chapter 8
ZenTai — the Value Equation

8.1 価値 — Value

This and the next three chapters are complementary; they help you understand the problem; they help analyse the issues to be found in the problem; and they help design the solution. 価値 (kachi) is used to represent the concept of providing value to the user. When you are trying to understand the problem and you are interacting with the users, you need to be collecting information about the user's value equation and think about it. Your job is to provide value to the user and that is **YOUR** problem. Of the four *ZenTai* pillars, value is the most important to deliver on. If you do not provide value, you will have a bigger problem to solve later, assuming that you do not jump ship or leave the scene of the crime like a hit and run artist.

> *Aside: be very wary about hiring anyone who keeps moving jobs every 12–24 months (or thereabouts) as this individual is probably not hanging around long enough to see the results of their judgment or decision making, and it is possible that they keep fleeing the scene of the crime. They could just be a lot of hot air with no proof of long term results!*

In information technology, it is a non-trivial challenge to understand and deliver sustained value because it is a slippery, non-static thingy. The concept of value is very important because obtaining value is why someone is using the technology in the first place. They want something out of it! If the technology does not give value and sustain the giving, we should not be surprised when people do not use the technology or barely use it. It is also a cost-benefit relationship as the value must balance out the cost and effort needed to obtain the value. If it takes longer and more effort to do something with the system than grabbing a pen and piece of paper, guess who wins.

There are many symptoms of poor value. Are there extra spreadsheets kept outside of the system? Is the system the tool of last resort? Are there many workarounds or perceived misuses of the technology? Do the users only talk about the negatives during an analysis without ever mentioning good bits? Do the users talk about how the system gets in the way and actually prevents them from getting the job done? Do the users consider the system a joke? Do the users actively suggest and advocate for a new system? Do the users rejoice when they hear of the system being replaced? There will always be some dissatisfaction or some minor bits that can be improved, but in general, the users will be happy with a system that actually helps them get the job done while saving them time, minimizing brain-dead tasks, helping to improve quality, reducing duplication of effort, and so forth. If the user fights or defends a system and does not want it turned off or removed, it is likely giving some type of value to the user but there is a caveat: it might not be the value desired by management ☺.

This chapter develops a systematic approach for thinking about the value of information technology. The approach can be used for evaluating an existing bit of technology, or for designing something completely new. The first step is to think about information and what is an information system. An information system is something that obviously deals with information. But, there is more.

If the Concise Oxford Dictionary or Concise Oxford Dictionary of English Etymology is checked, *information* is the results of *to inform*, something that is told, and items of knowledge. *Inform* is a verb and means giving form to, to furnish with knowledge, to form an idea, to describe, and so forth. In turn, *knowledge* is a fact or result of knowing: perceiving, recognizing, distinguishing; the theoretical or practical understanding of a subject; a certain understanding as opposed to opinion.

A system implies various components that come together, and an information system thus informs, forms, describes and furnishes ad nauseam knowledge. An information system can inform humans and/or technology about knowledge or provide knowledge about what to do. Information systems are quite varied in their manifestations:

- Medical — devices, patient records, institutional operations.
- Military — supply chains, devices, analysis.
- Macro-script — small projects, personal information systems.
- Scientific — packages, devices, statistical tools.
- Business application — payroll, accounting — records, input, reports.
- General applications — word processing, spreadsheets, databases.
- Tools — compilers, interpreters, web design tools, animation, photography.
- Systems — operating systems, servers, drivers.
- Telecom — communication switches and hubs.
- Realtime systems — imbedded systems, machine control, consumer electronics.
- Education — content, classroom and student support.
- And many others…

In some of these, the information system reveals something that is not obvious (through processing of raw inputs). Or, the information system automates information processing that could have been done by hand but is done by technology faster, facilitating higher volumes, and with better quality. Or, the information system allows something to be done in a way, or at a place, or at a time not possible without the information system component. For example, an optical stabilizer on a digital camera involves software at some level and processes some form of information that reduces the impact of camera shake on the captured image. This allows the camera to be used in different situations without a tripod and/or without the need for auxiliary lighting.

While we think of information systems and technology being associated with computers, the value of information processing via technology has been around for a bit. Consider a partial list of information systems that date before 1850:

1. Sun dials — 3500 BC
2. Abacus — 3500 BC
3. Calendar wheels — Aztec — 1000 BC
4. Compass — Chinese antiquity — 800 BC
5. Gauges and measurement tools — 600 BC
6. Torch signals, decoding — Greek — 500 BC
7. Mirrors — signaling — 405 BC
8. Flags — signaling — 400 BC
9. Sundial — Chaldean, Greece — 320 BC
10. Wind rose, astrolabe — 250–200 BC
11. Maritime chart — Carta Pisana — 1275
12. Clocks — 1286
13. Navigational compass — 1300
14. Slide rules — 1632
15. Calculators — Pascal — 1645
16. Sextant — 1731
17. Marine chronometer — 1759
18. Semaphores — 1791
19. Jacquard loom — 1801
20. Difference and Analytical Engines — 1822
21. Telegraph — 1838

All of these involve some form of information processing that provides value to the user. In terms of relative impact, some of the above can be said to rival or exceed the impact of computers. For example, the invention of the navigational compass around 1300 had significant impact. A compass reveals information that is not otherwise observable. It provides knowledge by processing the magnetic forces via a magnetic pointer and this pointer informs the user as to which way is North. Within a couple years of its invention, the commercial trade of the Venetian state had doubled (able to sail twice each season instead of once) and had also altered who the users were. More people could use the compass than could use the astrolabe, and it was possible to use the compass 7×24 in almost all conditions and not just when the sun was visible! People were then able to combine the use of the compass with maps and navigate great distances facilitating the exploration and settlement of various regions. The ability to sail around Africa and around the world impacted many things. For example, the price of pepper dropped to about 1/20 of its former land-travelled price. Think it through. Computers have had impact. But, have computers really impacted the world in the same fundamental ways that inventions like the compass or clock did? This is a good debating topic.

The point is that we need to understand value and what value means to the user. The compass had value; the telegraph had value; the telephone has value; ERP systems have value; imbedded control logic in a consumer electronic has value; ... Understanding value helps us understand what the information system (hardware and software) means to the user. If the system, whatever it is, is designed with keeping value in mind, then we have a better base. Consider educational software.

In the 1980s and 1990s, there was a flurry of activity to create CDs with multi-media content and teaching goodies for textbooks at all levels of the educational spectrum. In discussions with publishers in the late 1990s, several noted to me that they had never made money on the CDs packaged with the books and some of the editors were disappointed that teachers and students did not seem to value or use the packaged bonuses. Teachers I talked to also noted the useless CDs shipped with the books. At the same time I had some educational software being used in several different situations to enhance teaching and one of the teachers said that she did not want to teach the subject again without the software. What was the difference? Value!

Based on my own experience, almost all of the CDs shipped with books provided little or no value over what I was doing or could do in the classroom. Nor, did they give students any benefit. The software and tools did not seem to tackle the hard to teach or the hard to learn bits of a course. If there was value, the incremental cost of learning the new tool and using it exceeded the benefit and at the end of the day, the old way was close enough. Many of the publishers were basically just putting the printed word on the media and had software tools that mimicked the blackboard. When I was working with instructors, I did not do that.

I asked the teachers what was hard to learn and hard to explain. What caused the students' eyes to glaze over? What caused the teacher's concern and headaches? This is where I suspected I would find the best uses of technology and the high value stuff. If a teacher is going to change lesson plans and their status quo, it better be for a good reason! So, long story short, an Agile/Extreme approach was used to craft high value, high impact teaching modules focusing on what the teachers could not do otherwise. I did not start off asking them what they wanted software and technology to do. It was my job to invent that stuff. I needed to know their problems and where they struggled the most.

Once you know the main value drivers, you can do a number of things to make the user experience optimized for value. I will come back to this point. The next section delves into value like a root canal visit at the dentist. There are two aspects that will be developed: the analysis of value and a framework with which to do this. Analyze comes from ancient Greece — analusis: meaning to unloose, undo — and hence resolution into simpler elements. A framework or taxonomy is a structure, order, or plan. It is a structure upon which contents can be put. The purpose is to provide a systematic, efficient and effective way to think about the value of IT; when doing a requirements analysis, when thinking about an existing IT entity, when thinking about the potential benefits of IT.

8.2 The Value Framework

Imagine yourself performing a post-implementation audit on a computer-based information system that just cost/many days of effort and a few million dollars to develop and install. Did the company obtain the desired results? What was the expected impact? What impact was realized? What does impact mean in this situation? Could the impact, or part of it, have been predicted? Furthermore, how can your analysis of the impact be communicated to others? Or, imagine that you are purchasing a system and need to document and rationalize the investment? Or, you are thinking about creating a system and need to convey the value to potential investors or users. Or, you are thinking about how to enhance a current product and are wondering what the best features are to go forward with?

Traditional approaches to analyzing the value of IT have been focused on a cost-benefit view related to a single factor such as manpower costs, or have been anecdotal case studies using examples to make their point. The traditional models and concepts have also focused on the acquisition of the IT (why people buy it, how to sell it), and have not focused on the value that is obtained after the IT has been obtained. Furthermore, there are few, if any, frameworks discussed in the literature that focus on the socio-technical facets.

To address these shortcomings, I created a socio-technical model for organizing and structuring the issues and topics that relate to computer-based information system impact. As a socio-technical framework, it provides categories, themes, and identifies possible relationships between the framework's main constructs. The framework is high level and is not intended to measure or evaluate any specific impact. It is not intended to be a technical assessment tool (e.g., impact of radiation from cell phones), but is intended to focus on the direct behaviors and actions that relate to the information system and that define value or impacts (positive and negative). As a tool, it is meant to help the analyst quickly get to the ballpark of relevancy, and to help create a context for decomposing the value problem. Remember the ramblings about questions and knowing what questions to ask? Especially when you are in a time crunch and do not have all the time in the world to think? This is where a framework or systematic structure can help you. It can help identify potentially relevant areas where you should invest your time and effort and help identify the low hanging fruit. It also helps you avoid wasting time on work that you should never have thought of in the first place.

How to use it? Well, the framework can be used at the aggregate level of an information system in its entirety or can be used for subsystems or individual features or functions associated with an information system. It can be used to think about the value or meaning of a digital camera to the consumer who purchased it. The framework can be used to think about vendor claims for impact and it can help size up the potential for technology adoption.

In a sense, the framework is an "instance" framework and can be used to look at an information system in a specific instance at a specific point in time. For example, in a specific instance, the use of an information system might be used for leisure or it might be used for the person's job. The framework can also be used to profile a context (e.g., primarily leisure) versus the use and impact of the information system and identify possible conflicts (e.g., use of email at home for work purposes).

The framework is intended to be an analysis and organizing tool. Thinking through the framework can also be like a checklist and can assist someone with validating what might be relevant, what can be claimed to be significant, and what is reasonable to identify and talk about — while providing some structure to the argument. For example, knowing that a product is mature implies that data and historical writings are likely to exist and that it is possible to think about prior claims, current claims in light of evidence, and to speculate based on past trends and evidence. Knowing that a product is being used for business purposes helps to focus on the interactions and issues relating to business, and if the product can be used for both leisure and business, it is likely that two views or user models are needed.

However, the framework does not specifically define value or impact. This depends on what the IT system is. For the purpose of the framework and its use, value or impact is whatever is perceived or voiced as being the impact. Specific submodels or frameworks can be constructed for subsets of information systems (e.g., Consumer Electronics) that define in crisper terms what the value or impact might be, but at the high level, we can guide the discussion by using an abstract view of impact and value. That is, for whatever impact is claimed, we can talk about things like:

- Who is involved?
- What interactions are involved?
- When do the interactions occur?
- What initiates the process that yields value?

As a high level tool, the framework is intended to view the user of an information system from multiple perspectives: individual, small groups, subsets of organizations, organizations as a whole, and societal elements (decomposed or aggregate). The concept of a user is at multiple levels and implies the appropriate level of analysis for the information system being analyzed. A user is anyone or any grouping that is being directly or indirectly interacted with by the information system.

The proposed framework has five major constructs to help decompose and organize the issues related to information system impact. The five major constructs are:

1. Life cycle positioning.
    - There are multiple life cycles — a life cycle for proposed functionality or value, and a life cycle for the technology or way that the functionality is delivered.
2. Society or organizational structure affected.
    - The IT exists within a social situation or context. The IT can be used as part of the user's formal role in society, for leisure, and so forth.
3. Interactions between the structural components.
    - An IT system involves various components, including the user, and the interactions between the components delivering the value — either by providing information or conveying it. The interactions exist on a topology of interconnections and you cannot get from A to B if there is no road or connection.

4. Information conveyed via the interactions.

    ○ Once there is an interconnection or path from A to B, it is possible to think about the actual information or data being transported. The information has certain behaviors and characteristics that can be used when discussing value.

5. Physical, social, or personal impact (value).

    ○ The first four constructs create the tableau or focus for the actual value analysis. As such, the first four can be reasonably quick to sort out. The fifth construct is the actual analysis of *value or impact arising from the information that is conveyed on the topology via the interactions used in a specific social situation at a point in time*. Got it?

In the literature, you can find some frameworks or descriptions that include a life cycle concept. And of course, everyone talks about impact. However, I have not found a taxonomy that addresses all five concepts, and the existing taxonomies do not attempt to further breakdown "impact" in any consistent or organized fashion. Of course, my model is not THE model or final answer either. Any taxonomy or framework should be challenged — for example, are all five of the major components necessary? Are they the right five? Are the sub-categories and sub-topics within each necessary and correct?

The following sections will explain each of the five major categories and their corresponding subdivisions. An attempt will be made to rationalize why the category is important and what the category contributes to the understanding of information system impact. I feel that the five categories are necessary to address the more complete and larger picture of what is going on when an information system is introduced and used. I do not believe that focusing on one aspect at the exclusion of the others really helps in understanding the potential and realized impacts. An information system does not exist out of the context of when it is being used, who uses it, where it is being used, what the information content is, how the information is transferred and transformed, and what the information is ultimately used for. By answering or thinking about these aspects, it is possible to consistently and thoroughly analyze a past, present, or future information system situation. Such an analysis can aid in understanding what pre-conditions, facilitators, barriers, and side effects might be relevant. A structured analysis can also aid in separating symptoms and contextual aspects from the underlying issues. For very small situations, a structured framework and decomposition can be excessive. It is still important to think about the same types of things, but a formal or orchestrated analysis is not needed. However, as the information system situation increases in size and complexity, it is difficult to describe or think about the potential or existing impact without some form of organization for tackling the problem. What is the impact of introducing personal identification cards? What is the impact of the federal government's privacy legislation? What is the impact of the latest MRI technology? What is the impact of Napster? What is the impact of going wireless? What is the impact of introducing ERP? What is the impact of requiring every student to have a laptop?

In essence, the framework is an aid in viewing or understanding a situation in a consistent and organized fashion:

- Where in the life cycle is the discussion taking place or is focused on?
- What is the societal or organizational aspect that is being impacted or possibly impacted?
- What types of interactions are involved when the IS is in use or active?
- What is the information content or value or element that creates or facilitates the impact?
- What are the actual direct and indirect impacts?

8.3 Life Cycles

Information systems, especially computerized, go through the typical steps associated with inventions and other forms of technology. Someone has an idea and from there it might ultimately become something that will become real. Understanding the past, current, and potential steps or stage in the cycle helps you to understand what other issues or facets may be important. Each stage in the life cycle has dynamics internal to the stage and dynamics that allowed or facilitated the stage to commence and dynamics that eventually force this stage to be vulnerable and another stage to take prominence. A life cycle of an information system can be relatively short and have very few stages. That is, it never becomes more than an idea or prototype. The life cycle can also have rejuvenation and multiple releases or new capabilities fundamentally alter the information system.

For any given information system you can ask the following:

- What are the appropriate stages or cycles to consider?
- What is the current stage and where in the current stage are we?
- What are the life cycle dynamics that are relevant?
- What does the stage imply?

How do these questions help you? Speculating about the life cycle can help you think about what kinds of impacts might arise. For example, if the information system is based on emerging technology, there might be impacts on the emerging technology itself, impacts on governmental regulations, and changes in infrastructure. Whereas, if the information system is relatively mature and possibly in its decline, the impacts may focus on the installed base, issues of moving from a legacy system, compatibility with new systems, and so forth. Understanding the stage can also point you to the types of players that might be involved, possible issues with information being conveyed (e.g., standards), how widespread the impact may be, and so forth.

With computer-based information systems, there are often **two** life cycles to consider. First, the value, impact or result of the system might be a new invention itself or might already be in existence and be mature. Or, the information system under analysis might be a new invention for generating a mature result. For example, in a faster, higher quality, more pervasive way. The life cycle of the impact needs to be analyzed or considered in parallel with the life cycle of the specific information system.

For example, remember the introduction of the maritime compass about 1300? The maritime compass allowed mariners from Venice to find North any time during the day and almost under any sea condition. The concept of North was not new and the sailors knew what to do with North when it was known. The maritime compass was new technology for an old concept. The new technology changed many aspects of sailing and commerce; with the compass a law was passed that forced the fleet to sail twice a year.

X-Ray technology, and the areas of nuclear medicine would be examples of new technology and new functionality. The ability to look inside the body without invasive procedures was not possible before. The machinery allowing this did not allow more of something or more people to do it or to do something faster. Science did not even know what to do with the information initially and new skills had to be developed to study the output.

The hardest situation is when there is new technology and new functionality. In this situation, there is little historical data and little evidence to rely upon. In contrast, the easiest situation to analyze and to work with is when the technology and functionality are both relatively mature. In this latter case, you can expect to find existing data, studies, and situations that can be used to understand the existing situation — creating a more solid foundation for predicting the future.

There are three topics within life cycle positioning that I will expand upon:

- Cycles.
- Dynamics.
- Issues.

Cycles

Many inventions and technology-oriented products go through six phases or stages of evolution. They start out as someone's **idea** or dream in response to a problem, or a stated need. They can also start out as a solution looking for a problem. We already have it; what else can we do with it?

The next phase is marshalling or bringing together the **science and technology** necessary to start moving the idea into the realm of reality. It is possible to have the idea and then have a delay before science and technology is ready. For example, the idea of shopping from home and watching live performances from the comfort of your own living room existed in the late 19th century but could not be developed further until science and technology had also developed. It might be possible to explore part of the idea, but it is not possible to make a working prototype until the science is ready. This can lead to other investigations into the underlying sciences in order that the idea can be realized.

The third stage is creating a pilot or **prototype** that might have some limited usage but is not on the open market and the public cannot easily acquire it. For example, the early ARPANET was a prototype system connecting several universities before being adopted by other schools doing DoD research.

If an information system is in one of these early phases, it does not make sense to consider current, widespread, direct impact on society, organizations, or individuals. However, there might be impacts of a fundamental nature that will ultimately result in impacts not associated with the main system under consideration. For example, a tool may be written or created to help with the one project that ultimately turns out to be more important than the original project itself. There might be impacts on governing structures, standards, barriers, and such during this period as well.

Once something is out of the prototype stage, it is more generally available and there will be **early adoption** by certain parties. This might be a long period or relatively short. There are some things that are slowly adopted over decades until it can be considered that they are in widespread use while there are other things that are rapidly adopted in short order. In the early adoption stage, it is important to understand the rate of adoption, drivers, barriers, and facilitators. It is now possible to talk about the direct impact on the early adopters and the indirect impact that also results. The impact will be limited and there is a danger of extrapolating from observed impact on the selected few who feverously adopt new things to what might happen in the future.

Widespread adoption or deployment is where the bulk of the impact will be seen. The system has been fully rolled out to many of the targeted sites, or average consumers are now buying and using the system. How soon this stage occurs will often determine what product or system will become the de facto standard and it also determines the profit or value obtained justifying the initial effort and costs.

Following widespread adoption and success is inevitable **decline**. Sooner or later the information system will be replaced, set aside, forgotten about, or be relegated to back up status. This ultimate stage can be brought about by many possible sources. The impact is definitely limited, may have switched to a negative state, or be non-existent.

It is common to picture this life cycle as an S-Curve(see the figure below). You can map the phases to plateaus, reasonably straight sections, and the bends.

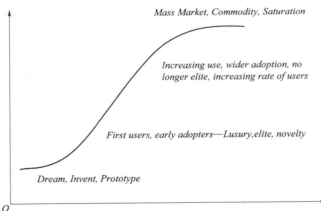

The horizontal axis is time and the vertical can be mapped onto various elements, such as market penetration and system usage. Concepts like adoption are context dependent and refer to the possible population that is relevant. For example, if an information system is intended for use in the airline industry, the adoption curve relates to the airline industry and not the possible adoption of the same software in the electronics industry.

You start out with the flat area of the idea, encounter the first bend as you get to the prototype and early adopters, hit the main slope of the S as users start using or adopting the system in droves and then you hit the top of the S with maturity and saturation. The overall shape or slope of the S indicates the rate of change in each stage. In the field of science, technology, and information systems, the S curve slope has steadily increased for the past two–three hundred years. Systems and technologies in the 1700s, 1800s went through the same steps as those in the 1900s but we are now moving quicker in each area.

There are also deviant S-curves that relate to bubbles (e.g., e-stores), fads (e.g., PDAs), early failures (e.g., beta format in video tape), disruptive technologies, and revolutionary (and evolutionary) developments.

The main stages are:

- The idea or dream.
- Invention — science and technology.
- Prototype.
- Early adoption.
- Widespread usage.
- Declining — set aside, replaced, unused.

Dynamics

Within any cycle, there are possible dynamics. For example, changes within a stage may be evolutionary or revolutionary. The evolutionary changes are incremental and are often small changes (relative to the whole). For example, Ford's Model-T had many small changes during its product life as did MS-DOS. A revolutionary change was the Model-T going to the Model-A and MS-DOS going to Windows. A product and its impact can diminish and increase as competitors develop new versions and new releases are produced to counteract the competition. Another concept related to dynamics is that of disruptive technologies.

Another dynamic to watch for is the move from novelty status for the rich and famous to commodity status for the working class. This move has a clear relationship with impact. The initial home computers were a novelty as were the Macintosh computers when first introduced. Hard disks, color monitors, pointing devices, and full-screen editors were also novelties. The initial cell phones were novelties as was email. All of these have passed from the small user base to that of commodity status. They have also moved from the rare status to ubiquitous. They are everywhere and no longer cause heads to turn. This transition from novelty to commodity and rare to ubiquitous is a major factor for marketing, management, and many other aspects. A book by Moore (*Crossing The Chasm*) is a suggested read for anyone interested in being in the game.

Another interesting dynamic associated with the impact of information systems is that of institutionalization. This is when the information system becomes one with the organization or societal structure that uses it. It is safe to say that email as a form of communication has become institutionalized. There are policies for its use, standard forms of etiquette, a specialized vocabulary and style, and many expectations about its availability. It is taken for granted.

Thus, when looking at a life cycle and a particular stage, it is possible to ask questions such as: what dynamics are present, what dynamics might be developing, and what actors and aspects are related to the dynamics. The main dynamics are:

- Evolution – rejuvenation
- Novelty – commodity (e.g., Crossing Chasm by Moore)
- Rare – ubiquitous
- Institutionalized – taken for granted

Issues

What are the issues related to cycles? In addition to the basic cycle of technology development and the dynamics within, there are five key issues to reflect upon. Depending on the information system being considered, the issues may or may not be important enough to discuss, but they are things to consider while thinking about the topic.

Before I describe the five issues, you have to think about the context of cycles since the issues relate to the context. Cycles and stages are not random events. Is the technology or industry pushing the cycle evolution or is the system being pulled by other forces: the users, government laws, etc.? What starts or extends a cycle? If a cycle is close to its end or beginning, it is important to understand what facilitates or hinders the rate of development. If a cycle appears to be mature and stable, what is sustaining it? What are the threats that can reduce its positive impact?

As noted, not everything is random. In reflecting upon an existing situation, the question needs to be asked about what was predictable. If something can be predicted, it can be planned for and might also be controlled for. If a system seems to be in turbulent times, it is worthwhile to ask these types of questions.

The five issues to possibly address or reflect upon for any major information system are:

- What starts or delays each cycle (or part of a cycle)?
- Are the cycles pushed or pulled? What are the needs? What is doing the pulling or pushing?
- What makes it easy or hard for each cycle to start?
- What makes it easy or hard to sustain?
- How much can be planned, anticipated, or controlled?

8.4   Society or Organizational Structure

In our quest for sorting out the impact of information systems, it is also useful to consider who and what is involved. The *who* and the *what* can identify the system's role or position within the larger scheme of things. For example, if the information system is targeted towards an individual in the family setting, then it is possible to identify a range of possible topics to further explore; these facets not making sense to explore if the information system is being used on the job for controlling the actions of a robotic arm. I have identified four major role structures within which an information system may be used by an individual:

- First, there is the individual's formal role within society: student, employed, retired, state-supported, etc.

- Second, there is the role and the activities within the household at the personal and family level. This is the nuclear family level and includes direct personal use.
- Third, there is the leisure role: what you do when you are not engaged in your formal role or performing functions within the household or family.
- Fourth, there are external interactions undertaken with institutions or formal entities such as companies, services, retailers, cities, provincial government, financial institutions, and so forth.

Pictorially, the four can be visualized as:

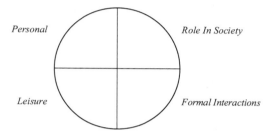

- The *personal and family* quadrant considers the information systems associated with day to day activities of survival and basic functioning. Included are those that relate to food and shelter. Any aspect that focuses on family and companionship would also be affiliated with this quadrant.
- The *role in society* can be many things. Five possible ways to view an individual might be a) child, pre-student, b) student — yet to enter job market, c) job within community (outside of the house), d) retired, and e) dependent — including at-home spouses, unemployed, disabled, and welfare supported. An individual may have a primary role (as employed) and secondary roles that may be concurrent (as part-time student). The role in society can also be defined and interpreted for organizational structures and entities — retail, service, and so forth as each might have interactions with other retail and governmental agencies.
- The *external or formal interaction* includes the civic, industrial, and retail interaction in addition to direct dealings with retail, factory, and other services. It also includes medical and health, education, roads, power, fire, police, military, etc. This includes the retail sector in which the individual interacts with to purchase or acquire merchandize. It also includes services provided at the civic level of government focusing on the immediate community. The public backbone or general services supported or supplied by the provincial and federal governments are also in this category.
- The *leisure* component groups the systems into a) ones in which the person is an active participant, b) passive observer situations, and c) vacationing type activities. In an active role, the individual has to actually do something to obtain and sustain the impact or value. In a passive situation, the individual might initiate a process (e.g., turn on TV and select channel) but then has no active role in creating value. In vacationing activities, the use of

information systems might be in the planning of the vacation, during the vacation, or after the vacation has transpired.

This categorization into four parts can help focus the discussion and think about what might be implied by information, responsibilities, liability, ongoing support, and so on. The role structures do not necessarily imply or create constraints as to where or under what conditions the information system is used. There are logical or reasonable contexts that can probably be identified as dominant or logical, secondary contexts where a system will be occasionally used, and other contexts that do not make sense for the information system to appear in.

The categorizations can also be viewed from either end of the interactions or usage patterns. For example, an information system connecting retailers and consumers can be looked at from the retailer view (what impact does the retailer get) or from the consumer view (what impact does the customer get). What impact does the province or country get as a whole from providing a public service or system versus what individuals get from using the system? Similar arguments can be phrased for each quadrant.

There are often legal and government requirements for external or formal interactions. Information systems used for leisure have different requirements and intended impacts compared to the others. Similarly, information systems designed for preschoolers would not have the same issues as would systems for the unemployed. One example of this reasoning is the concept of privacy legislation. Laws for businesses and government agencies may not apply to what you do yourself when using your home computer.

The categories arise from a simple observation: you are either at work, at home, out/in doing formal interactions, or out/in trying not to do work or normal house stuff. You then follow through with the reasoning: what does it mean if an information system is targeted for personal use versus commercial use? Over time you will develop lists of topics and issues that are relevant and the ones that should be dealt with first. In efficient and effective analyses, it is important to know what to do and what to focus on, not wasting time, effort, and energy on things that are better left unsaid or explored. In each sector you can talk about expectations, status quo, conventions, acceptance, legality, costs, return on investment, and so forth.

There might also be complementary systems within a sector where the whole is greater than the sum of the parts. It makes sense when analyzing an information system to reflect on possible complementary or orthogonal systems. Are there interactions or joint issues? Is one system co-dependent upon another for impact? Total impact or partial impact? There might be interaction and linkages, but are they important? Again, not every written up analysis requires this to be explicitly addressed, but the analysis leading up to the final report should have consciously asked these types of questions.

8.5 Interactions

In sections 8.2 and 8.3 I introduced the higher level constructs of life cycle positioning and social structure. These constructs put the information system under analysis into a ballpark for further analysis. They can be thought about rapidly in most situations and they create the context or background for the analysis. It is like undertaking a character role in a movie and the actor has to get into character to give a convincing performance. The actor becomes the old, aging king with physical deformities even if the actor is healthy and only middle aged.

If the concepts in 8.2 and 8.3 have been applied and thought about, we now know the types of players, general issues associated with general field or topic, and have an idea about the scope and domain that the information system exists in. Remember, the goal is to create information systems that provide positive value to the user! It might sound hokey and lame to think through these aspects, and you might want to just jump in and start creating as that is the fun stuff. However, if you do not consciously think about these aspects, how can you create a good user model? A usage concept? Realistic and useful personas?

Creating a good design and being able to create good designs consistently is hard work, and it is also sometimes quite boring. If it was easy and it could be done without much thought, we would have lots of great systems and successes instead of the other way around. Now that we have the basic social situation, the next task is to delve deeper and look at detailed aspects of the information system. Eventually, you get to think about the GUI and the fun stuff, but first things first and this is still about understanding part of the problem. If you do not understand the user's value equation, you are going to waste a lot of people's time and energy!

Since we know the players (computer/computer or computer/person and their social situation), the next logical topic is **interaction**. Information systems involve interactions by definition. Information is generated, transformed, conveyed, or presented in from/to interactions. Is information that exists in a computer and is generated by the computer itself and not shared with anything or anyone else truly *information*? Does the tree make a sound when it falls if there is no one in the forest to hear it fall? For convenience, I will assume that some form of interaction exists for an information system to be an information system. There is of course data and information within the bowels of a system used by the system itself or on behalf of the user, but the system must still exist for a greater utility than gathering data about itself for its own self-serving use.

There are three major aspects of an interaction to consider: the form, the purpose, and the characteristics.

Form

The form or style of the interaction refers to the underlying flow of information: who initiates it, who is in the sender/receiver dominance or submissive role. You can consider this the topology of the information system. For example,

- Is the interaction a broadcast or communication form whereby the sender sends information to one or more receivers?
- Is the information transferred from the sender to the receiver and is not retained?
- Is the interaction related to multiple users or systems sharing certain information or having a

common view of something?

- Is the form of the interaction having two systems co-operate and work together to achieve some outcome?
- Is the form of the interaction having one system do something on the behalf or for another system?

In each of these forms, there can be a 1×1 $1 \times N$ $N \times 1$ or $N \times M$ mapping of senders and receivers. Furthermore, the interactions can be parallel or serial.

Specifically,

- *Communication*: E-mail is an example of an interaction form that is intended for communication purposes.
- *Transfer from one to another*: Financial institutions transfer information about checks and other transactions between each other and when you request an account update at an ATM, this is another transfer of information.
- *Share with another or others*: Financial institutions share credit information with each other. It is not operational data like account transactions, but it is information that is used by many parties. Web sites that exist as information repositories or the definitive source on something are also of this form.
- *Do things with another*: Coordination of supply chains involving multiple firms is a collaborative application. Agent-based systems are another. Online gaming with multiple individuals playing together is also a special form of interaction.
- *Do things for another*: If you are at the functional end of something like a help desk, the form of the interaction is one of doing something for someone else. Software systems that take in a bunch of numbers or parameters and prepare forms (such as personal tax) are of this type as well.
- *Have another do things*: A reverse of the do-things-for-another category is when you are on the requesting side and logic or functions exist to reach out and seek the help of other software or systems. For example, web-bots or search engines that request other search engines to perform special searches are systems that explicitly rely upon other systems for some of the main functionality. In a way, all applications are of this form as well since they use the operating systems to perform the system level functions. In general, I will use this category for higher level functionality and not include the operating system analogies.

The form is interesting because it helps to start identifying what the information system is really doing. Is it the controller and instigator or is it the passive system responding to other systems. Active or passive is important for understanding impact. In active form, the impact is more under its control (potentially) while if it is passive, its impact will only happen if the controller or instigator comes and uses it. This helps to identify the sources and sinks, and the control hierarchy.

Purpose

The form leads directly into a discussion of purpose. Within each of the form categories, it is then possible to talk about or investigate the more detailed purpose. Why not jump immediately to purpose and why have a level above purpose called form? Having such a category is useful because it allows you to more easily identify similar or relevant systems for comparison, benchmarks, or probing. It is like grouping animals by their number of legs. Knowing something has a four-legged form versus a two-legged form can inform you about the issues you might expect to explore and consider.

Once you know the form, you can ask *why*? Why does the communication channel exist? Why is each communicated piece of information communicated in the first place? Why is it this information and not other information? What will happen if the interaction does not take place? What happens if the interaction is not available? These are all potentially important questions for thinking about impact.

There are other questions. Some of which are the following. Why is information being transferred? What missing function is fulfilled by having a co-dependent system interacted with? Does an individual interaction have meaning or do many interactions and connections have to be considered for the meaning to be identified or goal attained?

- *Why communicate?* Communication must match the situation that encompasses it. If the situation is rapidly changing, the ability to rapidly communicate is important for organizing, responding, and dealing with the expected levels of performance. Today's systems have changed the *how*, *who*, *where*, and *when* aspects of communication. Add in phones with cameras and computer hook-ups, the *what* has also changed. The bandwidth of information has changed and this can drastically alter the uses and value of the information that is conveyed.

- *Why transfer?* Business-to-business and other forms of information transfer are associated with many objectives. In some cases, the firms are building up a better profile on the consumer and in others the transfer is necessary to complete a transaction. Information is likely shared by various parties when you visit certain web sites, allow cookies to be active, or when you click on banner-ads. This type of indirect and non-transparent sharing raises many issues.

- *Why engage another?* Do you know everything? Might you not need help from someone about a problem or need get a different opinion? Sometimes it relates to the where, when, why, what, and how dimensions: someone telling you where to find something, when it is there or when to do something, what to do about a strange medical condition, and how to change the valve on the water line. If the interaction is of this form, you need to understand frequency of engagement, any patterns, and if similar engagements occurred in the past.

- *Do groupings of interactions have meaning?* Groupings may be horizontal or vertical. Horizontal groupings would be interactions occurring at the same time — somewhat similar to a number of error messages suddenly popping up at the same time. Vertical groupings would be trends over time, perhaps increasing communication or system usage by several departments, or many emails to one department every Tuesday and less on any other day of the week. Why were the interactions increasing and why were there more transactions on Tuesday?

These and other questions related to **WHY** help understand the issues and criteria surrounding the interaction. What is the driver or initiating force? What does it mean or matter whether the interaction takes place or not? Will your life change? Do your internet interactions affect what you eat for breakfast or lunch? How you get to work? Ask the *whys*. For example:

- Why were you emailing a friend? To share information.
- Why did you want to share information? Because your friend needed to know about the party.
- Why did your friend need to know about the party? Because your friend did not know that a party was on the go.
- Why did the friend need to be told about the party? Because the friend was not part of the inner circle.
- Ad infinitum.

But at this point, something is clearer. If the friend was not part of the inner circle, would your friend be welcome to the party and if your friend was let in, would he or she feel part of the party? You should always try to ask multiple *whys* and to force the symptom down to the real problem.

Characteristics

Thus far I have avoided specifics about what information the system actually deals with. This is abstraction and helps us with generic questions. Regardless of what the people or systems are working on or transmitting, we can ask the questions about cycles, social structures, and interactions. Thinking abstractly helps you to identify the essential or indispensable facets. *All that makes it what it is*. For example, let's re-consider the pizza making example. The essential aspects of pizza making are i) some trigger mechanism to know that a pizza should be made, ii) preparing dough in advance, iii) preparing toppings, iv) combining dough and toppings, v) applying heat, vi) dealing with the pizza once cooked, and vii) getting the pizza to where it should be. Those are the essential aspects of pizza making. Whether or not you use two ovens or one, if the oven is a hearth vs. conveyor, if there are olives or not — these are not essential and are not the essence. Before we get closer into the details, we have one final generic concept to apply to interactions: basic characteristics to think about for any interaction.

- *Norms*: It is likely that any interaction will have norms associated with it. These are the normal behaviors, baselines, conventions, and so forth.
- *Expectations*: There are the norms for interactions, but there are also the expectations for this given situation or structure. We have expectations about what will happen, what will not happen, when something will happen, when something will not happen, who will do what,

etc. These expectations are forms of assumptions and are one of the keys for understanding and analyzing the impact. The expectations may or may not be the same as the norm.

- *Exist at the moment*: Interactions exist in a moment of time and may have relationships to what happened before, what is happening currently, and what might happen in the future. Time has meaning. There is meaning before an expected event (e.g., email activity before a scheduled maintenance event), during the event, and afterwards. Time in the real world is rarely a simple case of t_i, t_{i+1}, t_{i+2}. Will the interaction have varying issues and impact depending on when it occurs?

- *More than one party involved*: Always remember, there are multiple players involved in any interaction. Information systems have no meaning without some form of interaction (usually multiple forms exist at once) and interactions require two parties to be an interaction (e.g., the essence of an interaction is having two players). There are likely to be different social structures involved (e.g., a single health care system may involve private practitioners, pharmacies, hospitals, government agencies, etc.). It is important to view and consider each one and not assume that all will be the same or have the same requirements.

- *Closing the loop*: The information system is a *system*. You need to think at the holistic level and include all of the key elements from start to finish. This is why I include the order trigger and the delivery in the Zen of pizza making! The delivery can implicitly include the consumption if you want, or since there is no right or wrong model make an eighth element for the pizza model if it is important.

8.6　Information

Each component of the framework progressively moves closer to the actual information and the impact. I have gone from the life cycle of the total system, to the major parts of society involved, and to the general types of interactions that exist between the players. Thinking in a systematic way helps us to identify key aspects at each level and these key aspects will likely be associated with assumptions and the ultimate impact of the information system under study. In this section, I will focus on a more detailed look on what is being handled by the interaction. An interaction requires something to be interacted upon (e.g., another essential part of an interaction).

I will look at the purpose of the information more closely, attributes of the information, introduce a life cycle view of information, and discuss information control.

Purpose

- *Information can be generated from inputs, captured and then later used*. For example, stores can capture sales information on customers, merge this with historical information associated with your client card and conduct targeted marketing.

- *Information can be transformed, stored, and either transmitted or retrieved*. ATM machines will transform your keypad entries into digital data, encrypt it and transmit to a server farm where your pin etc. will be validated, account information retrieved, and information sent back to the ATM. Information from an order entry system will eventually make its way to the billing and shipping subsystems and possibly to a collection agency.

- *Information can exist in a temporary state until additional information or processes are applied.* Certain processes such as insurance claims have partial information, and the various pieces are put together (medical claims, police reports, automotive body repair estimates, other parties involved, etc.) and this information may be quite fluid until all of the pieces come together. Pieces of the data may change multiple times as additional data is obtained.
- *Information can be reused for different purposes.* You supply information for shipping, but the store also uses the information for direct mailing. You get your car repaired and the vehicle maker and dealers use the information to track loyalty, trends of vehicle problems, etc.
- *Information can be presented in different ways for different audiences.* Information may exist in a digital way and be displayed in different ways to make it useful or to achieve different impacts. That is, the same information can have different impact depending on how it is presented and who it is presented to. Information can be presented in a variety of visual ways: text, tables, graphs, animations, etc. If the purpose of the information is for a scheduler in a factory to use, it better be in a form that allows the scheduler to see his or her factory.

Considering these and possibly other ideas, it is important to understand what the information system is directly and indirectly used for. You need to understand where the information goes and what the information is used for.

Attributes

Information is not just information. One item of information is not the same as another item. An instance of information today may not be the same as tomorrow's instance. Information has attributes that are related to information quality. The following attributes of information can be discussed:

- *Information can be too little, just right, or too much.* Imagine that you are doing shop floor tracking in a factory and the operations typically take one to two hours to complete. If the factory tracking system generates data once a day, this is not enough information. If the factory tracking system reports every millisecond, this is overkill and will generate extreme amounts of useless information. Probably a system reporting at a minute interval would be sufficient. If you ask someone their age, knowing to the minute is too much information, and an age rounded to the nearest 100 is also not useful. You have to match the granularity or information detail to the problem. If the granularity is not matched, the impact will be handicapped.
- *Information can be timely or tardy.* Several days after a mail option was updated, a message

was issued about the option changing. This is not a timely notification. The assessment of information quality must include this type of consideration. A warning that the server reached its maximum load is not as useful as getting a message in advance of the increasing load and its close proximity to a warning level.

- *Information can be ambiguous.* Too many users were signed on. What does too many mean? Many people did the function at the same time. What does many mean? Was it the majority? Was it above the expected norm? Ambiguous data and statements are next to useless. Does the information system provide sufficient information and detail so that the information can be interpreted in a clean and unambiguous fashion?

- *Information can be inaccurate.* Not all information in a computer is accurate. Sensing devices can fail and people can enter data incorrectly. There are many links in an information system process and any one can generate inaccuracy. Are there ways to prevent, detect, and correct inaccuracies? What is the impact of an inaccuracy slipping through? A common example of an error followed by a correction is a banking transaction. Look at your account records and on a regular basis you will see a deposit or withdrawal entered and then immediately cancelled. That was an oopsy. Not all errors are caught.

- *Information can be incomplete.* Information may be entered in an incomplete fashion or the information system does not provide sufficient options or features to capture the information completely. For example, people may have an address where they live and another address where they want the mail delivered. A mailing address may have the street, unit within a strip mall, and a box number for the unit within the strip mall. Some systems cannot handle this. They end up calling the small metal box in the wall a suite or an apartment. Incomplete mailing addresses can result in interesting problems. There are many other examples of incomplete data and issues that result. How does the system deal with incomplete or unanticipated data? What is the impact if the data is incomplete?

Information quality is a hard aspect to define. You can discuss what might happen if data is ambiguous, not at the right level of granularity, incomplete, late, or inaccurate. It is hard to put numbers on it. Is the information quality 65%? Who knows? It is important in an impact assessment to think about this. If the cost or risk associated with poor information quality is high, there needs to be a conscious and careful analysis on the impact. Can things be done differently and better? How can the risk be mitigated or controlled? In any system involving lives, health, and finances the issue of information quality is very important.

Information Life Cycle

Information does not just drop out of the blue (except from satellites). If information is viewed as a static entity, it is possible to miss many relationships and important facets. There are a number of issues related to the living or organic element of information that can be reflected upon in any analysis. Any important or essential aspects discovered through this inspection process become candidates for further discussion or for noting.

- *Information comes from somewhere or someone.* Who supplies the information? Information usually starts as raw sensory or generated data somewhere, is turned into information, and then hopefully in some cases turns into knowledge or a useful part of a larger whole. So, what are the primary sources and what are the intermediate transforming or conveying sources before the information is actually used for the various added value tasks or purposes? What are the sources of error and are there any special conditions related to the sourcing of data that affects quality and final impact?

- *Information is used somewhere or by someone.* While there might be write-only memory or storage devices, most information goes into mechanisms that can read or recover information that has been written or stored. Of course, some information may be collected and not used. It could have been used, might be used at sometime, and the potential for retrieval exists. Assuming that retrieval and use is possible, who uses it? Another software system? Hardware presentation systems? Who sees it? Who hears it? Who reads it? When and where do they do this? Does anything about the user or situation in which the user exists affect the impact of the information system?

- *Information is created, archived, and disposed of.* Very little information or data is permanent. Either through design or by accident, data is destroyed and lost. There are laws for retention duration in certain fields (e.g., tax information) but information can be filed and later disposed of. Does this process create any impact issues? Direct or accidental impacts? Immediate or long term impacts? On who? Why?

- *Information has currency — applicable or out of date.* The value or currency of information is not always the same and it can vary over time. For example, dental records of a child may have little value for cavity detection when the person is in his/her fifties. An MRI scan from five years ago may have limited value for detecting presence of cancer today. A bank account balance of last year might have little bearing on what is in the bank account today. Addresses change. Marital status changes. Family make-up changes. There are many things that change with time. If the information validity and value changes over time, what is the impact on the information system? Do past states matter? Does historical data provide clues and value for the current or future uses of the system? Sometimes yes, sometimes no. What happens if stale data is used? How is stale data detected? How is information kept up to date? What happens to stale data?

- *Information can be solicited or unsolicited.* How does the information or data arrive at the user? Is it pushed or pulled? Pushed information is unsolicited (think spam). Pulled information is solicited. Someone or something asked for information to be prepared or transmitted. It is important to understand what transactions are pushed or pulled and when they occur. Are there patterns to the events and triggers? Are these related to the potential or realized impact? If it is unsolicited, how does the interaction know when, who, how, what? If it is solicited, how does the solicitor know who, how, when, what? What are the costs or mechanisms for the solicited or unsolicited transactions?

- *Information can be transitory in nature and exist in the moment.* Sensory data in a control system exists for a short while and not all of it may be stored for later use. Some detailed data may be aggregated and the aggregation is stored as information. For example, the average temperature for the month may be stored on a tourist site and individual temperatures may not be. Other data such as prices for a sale may exist for the time of the sale and then simply be deleted when the sale is over. If information is transitory, does it have an effect on the impact?

- *Information can be imbedded (e.g., no active input sources) or dynamic.* There is now the idea of imbedded information sources. For example, what information is on the magnetic strip or chip in a credit or security card? This information is imbedded and some of the data is static and will never be changed or updated. Other information is dynamic and may change with every interaction or has the potential of change with an interaction. This can happen when the information is accessed or modified. What is the impact of having imbedded, static, and dynamic data in information systems?

If information is viewed as an organic and living entity, it is possible to follow information through its life cycle of creation, evolution, renewal, and eventual demise. The impact and issues related to impact may or may not be different at various points. If the impact is not affected, that is important to know. If the impact is significantly affected, this is also important to know and the relationships brought forward.

Control

The control over information is becoming an important topic. Identity theft, invasion of privacy, fraud, sharing of personal and private information, etc. are important topics in the analysis of any impact.

- *Can the information be stolen?* How secure is the information or data? Can it be hacked? Is access monitored? Who or what is the safeguard? What happens if the data is stolen? What is the liability or risk exposure? This can affect where the information is stored, design of security walls, and so forth.

- *Can the information be altered?* Or, can the information be copied or intercepted in transit? Is your password or credit card information secure and hidden from every internet server that sees the packet? If the information can be legally seen, who can change it? What changes are allowed? Any restrictions on timing? Can superiors override or change employee entered data? Are any changes tracked? Is the previous data also saved? Since people are people and machines do fail, is it possible to easily fix real mistakes? What is the impact if data that should be altered is not or cannot be altered? What happens if data is altered by accident or by intent? Can the alteration be detected and tracked?

- *Can the wrong information be used?* It is possible that the wrong information can be entered, generated, or accessed. People are not perfect, neither are algorithms, and neither are software applications. In some cases, the information is known to be wrong, but something is better than nothing and close is good enough. In aggregate form, little errors can cancel each other or be buried in the decimal place. To understand impact it is important to think about the degree of accuracy needed and know what the impact is of wrong information. Assuming that information will always be accurate and right is naïve. What controls can be put in place to minimize or recognize or correct wrong data? Can tests be done to detect silly results?

- *Can old or stale information be used?* We have already mentioned stale and old data. It is something that should be consciously considered in an information system if the use thereof can wreak havoc. Can it be detected? Can change control mechanisms be put in place to track and monitor data as it changes over time? Are there ways to know what data was used in any report or analysis? For example, it is common practice in writing to note the date and revision of material referenced and used: the edition, when last updated (even for websites). Old data can be useful to some people and useless to others.

- *Can the information be subject to the introduction of errors?* Shigeo Shingo is famous for his mistake proofing (Poka Yoke) techniques that are a key part of the Toyota Production System. He pioneered the ideas at Matsushita Electric Industry (aka the Panasonic brand) in the 1960s. Most people I know associate both manufacturers and their products with high quality. Shingo assumed people and machines are not perfect and that mistakes are natural. This is in contrast to some managers who have the idea that "we pay them enough, they should not make mistakes!" I have actually heard managers say this. In contrast, Shingo suggests that you analyze the process or system and look for where mistakes are natural. You then design simple little processes that either prevent or detect mistakes immediately. These ideas can be used effectively in information systems. Usually, they cannot be added in later and must be thought about in advance.

- *Can unsolicited and solicited interactions be controlled?* Spam. Tele-marketing. These are unsolicited interactions. So are denial of service bombardments or hacking activities. There can also be initially requested and then unsolicited interactions. For example, you might join a user group and then get sporadic emails telling you about various forums. Or, you might provide your email on a website or paper and invite people to contact you for further information. There are many forms of unsolicited and solicited interactions associated with information systems; e.g., error messages, pop-ups, warnings, update notices, etc. Therefore we need to distinguish between wanted and unwanted unsolicited interactions. Are there ways to reduce or eliminate unwanted interactions?

- *What happens when information holders are terminated?* Someone quits, retires, goes on sick leave, dies, or is terminated. What happens to the information that the person was responsible for? How much was in a system? Do others know what it is, what it is used for, what needs to be watched for, how updates are handled, what types of errors are common, ad nauseum? What is the cost or impact of losing key information? What controls are in place to protect information in the event of such problems? How much effort and cost is involved? What is the impact if done, not done?

- *Who plans or manages information and information system control?* You cannot have control without a control mechanism. Is control automated, hybrid, or manual? Who is in charge of keeping things safe and secure? Who controls the sharing?

Control is an important aspect. Too much control can negatively affect the desired impact, as can too little. Is there the right amount of control?

8.7   Impact, Value

Finally, I can talk about the V and I words — **VALUE** and **IMPACT**!!!! Perhaps I need a life or perhaps I have too much time on my hands, but I think that there is a lot to write about when it comes to understanding value and making computer systems useful to the user. It seems that many people use the words value and impact interchangeably. Check a thesaurus. There are differences, but there are a few meanings that do overlap and for the sake of not splitting hairs, I will use them interchangeably in this section. Impact can imply the generation of an effect; a positive effect can imply the generation of positive values. One could consider the impact as the instantaneous event or function and the value as the effect that occurs after the impact. In any event, think about the good stuff we want to have as a result of using the information system. If you are still confused about value, think about a lowly door stop. The *what* or purpose is to stop the door. The value of the door stop is in avoiding the many dollars needed to fix the wall if the door is slammed open.

I have not seen too many books that talk about value indepth. Most skip over this touchy topic because it is non-technical, involves humans, and other ugly factors. It is a subject that requires brain work. Perhaps many people think that understanding the user's value and requirements is the easy part, that anyone can do it, and that the real job is writing the code. Or, that the most important part about understanding the requirements are the pretty pictures and the diagrams that arise when specific methods are used. Or, that if you use the method or draw it like the book, the value will be delivered. Personally, I think that the hard part is not the actual software writing, and it is certainly not the thoughtless following of a recipe that ultimately gives nice diagrams without value. The hard part is delivering software that gives value. There is no magical or easy way to do this.

The other parts of the framework help you establish context and structure for the analysis. They will help you focus on important aspects and avoid others. They will help you identify various relationships that might have been overlooked. At this point you have information being conveyed via interactions between one or more players. So what? What is the real impact or value of the information system? What difference does it really make? Something is different, but does it make a difference? What will happen if the information system failed completely or was turned off? What are the big values and what are the bits that are needed to be done by the user and the system to deliver the value? What are the little values?

In this section, it is important to keep functionality separate from impact or value. When I teach or mentor, this always seems to be a hard concept for the pupils to grasp and keep a firm grasp on. What it does, is not the impact or value!!! Remember the door stop. A door stop stops the door at a certain point. If the door stop did not exist, someone opening the door quickly could create several hundred dollars of repairs. The *what* is the stopping, and the *so what* or the *value* is the prevention of damage. A door stop might cost less than several dollars. It prevents damage that is much greater. To help understand value better, there are seven topics that can be discussed:

1. Initiation.
2. Facilitation and continuance.
3. Source of value.
4. Who and what is affected.
5. Potential and scope.
6. Dynamics and control.
7. Dependency.

There are many possible points or questions within each topic and **not all of the points are relevant in all situations**. I cannot think of too many situations when I had to think about all of the questions or issues. You should be aware that the issues may exist and that there are many possible issues to consider. That is the point of the following subsections. When you first start analyzing information system value, it might be necessary to consciously ask "is this point or question relevant?", "do we know the answer?", "do we think we know the answer?", and "how confident are we that we know the answer without doing more work?"

Eventually, you will know what questions are and are not relevant. You do not mention everything in a report or study. You mention the stuff that *makes a difference*. If you make a point, ask: so what? Who cares? If it is minor and is not critical, do not raise it. Minor stuff creates noise and distracts the audience. You want to focus on the essence or the indispensable aspects of the system. What really provides the impact? What is the **real** value? The following explanations and examples borrow liberally from course notes (McKay 2003). Louise Liu was one of my graduate students when I started to formalize the value analysis framework in 2003 and she helped review and critique the course notes. She was also the first one to try to apply the framework (other than myself) to a real problem (Liu 2004) and as a result, she straightened out some of my ideas and fleshed out my points with good illustrations. I will try to bring in additional examples and illustrations from the other components to explain some of the points. Remember, not all of the following are appropriate for every case. Determine the important ones for your situation and then explore the phenomena.

The following sections are written from the perspective of analyzing an existing system — hence the tense and style. You can also take all of the points and turn them into present or future tense if you are considering the design of a new system. The points work both ways: to review or to design. I chose the style on purpose. It is probably wise to learn how to analyze and look at an information system before you actually build one. You should practice looking at different information systems and critique the value element (and the other *ZenTai* components). This will help you understand the value dimension and prepare you for designing and building systems.

For the following sections, it might work best if you imagine yourself sitting in front of a system that someone has hired you to critique…

8.7.1   Initiation

Why was the information system introduced and used in the first place? What was the situation and set of assumptions that launched it? Were there any conditions that facilitated or helped get value from the system?

a. *What perceived value was needed for the system to be authorized?* The driving forces for developing and implementing information systems can be economic reasons, survival reasons, or it can be that organizations seek to gain a competitive advantage by investing in information systems. For example, the manual customer files are too time consuming to keep current and to get consolidated data out of. A computerized system will address this.

b. *Was pre-existing technology or system structure needed?* Sometimes, pre-existing technology is needed; other times, system structure is needed. It's often that both are required. For example, in the days of dial-up access before high-speed connections, web-sites had to keep in mind the performance issues. Streaming video was not very practical and neither was voice over internet appropriate to consider. For system structure, think about what is being used. If you are looking for complete redundancy, you might need two different providers hooked to two different fiber systems and if your area is only supplied by one provider you will still have a single point of failure.

c. *Who were the groups or individuals involved?* It is useful to identify the related persons or groups for the developing and implementing of the information systems, as well as the actual users and the customers of the users. Each group is important for understanding how value is ultimately obtained (or not).

d. *Who pushed or pulled the information system development and use?* Was it technology driven or market driven? In the 1970s and 1980s, a number of companies simply invented new computer chips and other companies lined up to exploit whatever was invented. Here it is — do something useful with it. This is different from the market demanding something smaller and lighter, or with certain features and functions.

e. *Was the impact accidental or conscious?* The actual impact may not have been predicted by all participants. Was the impact documented clearly in the beginning? In some cases, the impact will match. In other cases, the impact will be different from that initially thought of. As an exercise, go look up the origins of the Xerox Star and study the user model and original expectations of the user interface that was invented. It is clear that the original design team did not anticipate what eventually happened to their ideas (hint: think window type interfaces and not text based).

f. *Was government, private, public involved?* To what extent, was the government or other parties outside of the firm involved with mandating requirements or imposing policies? The potential impact may be constrained or directed.

g. *What were the initial development and deployment costs?* Cost is an important component throughout the life cycle of developing information systems, particularly in the early stages. If the initial development costs are too high, the system will not get fully developed and the potential value may be marginalized.

h. *What was the shape of the S-curve and how long was each phase?* The S-curve can also be used to understand the past, present, and future of the information system. The curves can inspire reasoning and questioning: if something is in the very early stages of new technology and there is not a matching functionality, how confident can you be of claims of changing the world?

8.7.2 Facilitation and Continuance

It is important to know how the impact develops over time. Here are some of the questions that need thinking about if you are looking at an existing system:

a. *What is needed for the impact to exist at the present and how will the impact continue?* Is funding the critical factor? Is certain technology important? Is senior management support critical? Is the use related to specific people using the system? And, what might happen if different people use the system?

b. *Are multiple systems and structures necessary?* In a modern, complex world, there are normally several systems coexisting in the environment. When introducing a new system, it is usually necessary to reengineer the business process, which would often affect the organizational structure. Consequently, the impact on multiple systems and organizational structures may be interrelated. Does the value obtained from one system rely or depend upon another system generating special input data?

c. *Does the value arise from limited or extensive use?* Is the system used by a small subset of the potential users? Does it have sufficient (i.e., critical) mass to justify its ongoing use? Systems that are used by a small group are hard to support and sustain.

d. *If the system is used in multiple sites or by multiple users, how did the impact spread, what allowed it, what stopped it?* Is policy the critical factor? Quite often, when one organization starts to use a system, and the good word spreads, then several organizations in the similar industry start to use it. Eventually the system takes the dominant position for a certain period of time until another one takes over through competition. Was the system mandated by the head office and this is why every location is using it? If the system is public, why did the system only get a certain percent of the market share? Why not more?

e. *Is the impact regulated or controlled by policy?* It is possible that certain technology and ideas cannot be exported or used in certain countries. This is one form of government policy controlling the spread of potential value. Certain audit policies or strategic decisions can also regulate what a firm might do with an information system.

f. *What are the direct and indirect costs of ongoing usage?* Maintenance of an information system is an item to be considered during the planning for the deployment of a system. Planning does not only need to take into account on-going maintenance and technology refresh, but also on-going training. Maintenance often considers simple upgrades of operating systems and the hardware, but does it also include periodic development activities?

g. *Have there been evolutionary or revolutionary changes?* Evolution means no major infrastructure change, or major changes in technology. Revolutionary means there is a substantial infrastructure change or change in technology. For example, going from a Pentium II computer to a Pentium III computer is just an evolutionary change. But, moving a device like a camera from a film state to digital may be a revolutionary change.

8.7.3 Source of Value

What actually gives the value or impact? Get to the details...

a. *The value may be resulting from information that is new in its entirety.* For example, when multilane highways first introduced intelligent transportation systems, the displays had new information displayed. This information could be dynamically generated based on automatically detected traffic conditions.

b. *The value may be resulting from a different delivery of existing information.* One example is that of E-mail being used to communicate between two friends, possibly replacing the old form of mail. It is still text, but it is a different and usually quicker way to communicate compared to physical mail.

c. *The value may be resulting from a different use of existing information or transmission thereof.* Information can be interpreted differently and its value might depend on who views it. For example, quality numbers may indicate to production the number of defects being shipped, and if the same data is reported to Human Resources, the numbers might also mean that more training is needed.

d. *The value may be resulting from combining information in a new way, process or delivery.* For example, creating an integrated planning system that combines Industrial Engineering, Human Resources, production, finance and senior management oversight reports is new and different when all of the data is put together and correlated for management use.

8.7.4 Who and What Is Affected

The discussion of value has followed along the lines of why the system is brought into being in the first place, what conditions exist or should exist for the value to arise and to be sustained, and what type of information system manipulation is providing the value. It is now time to think about what is actually being changed or altered when an information system is used.

a. *The change may be to health, wealth, shelter, comfort, safety, or role.* On-line trading systems can impact a user's wealth. On-line auction sites can provide a seller with greater exposure to potential purchasers. Advances in nuclear medicine can help medical practitioners understand what is happening inside a body, and thus help the patient.

b. *The change may be operational, tactical, or strategic.* Operational is your day to day activities and decisions made each day. Tactical are decisions more related to policy or reasonably sized expenditures or efforts. For example, if there is a problem, some pain but the problem is not fatal to company. Strategic decisions are those that could cause bankruptcy if they are wrong or at least a major loss. Is the information system designed for strategic, tactical, or operational use? At what level is its impact felt?

c. *The change may affect where people are, what they do, when they do it, why they do it, and how they do it.* One simple example is the usage of online banking, which provides users with great flexibility in terms of when and where they can do their transactions.

d. *The change may be on the process and be visible or hidden from daily use.* For example, privacy laws have requirements for the destruction of data, security of data, and other functions and features — a normal user would not see the value of these safeguards until something bad happened.

e. *The change may be part of the end product itself.* With the growing prevalence of information services on the Internet, there are content providers, and functions such as search engines. In these instances, information becomes the actual product.

f. *Secondary changes may arise affecting other parts of the organization or process.* The use of an information system in the work place can result in changing the business process to meet the systems' requirements, such as the rearrangement of who records what information in what sequence. Or, when a certain task is done, moving a task from 9AM to 7AM. This could streamline the business function and could lead to more productive information flow from one party to another. Conversely, the impact might be disruptive and invasive, leading to inefficiency and contributing to ineffective processes.

g. *Secondary changes may arise changing the very nature of the problem.* For example, when online banking is used, customers move away from the bank counter or ATM connecting to the Internet. The impact on the bank is to save money on bank tellers and ATMs, but the bank needs to purchase computer software and hardware to establish the system. For the bank's customers, this requires them to use technology and that they can use it in the comfort of their homes. Online banking changes the dynamics and the nature of the banking system.

h. *The change may be uni-dimensional or multi-dimensional in what it affects.* Developing and implementing an information system could have impact on one organization on time savings, which consequently save staff time and money. But the impact could also trigger other areas of changes; for example, causing organizational structural changes, establishing the linkage with other stakeholders, or continuing the supply chain development, or gaining more market shares, etc.

i. *The change may result in interactions being created, destroyed, or modified.* Portable computers and telecommuting have created the condition where people can take their work with them anywhere and do it at any time. As a result, workers find their work is cutting into family time, vacations and leisure, weakening the traditional institutions of family and friends and blurring the line between public and private life.

j. *The change may be limited to certain societies, organizations, communities — creating real or perceived imbalances.* Access to a computer and information resources is not equitably distributed throughout the society. The access is distributed inequitably along racial and social class lines.

The changes can affect other aspects as well…

a. *Number of people involved — parallel or serially.* Introducing a new information system may require people to work together in different ways, and it might also mean more (or less) staff. If the system merges or combines work, this may imply better coordination and more parallel efforts between multiple groups. If the system progressively builds up value as the work flows through the organization, more coordination is also needed as one function is completed and another department must then engage and contribute its part.

b. *Skills or characteristics of the players.* With the demand of using new technical equipment or new techniques, users' skill levels are generally increased through relevant training and education. But, different skills might be needed and it might not be possible for all individuals to be retrained. There are also stresses (physical and mental) associated with change and the introduction of technology. For example, management might expect higher accuracy, more work done, certain tasks performed earlier in the day, all of which increase stress.

c. *The geographic placement and disbursement of the members.* Because of new technology, virtual teams, virtual offices, and distributed teams are possible.

d. *When the interactions take place.* Web-based systems make it more convenient to alter the time dimension. For example, it is more convenient for users to book an airline ticket 24 hours a day, 7 days a week by using online booking instead of visiting the agency during normal working hours.

e. *How long things may take.* Computer systems can certainly affect how long things take. They can take mechanical and manual processes and perform them quickly. A manual task taking one or more hours can be dealt with in seconds by hitting a button. Tasks that involve moving data from point to point are also affected by the speed of data transmission compared to sending physical mail to someone; E-mail makes it possible to exchange information instantly (almost) regardless of location.

f. *How information is presented and perceived.* Computer technology has also affected how information is seen and visualized. For example, presentations used to be conducted by using transparency slides or flip charts. With the usage of laptops and other software, presentations can be delivered in many different forms.

g. *What other interactions may or may not take place.* When new or different systems are introduced, prior interactions may be eliminated. For example, manual switchboards were made obsolete by automated switching systems.

h. *The influence or dynamics implied by the interactions.* For example, consumers often check the Internet for information about the goods they are going to purchase these days. When consumers go to a store and interact with the sales persons, consumers are no longer restricted to the sales person for information. The customer may even provide additional information about other products, sales in other stores, etc. for the sales person (such and such store has blah and is selling it for blah). The dynamics of interactions seem to be changed in many cases.

i. *The quality of the information quantitatively and temporally.* Technology can increase the amount of information available to a user or customer. In some cases, the information will be of high quality, but it is also possible to obtain copious amounts of poor or misleading information. Medical information on the web is a case in point. Caveat emptor.

I am not yet done though. There are still more issues to consider about who is impacted...

a. *Does everyone gain or lose equally?* When the ATM becomes popular at a bank, what happens to the bank tellers? This simple example reveals that not everyone gains from using information systems. After implementing a new information system, normally users get trained on how to use the system. How does it affect the performance of other individuals? Does it affect the performance of a group or an organization? Those should be assessed on a case-by-case basis and should not be taken for granted.

b. *Does the community, society or organization gain as a whole?* In some examples, it is clear, such as the invention of electricity benefited society as a whole (eventually). In other cases, this may not always be the situation.

c. *Does the entity gain at the expense of another entity?* When a firm starts to eliminate the intermediate channels and sells directly to customers via the Internet, the move can help the manufacturer. On the other hand, this affects the business of other competitors and also affects the traditional players in the distribution channel.

8.7.5   Potential and Scope of Value

An information system usually changes something (the impact) and the value arising from the change needs some further thought.

a. *The value may be neutral, negative or positive.* We've heard about a lot of examples of using computer systems and their benefits to business. But business also spends quite a bit of money to compensate and treat victims of computer-related occupational diseases (directly or indirectly through loss of time).

b. *The impact may be illegal, unsafe, unethical, unfair, and/or improper.* Computer crimes are the commission of illegal acts through the use of a computer or against a computer system. Computer abuse is the commission of acts involving a computer, which may not be illegal but are considered unethical. As with any other major new technology, computers have created new opportunities for committing crimes and have themselves become the target of crimes.

c. *The value may be negative by accident — a mistake.* This relates to liability and protection issues. It is now a common practice among many technology consulting firms to cover the liability issue with its clients early on to prevent any accidental events from resulting in litigation.

d. *The impact may change something, but does it make a difference?* Some systems may have temporary changes to a number of business processes, but it may not change the way that a company does business, or how the company profit would be affected; thus, the system doesn't make a real difference.

e. *The impact may create (absolute) or re-distribute (relative) benefits.* Sometimes, when a new information system was introduced, it created new benefits. However, there can be situations where the benefits are just simply redistributed from one department to another department.

f. *The impact may further create a substructure — societies, communities, industries, and organizations incestuously.* Since the World Wide Web became widely used, electronic commence was created as a new industry. Many e-groups were also created and eventually evolved to be e-communities.

g. *The impact may create options and further requirements on the societies involved.* As more and more information can be found from the Internet, with the files including images and video files getting larger, the demand for higher speed connections is also increasing. For society, it then becomes a requirement to build a communication infrastructure which involves various levels of government and communities.

h. *There is the latent potential and the realized potential.* Some information systems may have great potential, which will never be realized. It does not matter how great the software is, if the software is not used. It does not matter how great a function is, if the function is not used in a way that generates value.

i. *The impact may create greater good for some at the expense of others.* Some data mining systems produce an enormous amount of information, which could be very useful to a vendor but compromises the individual's privacy when the personal information is used without the individual's consent.

j. *The impact may have liabilities — short term and long term.* Consider business on the web that crosses state or provincial or country boundaries. What may be a legal activity in one jurisdiction, may not be legal in another. It is imperative that information system professionals become aware of how evolving Internet law will affect the medium they are charged with administrating, and be aware of what short term and long term effects this may have on Internet usage.

Scope of Impact:

a. *Broad or narrow.* When analyzing the impact of information systems, analysts can get attracted to a few areas that may seem obvious at the moment. It is more difficult to take a very broad view. It is important to ask various questions in different areas to access the broader scope of the impact.

b. *Deep or shallow.* Consider a consumer health web portal that aims to provide easy, one-step access to information on hundreds of health conditions, services and drugs. When there are not many users who know about the web site, it has limited impact and relatively shallow penetration. However, if more and more people find it useful, the impact to the community may increase.

c. *Temporary or permanent.* When using a new information system, redesigning business processes could potentially cause many clerical workers to lose their jobs. Will these loses be permanent? Can the workers now spend their time doing other more important, value-added services? There might be a period of time for retraining during which negative impacts will exist to be replaced by positive value after training.

d. *Hard or easy to achieve.* Whether it's hard or easy to deliver the value depends on perspectives. Some benefits of information systems can be realized quickly with enough resources; some may take very long time; some may not be able to be achieved at all.

e. *Immediate or delayed*: The impact to the society or the user base may not be immediate. Until a certain percentage of community users start to utilize a web site or information system, its impact may not be realized.

f. *Direct or indirect*: It is probably normal for a project team to consider the potential impact on certain users within the main client organization. It might not be as normal to consider the affect on other users who may not use the system but will still be indirectly impacted (e.g., using the output or affected by decisions made by the main client).

8.7.6 Dynamics and Control of Value

Value is not static and it can change like the wind...

a. *The value derived by a feature or function may change from positive to negative (and back) over time.* The dynamic nature of today's environment gives the possibility of impact to be positive and then switch to negative at times. This can be affected by when and where the features or functions are used.

b. *Negative value may be momentary and then disappear.* It is possible that when a new information system is first introduced to an organization, users will complain about productivity dropping down. However, these complaints may disappear as users learn more about the system, and become efficient in using it.

c. *The value may or may not be sustainable.* A positive value may be initially obtained when the system is first used, but will the impact or benefit be sustained over the long term?

d. *The value or resulting changes to the situation may lower the barriers to entry and alter the real or perceived imbalances.* For example, E-commerce would not have today's popularity in the 1980s. It cost quite a bit to set up a home computer and use the Internet. It is now easy to create E-stores and to access them. The ratio of on-line buying to brick-and-mortar store buying is changing as a result, and this has altered the customer–retailer balance.

e. *The impact may proceed from a luxury and exclusive item to commonplace and commodity status.* The personal computer was a luxury item when it first appeared on the market, but it is now considered a commodity.

Control of value generation:

a. *The impact may be incremental, continuous, or big bang.* Information systems are often tied very closely with the business strategies and processes. Big solutions will affect a business dramatically and need to be considered with caution. Incremental solutions will cause less change on the business world and consequently could be more controlled (positively and negatively).

b. *Can the value be predicted?* A wide range of people, including researchers and business people try to predict value, its timing, the nature of the value, the intensity of the value, etc. While past experience and knowledge of the unique situation can assist in forecasting, it is hard to predict the impact of information systems. Some people have done a good job predicting the future, but these are rare individuals and they have rarely repeated their winning performance.

c. *Can the value be measured?* There are different kinds of value. Cost-benefit analysis can used to assess the value in some situations. For example, projecting cost savings after implementing the system. However, cost savings can be hard to estimate because there might be hidden costs that are not initially considered. Performance impact, which is another cost to the business, can be included in the analysis. Sometimes the success of a project is measured by whether or not the objectives are met. Measurement can be quantitative and qualitative. Social impact and ethical impact are also hard to evaluate in quantitative ways.

d. *Can the value be controlled and dispensed?* It's important to understand the source of the value, and the driving force of the information system. If the scale of the impact is small and limited to certain areas, it might be possible to control. However, if the value generation relates to regulation and government, it might take a longer time or more resources to get it controlled. The Internet is such an example. The legal aspects of the Internet are still under development, long after its widespread usage.

e. *Can the impact be robust and fault tolerant?* Is the value sensitive to small changes in the culture, usage, demands, or user expectations? That is, would a minor change in one of these negate or neutralize all of the benefits? Would one small change in a competing system cause your users to flock to the competition? If these types of things can happen, the system value is not robust or fault tolerant.

f. *Can the value have a backup — alternative mechanism?* Depending on the importance of the information system, the user may require an alternative system in case of emergency. There are occasions when it is also necessary to provide alternative solutions for comparison purpose. If the value is generated by quicker and more effective communication and this is core to the business, is there a plan for alternative communications if the main system is down?

g. *Can discrete aspects or characteristics be controlled or changed?* When we examine the value of information systems, we could also ask this question to explore the possibilities of controlling various aspects of the system value. Is it possible to adjust or control various dials and knobs, or is there just one control mechanism for generating value?

h. *Can the impact of one change in the IS be predicted?* It is not unusual that modifying one part of an information system may cause unexpected effects on other parts of the system, especially when information systems become more and more complex. In some cases, it is possible to think through what might happen when a change is introduced. But, in other cases it is necessary to make the change in a controlled way and observe what happens.

8.7.7 Dependency

If you are looking at an existing system, you can also ask how dependent is the user on the system? Or, what is the value of the system dependent on? If there are easy alternatives, the situation is loosely coupled and the user is not at great risk. However, if the user has no real alternative, the user can be caught between a rock and a hard place.

a. *What if an information source changes?* If the source becomes unreliable, the information in the system will not be trusted; the end result from the information system may not be valuable.

b. *What if interconnections are changed?* Information systems often depend on interactions among various parties, such as in a supply chain. If the relationship between parties is changed, a contract ends, or one party goes bankrupt, the information system may not be usable since the continuity of the system is interrupted. The system may need to be modified when a new partner is found.

c. *What if turned off?* There may be a contingency plan to ensure that the business can still run — albeit in a less efficient manner. The system can be turned off logically or physically (i.e., license is not renewed or the power fails). If you cannot get at the data, convert it, or use it, what does this mean to the user? If you cannot generate that report at 7AM, what will happen? In some cases, it might be possible that no real impact will be seen, especially if the system is providing neutral value. Or, the situation might actually improve for some of the users if the system is giving negative value. It is always interesting to do the what-if and assess what would happen if the system failed or was not accessible.

8.8 Utility or Futility?

So, you might have waded through a few dozen pages on value and impact to get here. Over four decades, I have had to think about value a lot. I had to consciously think about it and not leave it to accident. As a result, I had to turn the idea inside out, upside down, and tear it apart. That being said, what on earth do you do with all of this?

As written, sections in chapter 8 can help you assess the value claimed or associated with an existing information system. If you play with the concepts, the sections can help you design and consider what you are doing or not doing while creating a system. In any event, the concepts can help you focus on relevant points. In the simplest case, you can use the points as a checklist to make sure that you have thought stuff through. But remember: there will **NOT** be a project where all of the stuff in this chapter is relevant at the same time.

You must use reasonable judgment about how much time and effort you actually spend on this. If the risks are low, the situation is a Class Ⅰ or Ⅱ degree of difficulty, and you can quickly adapt and improvise, you should not be excessive. However, if the system is critical or strategic, or a Class Ⅴ or Ⅵ, then you better do adequate brain work.

The biggest gain is in understanding what the IT means to the user. From there, you should align the interface and functionality with the value drivers and the steps necessary to obtain value. This is if you are the designer. If you are evaluating an existing system, you can identify the user's values and then critique the system to see how efficient and effective it is in delivering those values. From there, you can identify potential changes to the system to enhance value. You can do these types of things without thinking too hard or inflicting mental anguish on yourself. Having a systematic analysis of value helps you justify and defend investment decisions and rationalize the existence (or nonexistence) of certain features and functions.

If you have the user's value drivers squarely in front of you, you can easily identify the sequence of screens and functions needed to deliver the goods. You can move the fluffy stuff to the side and reduce the visual and operational noise experienced by the user. You can use this knowledge to provide information on the screen to the user — what value driver is active (what the user is trying to accomplish), what the user has done, what the user is doing now, and what remains to be done.

Having a value or impact analysis also helps with the user engagement. You can work with the user in a focused way. For example, you can have fun in discussions — "This is my perception of what you will get out of the system, is that right?" If you remember some of the stuff in earlier subsections, you will have identified various types of users, when and where they will use the system, and what the interactions need to be. This will help you document and explain the goals and objectives. Depending on the value, you can also quantify the value and set measurable objectives. Even if the concept cannot be quantitatively measured, you might be able to define qualitative measures or tests that could be done to ensure what you deliver.

I am sure that the previous sections appear as nothing more than a list of random questions and issues, and that you are not very clear on how to use them. One way to use the framework is to identify the main values to be delivered by the system and then use a fishbone diagram to think through how things are related to value generation. For example, you can use a line of reasoning: blah is the main value and to get this main value, these other bits are necessary. To get the other bits, you need these other things done, and to get these other things done, you need blah-blah. The fishbone can be used to visualize this. This will help with control flow through a system. Thinking about the interconnections and interactions helps you to remember who to involve as possible stakeholders and if they might need special functions. Contemplating the actual source of value helps with creating virtual machines or wrappers inside the system; to naturally fit the problem. There is not one clear and straightforward path through this topic or recipe. Just thinking about these various issues helps with understanding the user requirements and how the system should be designed to deliver value. Each topic is a bit of the puzzle.

These are the first and most obvious uses of the framework. Checklists and ways to organize the value chain on the user interface in terms of a work flow, and how to organize the inner software structure so that it matches the value generation. It has been my experience that a software structure organized on these principles is not bad. It is pretty responsive to user demands and requirements as they are usually driven by desired value.

It is also possible to use the framework for larger purposes and for bigger thinking and this may require the combination of tools. For example, if you have done a process model of the situation, you can work your way through the model discussing what values arise where — potential and existing. You can identify barriers to value generation and you can identify where certain values might be measured (or possibly measured). David Tse, as part of his Master's thesis, did this type of analysis for a hospital's cancer clinic and its web portal. He mapped the clinic's processes as a finite state model, overlaid the web portal functionality, and then worked the model with the value framework in a systematic fashion.

If you want to think a little bit harder about value, you can even use the concepts in the framework to create generic models of value. For example, for a class of consumer electronics, is it possible to create a taxonomy that helps understand value in a more formal or structured way? As part of a research assistantship, Sylvia Ng worked with me on such a task. The next section comes from a working document that we prepared and illustrates what can be done when value is thought about at a deeper level.

8.9 Value Analysis — an Example

Without going into all of the background, Sylvia and I derived a model of value that seems to capture the value equations for consumer electronics. We got there by going through the framework, listing, grouping, and thinking things through along the lines suggested by the framework.

For consumer electronics (CE), there are possibly eight dominating attributes associated with value. Think about digital cameras, MP3 players, DVD recorders, and things like that. The eight dimensions or attributes of value we derived were:

1. Utility or enriched value.
2. Results or usage.
3. Temporal dynamics.
4. Operator or customer.
5. Consumer dependency.
6. Spatial dependency.
7. Co-dependencies.
8. Concurrency.

Utility or Enriched Value

☐ First, value can be described as utility or enriched. Utility value does not have any conscious feelings of pleasure or enjoyment associated with it. The CE that supplies utility value is like a tool. Enriched value gives an added intrinsic feeling of pleasure, entertainment, enjoyment, or extra feelings of efficiency or effectiveness. The enriched value may be values such as memories, relaxation from listening or watching various forms of media, or just the awareness of using something that is cosmetically pleasing, is well-crafted, and feels good when being used. A CE product can be dominated by utility or enriched value, or be a balanced mix. The utility or enriched value can change over time, the context of usage, who uses it, and any of the other dimensions.

Results or Usage

- The basic value can be:
    1. value derived from results while in use.
    2. value derived through operation.
    3. value derived from permanent or semi-permanent results of use.
- It is possible for the value associated with a product to be primarily aligned with one or more of these points. Any CE product can be likely described by a weighted description of these three. For example, a product with a rating of 60-30-10 would describe a product supplying the majority of its value to a consumer from the immediate results, some value through its use, and a minor contribution from permanent results. Consider a camcorder being used for a live feed with a stabilizer active. The primary value is immediate from a stabilized image that people see. The secondary value is to the camera operator who does not have to use a tripod, carries less equipment, and has more flexibility. The third value is from better archived footage if it is to be viewed. If the purpose of the shooting was to create video for subsequent broadcast, the value model would be expressed differently.
- If the utility and enriched concept is combined with the results and usage, it is possible to create a number of similar product groupings in terms of value (as defined at a high level). The products in each grouping would share operational traits and possible expectations of the users. It is possible that a development for one product in the group could be easily exploited by other products within the group. Some of this happens naturally within research and development groups, but it can be harder to see when product development is widespread.

Temporal Dynamics

- Does the value change over time? Even if the same person uses the CE product, does the value perception and utility change? For example, consider a picture of a young child photographed by the parent. This picture will have certain meaning in the immediate future to the parent. There might be a different meaning and value several decades later. This is an example of the result value being different with time. The usage value can also differ with time as the younger person with strength relies less on the image stabilizer than he or she will as old age reduces strength and steadiness. For some products, the value will be reasonably independent of the temporal domain (usage or result). It is important to think about this axis during design, implementation, and roll-out. You want to make sure that the potential future value can be realized after the initial purchase is made. This is somewhat related to the ability to target and sell to different age groups, but it is a different dynamic. One is selling into the different age categories; the other is continued use and value across the different time zones. It is also important to think about the perceived or realized value increasing or decreasing over time through use or through factors beyond the immediate CE product.

Operator or Customer

☐ The operator of the CE product may not be the main target or receiver of the CE product value. For example, a father might be taking the pictures of a child, the mother involved with straightening hair, shirt collar, or wiping food off of the face. The father might get a nice memory for the office desk or wall from the many pictures taken. The mother wants ALL of the pictures that were taken, regardless of lighting, sharpness, or child's image. The mother, aunts, and grandmothers (and others) enjoy seeing all of the candid pictures "that is my Johnny. He has food on his face and tongue out, isn't that cute!!!" These are also the pictures that surface in later years during family celebrations. "Remember when he did this!" When the father starts to delete the pictures he thinks are rejects to save space or file maintenance issues, is the father thinking about the mother's value in later years? In a family, who has traditionally stored and organized the pictures after they were taken? The photographer or the interested party??? This is an example of different values for different people involved with the CE product.

Consumer Dependency

☐ Does the value change substantially depending on the individual who is the operator or customer? For example, would two photographers using the same camera have different values? Would two grandmothers have different values of the same photograph?

Spatial Dependency

☐ Does the result or usage value differ with the geographic placement or location? Would someone get a different value if it was used in a rural setting versus a city? House versus outdoors? In these types of ideas, the questions would relate to why does the value change?

Co-Dependencies

☐ Does the result or usage value depend on the presence (current, past, or future) of another CE product? Is the interconnection necessary for any or all of the value? Does the value result from a live connection or from a create-store-transfer-load-use connection? Does the co-dependency amplify, transform, or attenuate the value? If value can be derived without the connection or co-dependency, what is the new or changed value above and beyond the nominal values associated with each CE product? Are there multiple choices for co-dependency and leveraged value?

Concurrency

□ Does the value depend on the numbers of people involved in parallel or serial usage of the CE product? Is the value relationship linear or non-linear of one or more people participating at the same time in memory sharing?

By using these eight characteristics to categorize a CE product, it is possible to group and segregate CE products into value groupings. These groupings would not be based on media type, but based on the type of value and value dynamics. This major categorization allows direct commonalities and relationships to be seen. It also helps to identify potential, realized, and perceived values. Describing or measuring each dimension is problematic. Relative or qualitative values are likely possible to think about from surveys, field studies, or focus groups. A polar star type of graphic might be a way to represent the dimensions and the relative position.

Each radial edge measures value characteristics that a CE provides. For instance, for the edge representing concurrency, a CE may allow many concurrent users on one extreme, and only one serial user on the other extreme. Marking off the characteristic of each value on the radial edges and connecting the markings gives a shaded area that tells what type of value the CE provides.

For example (this is totally fake):

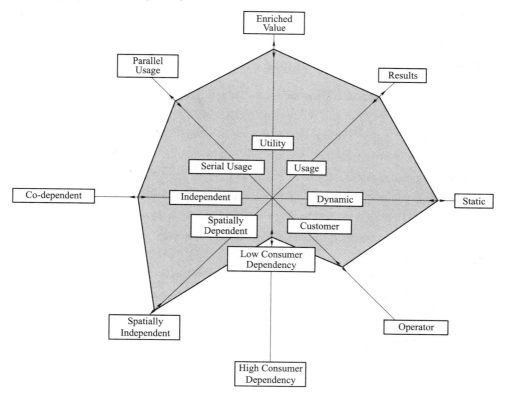

Value Representation for Consumer Electronics

The general shape provides the value map. This would be the composite value definition for one instance or one set of assumptions. A product would have different maps at different points in time. A single map could also be used that would illustrate the dominate weight or priority (e.g., pleasure is the intent, not utility). Consumer products could also be analyzed and grouped by similar shapes of the value map to identify complementary or potential opportunities.

Going through the categorization process as part of a product management process also ensures that each product's attributes are thought about. It is easy during a product development process to forget or otherwise ignore a key factor that will alter or define value.

Another use is to identify cross groupings of similar characteristics along one or more of each of the eight attribute types. For example, this could potentially show complementary products that are similar on the spatial dependency: two seemingly independent CE products that provide value in the same geographic space.

A systemized categorization can also be used in the analysis of competitive products in a gap analysis. This is how you could identify unique or innovative or better or worse value aspects.

Many of these insights are undoubtedly being obtained today through ad hoc processes and through the intuitive skill of the most experienced designers. The challenge is to make the practice and concepts more widespread and institutionalized throughout the organization.

The following sub-section interprets the general IT analysis framework and the high level Consumer Electronics model and speculates what an operational model of Consumer Electronics might look like.

8.9.1   Operational Model of Consumer Electronics

In this subsection, a more detailed systems analysis of consumer electronics will be undertaken. Section 8.9 started with a specific sub model for Consumer Electronics. Combined, the five general structures in the general value framework model and the eight dimensions in the high level CE model clarify and give a structure for viewing and talking about value. You use them together to think through the general strategic and tactical aspects associated with value. However, this is not yet enough to make a value analysis crisp. An operational model with more preciseness is necessary. For each grouping of IT based systems, there would be a high level value model similar to that presented in section 8.9 and an operational definition. In this section a preliminary concept for a CE Operational Model is presented.

In the operational model, a CE can be thought of as a system with various components. There are three main classes of components, and different component types within each class:

1. External Components.
   a. Environmental.
   b. User(s).
   c. Electronic(s).
2. Connection Components.

3. Internal Components.

    a. Interface.

    b. Input.

    c. Processor.

    d. Storage.

    e. Power source.

A diagram of the operational model can be found in Figure 8.9.1. The external components are any components that exist outside of the physical boundary of the CE. The connection components then connect the external components with the internal ones. The components and their affects on the value attributes are described in further detail below.

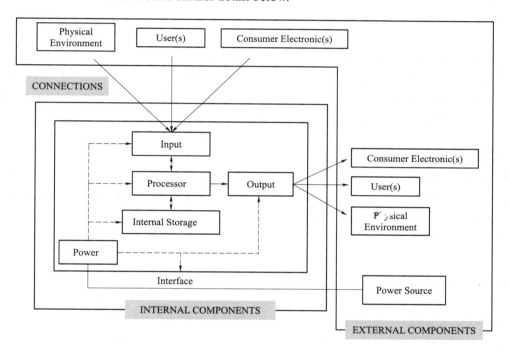

Consumer Electronic Representation

External Components

Physical Environment

The physical environment is the surrounding situation that the CE is used in. Characteristics of the physical environment include place and time, and can be described by variables that affect CE usage such as light and noise level, humidity, and available power outlets.

The physical environment is directly linked to the *spatial dependency* value attribute.

Depending on the environment, users may get different values out of the CE. For example, users who listen to music using an MP3 player in noisy environments may not get as much value out of the player as those who listen in less noisy environments.

User

It is important to define the type of user using the consumer electronic, as it is the user who is impacted by, and finds value in, the device. Users affect all eight value attributes, especially *consumer dependency*. Whether a user finds value in the *results or usage* of the device depends on the user's age, social status, culture, gender, and other characteristics. One CE may have different users, and some users may just set up the device while others may just use the device to produce output after setup.

Continuing with the MP3 player example, an athletic user may value a light weight portable player to use while jogging. On the other hand, a user who uses the player at work while sitting at the desk may prefer a larger player that is able to offer more storage capacity for songs and that might also have a larger display.

Electronics

Often CEs do not work in isolation and they need to connect to other devices to give the user added value. For example, an MP3 player may connect to headphones, speakers, a stereo, a computer, or flash cards. The external devices that a CE connects to affect the *co-dependency* value attribute.

Connections

The connections are what connect a CE's internal components to its external components, taking in input and power and giving output. The connections can be wireless, infrared, direct, or through cable, each of which affects transfer speed, capacity, error rate, security, as well as setup required.

The connections affect a CE's value through affecting *concurrency*, *co-dependencies*, and *spatial dependency*. For *concurrency*, users may be able to use the CE in parallel depending on the number and types of connections available. Different connection types such as wireless or cable connections may allow other users in diverse locations to use the CE simultaneously, either directly or through another CE. Users may use physically different connections for parallel usage, or share a connection. Value is affected by how many people may concurrently use the product, the quality of a connection as more users share it, and the type of usage that may be shared.

With the MP3 player example, perhaps the player has two headphone connections, and users may listen to the same music together. Perhaps the player is able to play different streams of music to each headset. Or, if the player may be connected to speakers, then anybody within earshot may enjoy the music. Each scenario changes the value of the MP3 player. DVD recorders provide another example: can you play back a movie, while recording a second movie to the hard disk, while dubbing or copying a movie from the hard disk to the DVD? Or, can you only do one or two of the tasks at the same time? What would be needed to do all the three at the same time?

Connections also affect *co-dependencies*. If a CE's value is dependent on usage with another CE, but a connection between the devices is not available, then the value is lost. As an example, an MP3 player may be dependent on a connection to headphones or speakers for users to listen to music. Similarly, connections affect *spatial dependency*; for example, the voltage rating on an MP3 player battery recharge power connection affects whether or not the player may be used regularly in North America or Europe.

Internal Components

Interface

The interface is the component between the connections and other internal components. The interface that the user interacts with can be broken down into two parts: controls for taking input, and displays for showing output. Controls may either be implemented in hardware or software, and can be analog or digital. They may take input in several modes such as speech or mouse movement, and vary on how much and what type of user movement is required. Intuitive or easy use of interfaces decreases the amount of learning required by the user to know how to work the consumer electronic. Other types of interface include any sockets or receptacles for storage devices, power connections, or networking.

The interface affects the *results and usage* of the CE, as well as the *temporal dynamics*, *co-dependency*, and *concurrency*. In terms of *results*, all CE outputs are passed through the interface before reaching their intended destinations, and therefore the way the interface displays and transmits output very much determines the results of the CE. For example, music is delivered to the user through the interface between the headphones and the MP3 player. A poor interface may result in noise in the music played.

In terms of *temporal dynamics* and *usage*, the way the interface takes in CE input affects value. As an example, adults may have trouble picking a song using buttons on a MP3 player the size of a thumb. On the other hand, youngsters may find the same player conveniently sized. As the youngster grows, he or she would find the player to be of different value.

The interface also affects *concurrency*, affecting the number of users that may use the CE at the same time, and what can be used. If two users are connected to an MP3 player through separate headphones, but there is only one set of controls playing the music, then both players would not be able to control what he or she hears at the same time. On that point, is it possible to have the one MP3 player provide two separate feeds at the same time to allow sharing? Would you even want to do that?

Lastly, if the CE interface does not provide the output or feedback in a desired form, dependency on another CE to provide value may arise. For the MP3 player example, if the user wishes fast forward between tracks, but the control is not externally available on the MP3 interface, the user may depend on headphones with built in controls to deliver value if the MP3 device recognizes controls via the headphone.

Input

The input component reads input, and includes sensors that read from the physical environment, readers of devices connected to the CE, and readers of user input. Sensors may sense light, sound, heat, or other physical variables; as an example, an MP3 player may sense radio waves to give radio output to the user. Readers of external devices may read digital or analog data; an example is a SD card reader. Readers of user input collect input through the interface; an example is a pressure sensor that registers input when the user touches a scroll pad to scroll through songs on an MP3 player.

The capability of the input component affects *spatial dependency*. For example, good radio sensors allow a user to listen to radio in areas farther away from broadcast stations. The input component also affects *results or usage*, the most obvious example of usage being when input is required from the user. In the pressure sensor example above, a poor sensor may require a user to press harder, or one that is too sensitive may cause overshooting in the scrolling result.

Processor

The processor is the brain of the consumer electronic, and can be classified by processing speed, cooling requirements, the type of information it processes, and the type of processor it is.

The processor of course affects the *results or usage*, as it generates all output of the CE. Noise produced while processing as well as time delays in processing affects also the usage of the CE.

Storage

The storage component is the memory of the CE, and some classifications include capacity, speed of storage and retrieval, type of storage (card, disk, tape, or other), type of information stored, and whether data is backed up in case of damage.

The storage affects *results or usage* as well as *co-dependency*. The *results* are obviously affected by whether outputs may be stored internally or not. An MP3 player that stores MP3s internally means that the user does not need a separate input connection to play music; music is simply read internally. If the storage type or capacity required is not sufficient, the device may then be dependent on external devices to add value; for example, users may turn to the PC for storing additional MP3s.

Power Source

The power source determines how the electronic device receives, stores, and uses power. The power source can be chargeable or not, and be removable or not. The capacity, voltage rating, and life are other characteristics.

The power source affects value through affecting *spatial dependency*, *temporal dynamics*, and *results and usage*. If the power source is chargeable, then the user may take the CE away from an access point to a power grid, possibly outdoors in a park. The value of the CE over time may change with the capacity of the power source over time, and the amount of power available affects the quality of output the CE generates. The shorter the battery life, the less time usage a user gets out of a certain amount of power, which can then result in lower CE value.

Output

The output component takes any output generated by the processor and sends it to the external components through the interface. Outputs can be in different modes (sound, light, etc.), can be written at different speeds, may require user responses, and generate value to the user by giving results.

The output, along with the processor, are the components that determine what type of output the CE is capable of producing, and hence the value of the CE's *results*. The output produced should be compatible with the interface to provide value; high quality music is wasted if it must pass through a low quality interface on its way to the user.

Overall Consumer Electronic

Overall, a consumer electronic may have some or all of the components described in this section. The size of the entire device would determine whether it is portable or not, and the components themselves may be dynamic or static. Other characteristics include the stage of the lifecycle it is in, what signals pass through (for example, audio, video, and still pictures), and the type of output it provides. One purpose of this example is to show that you can have a dialog about each component and how it relates to value. Does it extend or facilitate value or does it constrain it? Is there a way to engineer more value?

In summary, the overall CE may be classified by four functionalities:

1. An input device primarily used for capturing content.
2. A processing device primarily used for processing content already captured by another device.
3. A storage device for storing content.
4. An output device for showing processed content.

Any CE that encapsulates all functions would be less dependent on other devices than a CE that encapsulates only a subset of the functions. A comparison can be found in the figure below. *Co-dependency* ultimately is a function of what type of output is required by the users, and whether one CE may provide that output at all times.

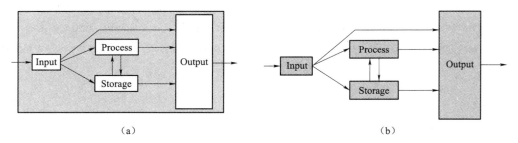

A CE that encompasses all four functions of input, storage, process and output(a)

is less co-dependent than a CE that encompasses less functions (b)

The lifecycle that a CE is in also affects the value attributes. Device A may be *co-dependent* on device B to add value, but if B is not being widely adopted, even though the connection between the devices is seamless, users of A may not get the maximum value. For instance, an MP3 player may include an SD card to save songs. If usage of SD has been widely adopted, then the user of the player can share songs with other users who use SD, and put the SD in other widely adopted devices for added value.

The overall CE also affects *temporal dynamics*. A CE revolutionary in design may not become obsolete as fast as an evolutionary design, giving more value to the user over time (but, if it is not widely adopted, it might have a very short life). Backward and forward compatible products are also more likely to give value over time, and the same for products designed to last long. Perhaps the batteries keep their original storage life over the years, or the processor's disk does routine self-cleanups. Or perhaps the user interface is flexible and may change text display size or button functionality to adapt to changing eye sight and hand dexterity as a user ages.

Discussion

Even if a formal structure is not used, a focused look at value and what constitutes value is important. For example, what about the father deleting pictures that the mother (and other family members) may value later? Should a digital camera equipped with adequate memory keep 4×6 inch quality snapshots in storage, when the larger image is deleted? Should a separate function be needed to delete the images that might provide memories?

8.9.2   Considering the Elderly

The value discussion forces you to explicitly think about the user and assumptions being made about the user. From my casual observations about electronics and the toys, most of the goodies seem to be designed by and for the 20-something's or 30-something's. This is interesting because I remember a newspaper column noting that people over 50 control over 70% of the net wealth in the USA, people over 50 have over 50% of the disposable income, but only 5% of the advertising budgets are spent on people over 50. If the advertising budget reflects the research and development focus, then it is possible that the meaning and value of Consumer Electronics to the aging population is under-emphasized. This suggests opportunities to exploit the potential needs and values of the older generation. Important point: older people do not want to buy products labeled "for the old and feeble," but it is possible to provide alternatives and elder-friendly choices.

Think about this: do the people 40'ish and older have different value maps if you do the analysis? Do they use the product for the same reason in the same way, in the same places, at the same time as younger people? Are there physical differences between the young and the old that should be taken into account in the IT? Hardware and software? I spent a little bit of time thinking about this with an undergrad research assistant (Hao Xin) and there are definite differences and concepts to consider. There are the physical realities of sight, hearing, and joints that impact what can be seen, heard, and manipulated. The older generation may also have different experiences and comfort zones when they approach IT. I recently bought my mother a new TV. The concept of a TV was not new to her. She had used TVs for about five decades. It took me 1 1/2 hours to show her how to use the new TV and she had great anxiety about it. Why? I will return to this later. But, just the fact that she had anxiety and difficulty using the new TV is interesting from a user-product relationship perspective. Sorry, assumptions being made. I find it interesting, while you might not, but if you are going to be a designer or architect, I hope that you share my interest in such things. Hao prepared the following diagram based on her research:

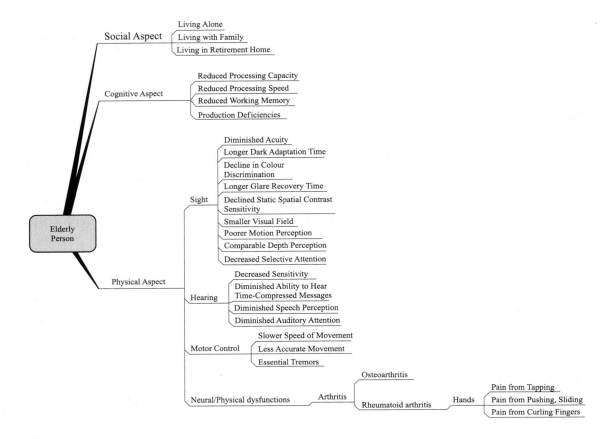

Based on our discussions, she then related this model of the elderly to Sylvia Ng's CE model presented earlier. The following is a slightly modified extract from her final research paper (reproduced with permission):

How does the model of the elderly interact with that of CE? How do the two models together generate information that will help designers evaluate consumer electronics and make valid design decisions? To arrive at the interactive model, the CE model underwent some refinement.

Firstly, how would consumer electronics be evaluated according to the CE model? Usually, a dimension has grades of membership; how would a CE be placed on each of the dimensions? It is difficult to quantify or determine the membership on the CE dimensions. For example, how can one decide if an electronic is 50% or 80% enriched on the utility-enriched value dimension? After some debate, it is determined that the dimensions are either binary or tri-nary. The value of a CE would be at one end or the middle of the eight value dimensions. This is a starting point and keeps it simple. It might be necessary to have more resolution, but you can start with a few and see what happens. In some cases, the scale can even be binary. Unless otherwise noted, the scale used in this description is tri-nary.

Secondly, how does the Elderly Model modify the placement of an electronic on the CE model dimensions? More precisely, without specific examples in mind, how can each of the dimensions in the Elderly Model *potentially* affect those in the CE Model?

The modifications of the CE model and the potential for modification by the Elderly Model are discussed below.

Utility-Enriched

A CE can be purely utilitarian, purely enriched, or a combination of the two. For example, a person might consider a classic desktop computer to be utilitarian (a working computer), an artifact of the first hard-drive ever made to be enriched (historical significance), and an Apple iPod to be both utilitarian and enriched (plays music and nicely designed).

How the cognitive dimension of the elderly modifies the utility-enriched dimension is difficult to say. The elderly usually experience reduced information processing capacity, reduced speed of processing, reduced working memory, and production deficiencies. But as long as the elderly still maintain normal mental and intellectual health, the individual can find enriched value in a CE. "Enriched value gives an added intrinsic feeling of pleasure, entertainment, enjoyment, or extra feelings of efficiency or effectiveness" (from the work done with Sylvia Ng as noted above). Creativity, intelligence, education, and ethnic background are just a few of the factors that will affect how a user will evaluate a CE. These factors are highly variable from individual to individual. What one person finds beautiful could be hideous to another. Production deficiencies here play a more important role than the other three aspects and negatively modify the utility-enriched dimension, because the elderly today have learned how to operate an older generation of CEs (e.g. microwave ovens, television). They have learned how to operate CEs in a certain manner, and have developed certain expectations on how CEs ought to operate — these are production rules that they have learned and try to apply, but might not be applicable to the new classes/generations of CEs. This is one possible reason why elderly persons might not be able to use a new TV even though they have used televisions for decades before.

The physical dimension of the Elderly Model has high potential to negatively modify the evaluation of a CE on this dimension. If a CE just works and helps a user accomplish a desired goal, it is utilitarian. However, if a CE is easy to operate, and does not place demands that exceed the physical capacities of the elderly, then it has enriched value. The physical limitations of the elderly place restriction on the design of CEs. What works for younger adults might frustrate the older user. For instance, the ever smaller handheld devices with small screens and smaller buttons are likely to stress the eyes and aggravate arthritis, thus decrease enriched and/or utility values for the elderly.

The social dimension does not have high potential to modify the evaluation of CEs, because the operation of CEs is independent of social environment. A CE operates the same way, for the same person, under different social situations.

Usage-Results

A CE is usage-oriented if the value comes from the results while using it or from the operation itself, and it is results-oriented if the value comes from permanent or semi-permanent usage of it. The results end of the spectrum has a temporal effect, because the value is derived from usage over time, or results that have been accumulated over time. A CE's value could also be a blend of usage and results. For example, a watch is usage-oriented (if it is used to keep time); an exercising machine is result-oriented (the benefit of the CE shows only after long-term use), and a video-game machine is both usage and results-oriented, because the user derives value while using it and from continuous play.

The modifiers from the Elderly Model that appear to matter for the usage-results spectrum are cognitive and physical. Cognitive factors have the potential to negatively modify this particular dimension. Certain types of CEs might require more working memory for its operation; for example, remote controls with 30-plus buttons demand more in terms of information processing and working memory. When long-term memory is corrupted by disorders such as Alzheimer's, the value is pushed towards the usage end of the spectrum for certain classes of CEs, mainly because the brain cannot retain the consequences of short-term usages and accumulate them into long-term results.

The physical factors could negatively modify the usage-results spectrum if an elderly cannot actually operate a CE due to physical limitations, but the same spectrum can be positive if the elder can enjoy the outcome of it when others use it and pass the results to him/her. For instance, grandparents receiving an album of pictures of their grandchildren taken with a digital camera, although they do not own or operate a digital camera themselves because they have arthritis or hand tremors.

Temporal Dynamics

Temporal dynamics is a binary spectrum, as the value of a CE either changes over time or does not.

The cognitive aspect of the Elderly Model has the highest potential to modify it, since memory is needed to compare the value now to the value before. When memory is only affected by the normal process of aging, how a person values a CE depends on his/her personal preferences. However, when memory is degraded by psychological or neurological disorders, the temporal domain becomes less prominent.

Operator-Consumer

A person can be an operator or a consumer or both. A person who owns and uses the CE and enjoys the output of the CE is both an operator and a consumer. However, a person who cannot operate the same CE because of ownership or physical limitations can still get the output of the CE from someone else and become a consumer. Similarly, a person who operates the CE for the sake of someone else is just an operator. For example, a grandparent at a family picnic becomes a consumer of a digital camera even though he/she does not own or know how to use one. The parent who takes pictures with the digital camera is both a consumer and an operator. Finally, a relative who is asked to take pictures with the camera is an operator.

The operator-consumer spectrum is modified by cognitive and physical aspects of the user. Cognitively, a person needs to have a level of rationality to consume the goods. Physically, a person needs to be able to use the CE (e.g., push buttons, press keys) to become an operator. For the elderly, the cognitive and physical characteristics might push the value of the CE more towards the consumer end of the spectrum, if they cannot or would not operate a CE due to physical limitations or frustrating design.

Another modification to this spectrum developed during the course of analysis — the operator-consumer spectrum became a sub-topic of the usage-results spectrum. Whether a user is an operator or a consumer depends on his/her original desire: does he/she wants to use it, or is he/she merely interested in the end result? So usage corresponds to operator; results corresponds to consumer, while in the middle is where usage/results and operator/consumer converge.

Spatial Dependency

Spatial dependency is a binary spectrum. A CE is either spatially dependent or independent. The value of the CE is modified by the physical aspect of the Elderly Model. Sight and hearing are good examples of how a CE's value changes spatially. Since elderly persons have poorer color discrimination and longer glare recovery time, the value of a handheld device with a reflective screen is different when it is used indoors versus when it is used outdoors. Similarly, since elderly persons have poorer ability to shadow audio streams in one ear while listening from both ears, the value of a television changes depending on whether it is used in a quiet setting like a residential home or a noisy location like a sports bar. The cognitive and social aspects do not really affect the spatial dependency spectrum, because the mental state does not usually change from place to place to affect the use of CEs, and what is spatially independent is usually also socially independent.

Consumer Dependency

The consumer dependency spectrum is binary, as a CE's value either changes according to each individual user or it does not. The cognitive and physical aspects/dimensions in the Elderly Model have high potential to modify this spectrum, because they modify the characteristics of the consumers, including their ability to operate and appreciate the CEs or their outputs.

Concurrency

Concurrency is a binary aspect, and the social aspect of the Elderly model has high potential to modify this spectrum. A big determinant is whether an elderly person wants to share a CE or the outputs of a CE with others in a parallel or serial manner. A CE could have the potential to be concurrent, but if the elderly user does not want to explore the benefits of concurrency, then the potential is unrealized. An elderly living in isolation or feels isolated is less likely to have concurrent use of a CE.

Co-dependency

The co-dependency spectrum is binary, as a CE could either be dependent or independent of other CEs. This spectrum is not modified by the Elderly Model, because the co-dependence of a CE on another is not altered by the characteristics of the user. The potential for co-dependence could be unrealized, but is not changed by the individual.

Hao's and Sylvia's combined example illustrate how the concepts of value and a contextual value analysis can be used to guide a systematic analysis. The choice of Consumer Electronics was used because most people have some exposure to this class of information system. A similar analysis could be performed for specific devices such as digital cameras and for different classes of technology.

From this CE example and the work done with Hao and Sylvia, it is possible to critique classes of devices and to think them through. I do not consider it possible to have a common design for all consumer electronics, but it should be possible to have consistent themes and principles guiding their designs. It should also be possible to develop specialized, common designs for classes of devices within a larger category of devices, such as MP3 players within consumer electronics. Or digital cameras within consumer electronics.

It should also be possible to consider the differences in value and usage when different types of users approach the products. For example, the young versus the elderly. Why not include a simple remote control with big buttons on it with the TV, along with the fancy 30-button model? You do not have to call it the Old Person Control, but it could be Quick Play Control! Why not provide enhanced screens and simple wizards on DVD recorders which are task oriented for the people who have trouble remembering the 23-click sequence? There are many things you can do or think about when you start thinking!

OK, I admit it. The above analysis of Consumer Electronics might appear to be a little bit of overkill. Just build the MP3 device and sell it. Who cares about a value analysis anyway? There are three reasons for including this long example in the text. First, I wanted to illustrate how the ideas can be used. Second, if you are the producer of something, you might have a MP3 type of situation on your hands and you need to identify possible areas for evolution or marketing. A systematic analysis will help you. Third, if you are eventually going to create a strategic or mission critical system, it is good to have had some prior experience in analyzing value. You learn by making mistakes in low risk situations before screwing up the big project. Doing a value analysis on what might be considered minor topics is a good way to learn how to do a value analysis and hone your skills. Think about it, you would not want to do this the first time on a big, expensive project! A fourth reason would be if you were ever challenged to do a value map of something. This methodology gives you a clue about how to do one.

8.10   Process Models and Value Mapping

The Consumer Electronics example was a nice combination of technology and software and was relatively easy to think about and relate to. What about a more software oriented situation? David Tse in his Master's thesis looked at the value of a web portal as part of a cancer clinic and one of the goals of his research was to identify metrics related to value and try to understand how such things could be quantified. As part of his work, he created a value model for the patient. Instead of eight dimensions like the CE model had, it had four:

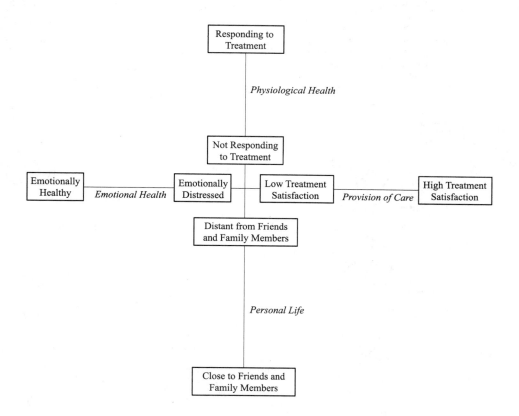

David then prepared a process model of the cancer treatment process(shown on Page 152).

David then used the two together to perform an information audit and analyze the value situation. The information audit looked at each state in the process and documented such information meta-data as the type of information, source, quality, and usage. The value was also noted. This was then viewed from the web portal interface: what information was available, what could be measured about value, where it could be measured, how it is measured, etc. It was also possible to then discuss what additional value could be obtained if the web portal was extended to cover more of the process.

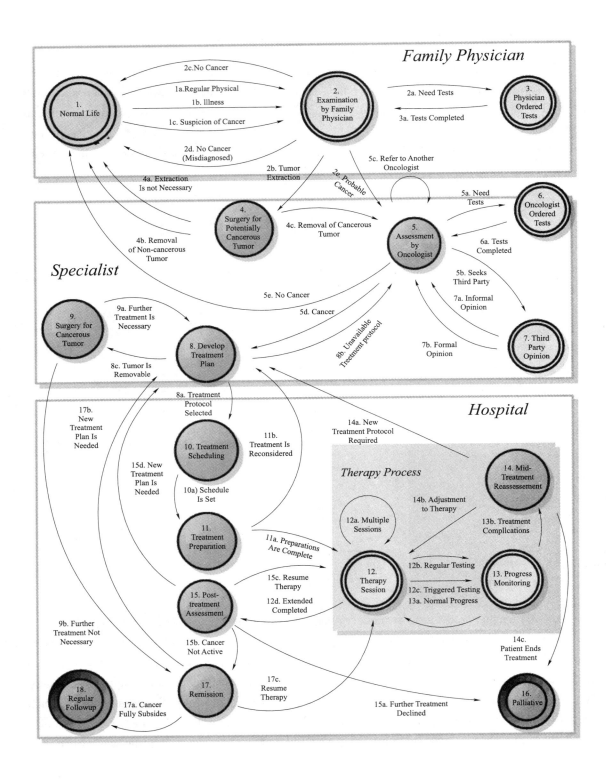

8.11   Exploiting Existing Technology

The world of IT based solutions is ripe for the picking. It seems everywhere you turn there is opportunity to create a better design and give higher value. Once you have a better handle on value, you can think about how you can exploit the base IT (e.g., hardware and software tools) to provide higher value.

For example, consider the DVD recorders shipped with a hard disk, the ones appearing around 2005—2006. The technology imbedded in the device was not exploited. Whether or not it is true, the devices gave the appearance of being simple, evolutionary progressions from the old VCRs. Instead of assuming that such a device was an evolutionary extension of the old VCR's of the 1970s and 1980s with no hard disks, limited CPU and memory, and very restricted display capability. Think of what you could have done if you stood back and viewed the situation as a revolutionary opportunity: what could you do with a TV screen as a monitor, hard disk space, and some smarter code! There are many simple things you can do to make devices like this easier to use if you think a bit.

Unfortunately, many of these devices seemed to have been stuck in the dark ages with a focus on the advanced electronics with an apparent disregard for understanding how non-engineers, and the non-young will (or could) use the products. As I have already commented, the folks over 40 are a great target for market share growth but only if they are understood.

Aside: To better understand the average user's experience with an IT product, I have suggested to some manufacturers that all of the senior executives get some goodies delivered to their home and then (without the aid of the company gofer's) watch members of their family unpack, install, and use the stuff. And, instead of going to the office, stay home a few days of the year and actually use what they build and ship. It is my hypothesis that if they actually watched their fathers, mothers, neighbors, spouse, children, etc. crack open a box, try to figure out the instructions, and then try to use the hardware or software, then a rapid re-design would result. It is not pretty for most things. But, you would never know if you never left the factory or office.

This suggestion also goes for those who write software systems: do you ever actually use the product yourself? In a real situation? Do you go and work with the users? When the software is first deployed? A year or two later to see how the users are really using the software? For a whole week and not just for a few minutes?

8.12   Conclusion

So, this completes the first of the four pillars: the value analysis. This is the necessary but not sufficient pillar for creating a really great IT product. Most of this material came from a course I regularly taught on the impact of IT on society. When I was initially asked to teach the course, I researched the literature and discovered that while there were many nice stories, I could not find a good model and description of IT impact and value after it has been purchased. There are also many adoption and acquisition models that try to understand why someone buys or adopts technology, but very few considered the situation after the user has spent the dollars. I was forced to think back through my career and look at other literature to create an organized theme for the course. This was the analysis framework and it based on socio-tech principles and concepts combined with a user-centered view of the IT universe.

At an early stage of science, the first types of models and theories are ones for organizing thoughts. For example, the atomic element table in Chemistry is simply a way of organizing things. It does not explain, evaluate, or measure. Once something is organized, clear relationships can be identified, common relevant traits isolated, and more detailed models generated. The general IS framework in this chapter is of this type, as is the more specific CE model used as a domain example. These models simply guide and help the analysis: what questions to ask, how to group, and what are the major attributes? As is shown in the specific MP3 example, this view can help generate a focused analysis on the details. It is not claimed that the framework ideas are the only way to systemize or analyze the study of deeper value, but the framework gives an example of what might be considered.

How do I personally use these ideas? I use the ideas to create a reasonably fuzzy view of the user space and I try to focus on why someone is using something or wanting to use something. I believe that the main interface flow must focus on the chain of events needed to obtain the main value for the user. If there are several different users or different objectives, I prefer to stream them and create task oriented interfaces instead of creating a confusing mess of features and functions that can be used at any time by any one for any purpose creating lots of noise and confusion for all. Think of the task-oriented or task-allocation interface as a form of imbedded guidance or wizard.

I like to highlight the path for the user and let the user know where they are along the path, where they have been (what has been accomplished), what they are now trying to do, and what remains. I like to make sure that whatever they do along the path to value anticipates the future while allowing changes to anything done in the past. It has to be easy to go back and change things along the path. My goal is to minimize the user's effort and to make the experience as efficient as possible. If the system can figure it out, then why should the user do it outside of the system using napkins and crayons? For example, if the user talks about scheduling 100 parts, or 1 hour of production for a part, or a percentage of a shift for a part, or "in whatever time is left on the line, make this part," then I build a system that allows them to express capacity or use in these different ways. How the user expresses their constraints and inputs is important for designing the value chain interface. You should match the semantics of the user and not the other way around.

I try to remember that they are using the system for a purpose and not just because they want to use the system to fill in time. Consider it purpose driven design. I do not have a DVD recorder because I like watching the display. I have the DVD recorder to record, play back, and to make DVDs that I can play on other DVD players. I really do not care what the color or shape of the LED display is. And, I really do not care about all of those fancy features that a power user or movie creator wants to use. Every time I use the device I have to press the same buttons, skipping over the same features, drilling down to the same point never doing anything else along the way, never using the power user features. Never: in over four years of use. All of this takes me time and introduces confusion and possible error. When I want to finalize a DVD and make it playable on another device, it takes many clicks on the remote control. Think about the buttons on your digital camera. How many of the functions do you really use and how often do you have to drill into the menu to do a function? Occasional camera users use a camera one way. Non-serious amateurs use a camera a different way. Serious amateurs use a camera differently, and professionals another. One of the cameras I personally use is one targeted towards serious amateurs, but the interface is not. This is not purpose driven design and it can be very frustrating and annoying. There are features on the body that a serious photographer would rarely (if ever) use and very common features buried in the beast's menu structure that are often used. Sigh. I once did some cherry blossom shots at night in Japan and it took two minutes per picture and quite a few clicks on the back of the camera for each shot. My host was very patient. If it was not so painful, it would have been amusing. The photos were of excellent quality and the technology was great. However, the usage model was not as great and could have been improved in my humble opinion.

When possible, I build systems that have small self-learning functions that will pick up frequently used activities or special cases and automatically adjust. Or, I work with the users and model what they do at 6am versus 7:30am versus 10am and 2pm — and what they do differently on Monday versus Friday and the beginning of the month versus the end. While the mass of functions etc. might be provided in other menus, the main screens should be uncluttered, and the most useful and frequently used functions are focused upon for the current context — the current purpose of using the system and what this implies.

Can I reduce typing? Can I reduce the impact of user confusion or error? Can I prevent errors? These are all forms of waste. Can I make it as easy and as fast as possible for the user to get in, do their work, and get out? If I understand the value equation or value chain, I can build what I call optimized GUI. And guess what, this is the type of interface that I have found users like and do not want to have turned off or discontinued. They will defend it and fight for it. They do not fight to get rid of it.

Chapter 9
ZenTai — the Comfort Zone

9.1 安心 — Comfort

This is the next important facet after understanding value — **Anshin** — 安心 — the comfort pillar. How comfortable will the user be using your software or hardware system? This covers stress, anxiety, fear, fatigue, frustration, annoyance, you name it. These are some of the negatives. You should at least aim for neutral and if you can, go for the positive side. If you are neutral or negative, the user will not likely have too much loyalty to your product and the barrier to exit will be low. That is, they will flee like rats from a sinking ship at the sight of a better, more comfortable system offering the same or better value. You might have slightly lower value, but if the comfort factor is high, you will probably keep the users as customers or clients.

One of the biggest obstacles for creating comfortable software might be you. It is difficult for someone in the IT field to purge themselves of the "I think" and the "I like" syndrome. This is the "I think that this is what the user will want," "I think the user should be able to do this," "I think that the user should do this," "If I like it, the user will too." Why do you think that your thoughts or likes matter? They might if they are based on some decent research and analysis, but if you are simply doing the "I think" without really thinking, your thoughts are probably not worth more than a gram of cow dung. And, it does not matter if you would like to use it, or if you think it is intuitive, or if you think it is super efficient and obvious. What makes you think that you think or behave like the normal user???? Look in the mirror, do you do what a normal user does? Do you look like a normal user? Unless you are writing software for another software creator, your assumptions might be a tad off reality. If you take the time to do the stuff recommended thus far in this book (or in similar books), your thoughts might mean something. Imposing your own arbitrary view on others is something that as an architect or designer, you should learn to avoid. When in doubt, shelve the ego, and ask the user!

9.2 Sources of Discomfort

There are a number of sources of possible discomfort. Many are fear driven and on any given project, you might see any, some, or all of the ones below. The list is partial!

- The fear of making errors that cannot be caught or seen till after the fact.
- The fear of making errors that cannot be un-done.
- The fear of making errors that cause many other things to happen and many more things to fix if the error is not caught.
- The fear of making an error that takes a lot longer to fix (in terms of steps to do) than the original operation.
- The fear of doing things that will disable or prevent other things from happening — by accident, and then not knowing how to turn them back on.
- The fear of using something periodically and forgetting what to do.
- The fear of using something periodically and the something has been changed on you.
- The fear of not finding the equivalent functions or needed functions.
- The fear of not knowing what the options are and what the options imply.
- The fear that your expectations about time and effort will be wrong and it will take you a lot longer and more effort to do something.
- The fear of losing acknowledged expertise with existing methods and tools — a lowering of status (and in extreme cases, a job re-classification down).
- The fear of more things to remember, details about what is keyed, what is the right way to do things, what not to do, what steps are needed to go from A to B.
- The fear of new names and terms for things and the need to know them and use them.
- The fear that the learning curve will not be taken into account in your work load.
- The fear that the IT is not stable or in its early life cycle and will lose work and will require stuff to be done twice (or more times).
- The fear that the IT will not be available due to installation issues when it is needed.
- The fear that when you start using something, the bosses' expectations will change and you will have more stuff to do, perhaps at a higher level of quality and quantity expectation, without the offset of other work being taken away or accounted for in your work load.

I have encountered all of these fears at one time or another. And, probably more that I do not remember. There are other things that also cause discomfort in the form of stress or frustration. You can see many of these if you just sit and watch a user try to use your system. For example,

- The user knows exactly where they want to go in the system, but must go through a long drill-down or menu selection process each and every time to get there.
- Calculations or paperwork must be done outside of the system and then input into the system in a non-transparent fashion.
- The user interface requires constant alternating between keys, mouse, and possible mouse movement; it is made worse when things are all over the screen.
- When you can see something that you just want to click on and drill down on, but the system forces you up and around and down again before you can get there.
- When you are doing a number of steps and you do not know how much is left to do, and/or you cannot remember where you have been and what you have done.

- When the user must keep an eye on different areas of the screen for subtle changes that are important and can be easily overlooked.
- Doing seemingly redundant things. For example, a button for logout or sign out, then another screen that displays another logout button and then another confirmation screen to really logout. One of these steps is not really necessary.
- When dealing with a single system and the system interface assumes that there is more than one thing on the screen and forces you to pick the only option.
- When you go to click on something and it moves on you before you click on it.
- When you get error messages that are so vague or poorly worded that they are useless.
- When there is visual noise or clutter (animations or flashing bits) that is distracting and takes focus away from the key part of the screen.
- When the system assumes that you are a novice or first time user and never updates its view of you; the interface has a one-size fits all view and the size starts at newborn.
- When the system uses special terms or short forms that are anything but common knowledge and requires you to learn their conventions — because it was more convenient for the programmer.
- When the system requires you to remember something from one screen (either that you typed or saw) and to re-enter or use the something on another one or more screens later.
- When you are typing away merrily and something pops-up taking control of the screen and you are taken away from your application in a heavy handed way and you lose your keyed input and you have to do things to get back to where you were. Worst is when your typing continues into the newly popped up screen. Lots of confusion.
- When you have to do a lot of keying from a relatively small set of options that could have been on a click or button basis and you keep miskeying.
- When you have to wade through a long list of infrequently used items to get to the most common or most likely item.
- When a feature is available at one point in time or one screen and after you do another feature, you cannot use the desired feature or go back — just because you forgot to do B before C and it really does not (or should not) matter whether you do B–C or C–B.
- When you use a system that is supposed to be used by a wide range of users and the system obviously favors one set over another either by group or by age or by profession.
- When something is to be used in a repetitive fashion with dozens or hundreds of similar interactions and the system assumes each is a brand new interaction or engagement; no concept of multiple items or objects being acted upon.
- When a system does not have predictable results.
- When the system is changed and it is not clear to the user what is changed.
- When there are known errors or shortcomings and these are not shared with the users.
- When an old function fails to work in the new version.
- When the new system is not as powerful as the old.

- When old parameters or preferences are not carried forward to the new system.
- When old data or information is not easily merged or migrated to another job, role, or system.
- When the system randomly or inconsistently resets stuff to defaults or earlier values and what you have set does not stick.
- When the system delivers up inconsistency in responses, response speed, feedback, interface styles (and feel), and so forth.
- When you have spent a great deal of time doing something and something causes a page or screen refresh or a re-login and you lose all of your work.
- When things that should be remembered by the system are not.
- When things that should not be remembered by the system are.

Got the picture? All of these discomforts I have seen and have had to deal with. Give me more time and the list would just get longer. This is a partial list to say the least. I have to admit that most of these bug me too, but I can cite chapter and verse of studies I have done where the users have voiced these sentiments about their existing systems.

If you have only one user (or a couple), you should be able to design and deliver a system that avoids most of the above problems. Remember, you will never make someone 100% happy all of the time. Forget about that goal. But, you should be able to satisfy one user enough that they are happy. Perhaps about 90% of the time? Beware of those people who will never be happy with anything, and will whine and complain about EVERYTHING and once you do what they ask for or suggest, they will whine about that. Based on my experience, you will hit a few of these in your career, but most users are not of this type. Users are humans and they will have bad days, and will — due to necessity — try to do things never anticipated or imagined. These situations will be in the minority if you have done your homework and have built the system well. They will still occur, but will be relatively rare. They should not happen weekly! The system will also have bad days and everything possible that could go wrong, will go wrong and it will be at the worse possible moment. This will happen and you cannot avoid it, but these situations should also be very rare.

If you have a whole bunch of users, you have to go for the classic 80–20 rule and bite off part of the bullet. You have to compromise, do what is consistent and try to hit 80% of the people 80% of the time. When possible, allow workaround and options that allow the 20% to get their job done, but you will often be wrong or in a problem spot with someone. What you want to do is give an overall good feeling and be in the general comfort zone for the vast majority.

Most users will accept some sources of discomfort and with a little bit of muttering will just put their heads down and put up with it. If you can fix this stuff, you should, but unless you are really there and know what is being muttered about, you will not know what to fix. Sometimes you will hear second hand from the user's boss or co-worker that something is annoying. They either forget, or do not want to bug you, or they might be afraid of the big bad programmer and are afraid of saying anything. So, while a little bit of 'less than optimal' user interaction is going to happen and will be accepted, if you deliver a system with many sources of discomfort, stress, and frustration, the users will be less than impressed. And, do not be surprised if they actively pursue the replacement of your system. Or, try to sabotage the system.

9.3 Increasing the Comfort Level

You should consciously consider a number of factors when designing the system. One very important factor is the existing situation. Is there a legacy system or is the user switching from a manual system? This factor suggests possible forces or gaps that you have to consider. For example, how long has the user been using the existing system? If the user has been using it for a long time, there will be a fairly large force against you; there are habits, comforts, and built-up experience. Various skills would have been created and reinforced. If you are computerizing a manual situation, something totally new and different will create other problems. In this case, you need to zig before you zag. That is, you might have to migrate in phases or put in features that allow the user to do the same thing they did by hand and then slowly get the user to do things differently.

You want to minimize gaps and possible fears. In one system, the old scheduler was close to retirement, barely computer literate, but had some (very) limited exposure to Excel. I built his new tool in Excel so that he would have some degree of familiarity and comfort with the new system. There were no macros, and all of the logic was in Visual Basic for Applications, but the old guy just thought of it as another Excel thing. It was not considered something totally new or different. This made the system more comfortable for him. I used Excel as the host and just manipulated it. For the user, it could not be a big, scary IT system if it was just built on Excel. He did not know that there were many thousands of lines of code buried behind the scenes.

I will continue using this particular system (see papers co-authored with Wiers and Black) for some examples of comfort. This old scheduler also had an interesting way of clearing out fields. He would put blanks in the field: if you do not see it, it does not exist. So… The system has an extended empty field check looking for one space, two space, three space, etc. Basically truncate for blanks left and right. If you can't fight them, join them.

The factory had also used a big metal board with magnets on it for scheduling. I made the system look like his wall and even called the things on the screen magnets. I used the scheduler's terminology — nouns and verbs — reducing training and discomfort. By using the various methods described for how to understand the problem, I had a reasonable understanding of the processes and how he did his job. Even better, I had the person assigned to back up the scheduler assigned to me for a few months. Having this person sit with me as I coded was key in making the system match the scheduler.

I reduced typing by giving most of the common responses as buttons, and I made menus and functions that matched the scheduler's tasks. If this old scheduler could do it faster or better by pencil, that is what he would do. I had to beat his manual processes (substantially). For his repetitive tasks, I made optimized GUI components that presented what information was needed (not more or less) and gave him specialized buttons and wizards. This reduced the noise and allowed him to concentrate on the task and not the system. The tasks were also organized to match the early morning rush, the period after the rush, mid-morning, afternoon, and so forth. This made it easy for him to step through the day and also made it easy for replacement schedulers (when he was sick or on vacation). The information used by the scheduler was also analyzed, as were the planning meetings, and the discussions held around the scheduler's desk. Reports were generated that combined data from a variety of sources so that for almost all decisions all of the information was available in one place without any hunting. If calculations were part of the discussions, the calculations were already done on the report, reducing the manual effort. There was one fourteen-page report in the new system that replaced close to three hundred pages of various reports generated each and every day by the corporate information system.

There was a lot of fear and skepticism in the scheduling area. This particular factory area was very dynamic and creative problem solving was an hourly event and it was not possible to assume that the computer database and logic was up to date or could represent reality. By using Excel worksheets for output reports, the scheduler was able to edit freely and make the paperwork match reality. The computer would take its best shot, but it would never be 100% accurate. Since the scheduler was comfortable with Excel, he felt OK with just going in and directly editing the report (aka a spreadsheet) to make it right. This was not just a case of tailoring or customizing the format of a report, but this was live editing of the report itself, changing numbers, machine names, part ids, etc. The paperwork sent to the workers had to be right and the scheduler essentially had to be able to lie to the computer and the computer system had to have faith that whatever the scheduler said or did, was indeed feasible. In this tool, the user was in charge and not the computer. This ability to change and control the output overcame many issues and created a nice, comfortable zone for the scheduler to work in.

I ended up building a number of systems for the one company and the same approach was taken with each system: a good initial study, followed by an Agile/Extreme development focusing on value and comfort. You will help make the system comfortable to the user if you match their vocabulary, semantic elements (e.g., what they talk about and how they manipulate them). This is not something that can be done repeatedly by accident. You have to consciously do it.

The users will feel comfortable if they feel protected from the unknowns and if there are few costs of errors. There are various ways to do this and the specifics will vary with the situation. For the scheduling system noted above, I made sure that the scheduler could restart or work from a safe place. The system also included its own backup system so that the scheduler could recover and reset at 6am or 7am by themselves without the need for the MIS folks who would turn up about 7:30am. Since the scheduler could recover and reset, there was little fear of messing things up to the point of no return. Further comfort was given by designing the system in such a way that many things could work independent of each other and while some bit might not work, the system could still give value and do something.

There is another technique I want to mention for improving the comfort zone: mistake proofing. I briefly pointed it out in an earlier section. Shigeo Shingo developed a technique at

Panasonic in the 1960s and then refined it at Toyota. This is Poka Yoke — mistake proofing. You should assume that humans are indeed human and not robots. They will make mistakes no matter how much you pay them or yell at them or train them. The Shingo approach is to study a situation and determine what errors are possible and either design the system to prevent them or to detect them. It works well, even for software.

9.4 A Comfort Analysis

To deal with the comfort question, you can do a comfort analysis for the system you are analyzing or crafting. Some of the key questions to ask are:

- Where is the user coming from, and what are they currently comfortable with?
- What are the user's general computer skills and how have they developed them?
- What are their fears?
- What are their expectations?
- What are the possible impacts of user or system error?
- What are the possible impacts of delayed or partial operation?
- How often does the situation change or require changes to the system?
- How close can the system model the problem and prepare output without further human, manual manipulation?
- Are there any times that the user has to use your system in a rapid, time-limited way?
- Are there any times that the user has to use your system in a repetitive fashion?
- Are there any times that the user has to combine or use information from various steps, parts of the system, or from different sources?
- Are there any times that the user has to use your system with the boss hovering, asking "do you have the answer yet?", "is it done yet?"
- What are the ways errors can be made or introduced?
- What has the user built skills in and how close is this to what you are thinking about building?
- How much training will be needed for basic equivalency?
- What are the terms and semantic units used by the client and how close are they to your system?
- What are the user's expectations about the system? Its flexibility, its robustness, and its ability roll with the punches without being down for the count?
- What is the impact if the user does not use the system?
- How is the organization supporting the training, learning, and migration from manual or legacy to the new system?
- What are the most frequently used functions and features?
- Are there any temporal or time based patterns as to when something (or how much) needs to be done?

- What is the impact if the user circumvents or works around the system?
- Who gets blamed or yelled at when things do not go right in the business — regardless of who actually decided things or did things?

Answering these questions will help identify the gaps, potential fears, and areas that need special attention and development. You can also go through the lists in the previous subsections. You can consider which are relevant and which are not, which are present and which are not. Then you can decide what to do. Not everything worth doing is likely to be done or is likely to be done in a pure way. You have to prioritize and decide what should be done when and how. **Not everything worth doing is worth doing right.** Some features and functions just don't need to be perfect. And remember, you cannot build a system that costs the same as a jet fighter and sell it for the price of a single-engine bi-plane. You can do that once perhaps, but you will not stay in business if you do it all of the time. Over time you will learn what trade-offs must (or could) be made and how to make the appropriate decisions.

If the project is important enough, you can perhaps take the time and build up a toolkit and a special interface that addresses many of the fears and annoyances that lower the comfort level. If you have a good toolkit to start with, it will make the crafting of a good system easier and faster.

Chapter 10

ZenTai — the Experience Factor

10.1 経験 — Experience

This is **Keiken** — 経験 — the third pillar. There are two aspects to the experience pillar. First, everyone brings experience with them, experiences with other software, experiences with terminology, and the related skills. Second, people by using the software and hardware have experiences. As a designer or architect, you should consciously consider these two aspects.

As you do the user analysis, you try to understand what experiences the typical user has had and what these might imply in the design and operation of the system. This helps you understand how you should or could address the comfort zone and the value equation. Someone with a great deal of computer skills and experience with a similar package will require different steps and help to achieve a certain value than someone who is a rank amateur and is a newbie. Someone who is in the domain or function area every day will have requirements different from someone who is there once every year or two. As I am writing this, I am wondering aloud if there are systems or times as a designer you can ignore the experience factor. It might not lower sales or impact your profits. For example, I do not think that many developers thought about this when I think about the software and systems I use on a daily basis. So, people will buy software and use software that ignores experience, I have. There is a caveat though, there are usually other pressures or reasons for buying these poorly designed and annoying software packages; just like people will buy software that does not deliver up any real value or provides comfort — *if they do not have a choice or have seen something better*. However, I do not think such software is anything to be proud of. It is not great software. It is not software that users will want to use and value. As designers and developers, we can do better.

You should also think about the experiences that the user will have with the IT. Did you consciously write this down the last time you designed a system? Under what conditions will someone use the software? For how long? How frequently? What functions will be used a lot? What will the user do repeatedly? These are the active experiences. There are also the passive experiences of what will be seen. In both cases, the experiences will i) create an initial impression,

ii) be measured consciously or subconsciously against expectations based on previous experiences and knowledge, and iii) become another part of their experience and either reinforce or challenge skills and abilities.

How people experience things is also varied. Some people will explore and poke about. Some will just use what they need and will learn it when they need it. This characteristic seems to be independent of expertise. What assumptions are you making about how someone will learn and explore your software?

10.2   Prior Experience

Prior experiences help to set expectations and perceptions and create a kind of force or momentum. People like to be able to use existing skills and experience to do things, especially when time is tight and there is little value to be gained from presenting the user something totally different. If you can match or exploit existing knowledge and skills in the user interface or set of functions, then you are explicitly recognizing the user's experience and what this implies.

To understand prior experience, I try to think about the following (using those ethnographic methods noted earlier):

- What is the normal work day (week, month, year) for the user?
- What do they when, for how long, and how frequently?
- What is the total time, per elapsed time interval, is the user doing things that the current or proposed IT will do or interact with?
- What kinds of things can happen between encounters that can interfere with memory and re-engagement with the topic area?
- What meta objects or semantic units are users employing or using? What do they use in conversations, on the phone, during discussions at the desk, at meetings, and in reports?
- Have they used manual or other IT tools for part or whole of the task? And if so, what and for how long?
- Part of experience is learning to deal with things that change, so it is also wise to look at how their experiences will change. Various change topics to consider may include types of change, frequency of change, and how they handle fixing and changing things. You also need to think about the change triggers, how change can be identified, how it can be anticipated, and how change can be responded to.
- When was the last time a new system was introduced? (this helps understand potential flexibility or challenges to change)
- What was the user's experience with the last change or new system? (this helps understand expectations and possible barriers)
- Does the user have experience in a similar domain where the skills and abilities might be transferable to the new one?
- What is the range of experience (or expected range) of the users? How many different

experience models will you need?

- Who will do the function if the user is sick, transferred, on vacation, quits, or fired? What are the assumptions about this person's experience?

These and other questions will help you think about the prior experience and the possible implications for the IT. You can deal with prior experience in various ways and it is unlikely that you will use all of the ideas in one system. Here are various ways that I have tried to deal with previous experience in systems I have been involved in:

- Multiple user interfaces matching the users' experience and expectations — with the user being able to select between the options.
- Objects and functions matching their basic terminology. Even to the point of some things matching the way they do it manually.
- Meta-objects and meta-functions matching the user's major units of work and manipulation.
- Task oriented menus and mini-wizards to lead the replacement or new user through the tasks — first this, then this, then that — all interlocked to prevent out of sequence or duplicate operation.
- Task oriented menus and mini-wizards reflecting the frequency and timing of when usage might occur. These might change depending on the day of the week, day of the month.
- If users (some or all) have experience with similar technology, interfaces that allow the user to drive the system with little guidance.
- If users (some or all) have little experience with similar technology or functions, interfaces that help guide the user.
- If change and creative-problem solving is part of their experiences, similar ability and support is provided in the system. Even to the point of the user lying to the computer and the computer documenting and reporting seemingly infeasible or impossible things. For example, in some factories, one machine can be temporarily torn apart into two machines by a creative worker with a suitable wrench and do two jobs at the same time although the computer system thinks that the machine is not capable of doing it. The report or instructions for the factory floor have to show two jobs at the same time on the same machine although the system does not think this is right.
- Multiple meta-objects or concepts that get resolved to a common form by the software. If the user expresses things in multiple ways, I try to support all of their ways instead of asking the user to do the conversion. For example, if the scheduler or planner uses x per hour, % of shift, nnn for shift, etc. in their normal communication and thinking, I support all three (or more).

10.3 Experiencing

An encounter with the system is an experience. If someone uses something once or twice a year for a half hour, it is not reasonable to expect the user to either gain skill in doing that task, nor to remember how to do things, nor value the challenge in doing so. However, if the person is going to use the software six to eight hours a day, five days a week, for many years, then the case is different.

If you have many user types and there is high variability within the types, you might need to consider a number of factors about their actual encounters with the IT. Here is my list of considerations:

- How often, when, for how long, and under what conditions will each type of users use the existing or proposed IT?
- What does this translate into — in terms of lapses between uses? What can be forgotten?
- Since experiences contribute to expectations, what can change between experiences?
- Is there any clustering of experiences around a time element or contextual element?
- Is there any pattern of experiences over time?
- Are there complementary systems being used at the same time that reinforce experience and learning?
- Are there non-complementary systems being used at the same time that weaken experience and learning?
- What is considered positive or negative experiences?
- Are positive or negative experiences on some form of continuum?
- How might experiences impact how the user will know where controls or functions are, and how to use features and functions?
- How fast does an experience develop skill and knowledge about the system and its operation?
- How much does learning the skill take away from the actual value-added function as the skill is being learned? Versus when the skill is developed? Does it take more time or more effort?
- Can the skills associated with the experiences deteriorate over time if the system is not used? How much is deep knowledge and skill versus superficial short term memory and motor skills?
- Will experiences affect the sequence and groupings of functions that a user might choose?

The key is to make believe and to pretend that you are the user and think about what you will encounter, how often, etc. and think about what all of this means. Always try to remember that it is not about you, what you do, and what you like. You have to project into the user space. From there, you can think about ways to help experience or exploit it:

- Task oriented menus and mini-wizards for infrequently done tasks and functions.
- Alternate menus or functions for experienced or in-experienced users.
- Mini-wizards for users who infrequently use the system and have no hope for remembering what is where and what to do.
- Functions to hide controls and noise if they are hardly (if ever) used.
- Functions to set preferences for commonly used items (and make them stick from session to

session).

- Functions to hyper-jump or move around the system at the experienced user level instead of having to drill down and around like a novice.
- Notes and information presented when a new system or upgrade changes an existing function or option. The dynamic notes and information can adjust a user's understanding and expectations, possibly avoiding problems before they occur.
- Simple learning mechanisms to pick up patterns of use and create macros for them.
- Simple learning mechanisms to pick up exceptions or special cases and to automatically create alternatives.
- Avoid changing experience related functions during peak times or when the user does not have time to change expectations and learn the new stuff.

Chapter 11
ZenTai — Evolution

11.1 進化 — Evolution

The final aspect of *ZenTai* is to consciously consider **Shinka** — 進化 — evolution. This has two issues and both end up being more internal than GUI external. You need to think about how the environment that supports the IT will evolve and make this evolution as transparent and as painless for the user as possible. You also have to think about how the problem space will evolve and allow for some software evolution that will be quick, painless, and transparent. While the user will see the direct forces of value, comfort, and experience at work, the evolution piece is more subtle.

Evolution is about change and what can change and how change is accommodated. It is not possible or reasonable to think about anticipating all change, but it is possible to think about types of change and what is likely to change. One approach I have taken is to think about what is easy to change in the real world and make sure that it is equally easy to change in the software. If the IT is expected to have a short life, evolution might not be an important topic. However, if the system is supposed to be used for 10–20 years, then evolution should be consciously thought about.

This chapter is the shortest of the four on *ZenTai*. But, that does not mean that considering evolution is not important; it is just quicker to think about than the value, comfort and experience pillars. Evolution is also very specific to the application being built or analyzed and as such; there are a few general ideas to think about, but the specifics will be situation dependent.

11.2 Environmental Evolution

This is the hosting or executable environment. What happens to the application and user experience when the lower level OS is upgraded or hosting tools are upgraded? For example, if the tool is built on a spreadsheet platform, what happens with a new Windows or Office upgrade? What options or preferences are reset? What happens to the visuals? These are questions that you need to think about in advance. Some of my considerations are:

- Will the underlying OS and hosting environment (e.g., development tool, execution base) be in synch or can they be out of phase? Do I need to consider a single vendor solution?
- If the system is mission critical, is it possible to minimize the scope of dependency or to

thinly wrap the environment interface to create a virtual world for the application?

- Is it possible to avoid using features that are rapidly changing and rely on mature, stable features?

- Is it possible to put in version checks so that the user knows that a change has occurred and that extra precautions might be needed?

- Is it possible to pre-design and consider ways of exporting, and re-importing into new and improved data repositories? Doing this from the beginning and ensure that upgrade and migration is possible?

- Is it possible to consider running new/old in parallel via macro playing and ensuring that the new system does not result in any data or logic surprises? That is, run a stream through the old system, through the new system and ensure that the outputs (service to service, service to user, and service to database) are the same. You might have to design this idea in from the beginning and create the necessary foundation logic in advance.

The list is not long, but each point can have a long list of side points to consider in a real situation.

11.3 Functional Evolution

The environmental evolution is simpler than the functional evolution. The environmental issues are transferable across the same host/operating system pairs and you can exploit knowledge and existing solutions. The functional ones are a lot more fun to deal with.

You need to have a really good handle on the user model and user situation if you are to address the functional evolution challenges. You need to know what is easy to change in the real world and what is going to be difficult. You need to know the Zen of the problem and what are the parsley-on-the-plate options. Internally, you have to make some decisions and have some design schemas; you cannot make EVERYTHING flexible or variable. So, the trick becomes knowing what to make firm like a stone, and what to make flexible like pudding.

Report requirements will change with management changes and the business situation. McKay and Black (2007) describe the ten-year evolution of a decision support system that highlights the drivers and issues behind functional changes. The direct users and the users of the output cannot predict how they will be able to use the IT for better decision making or value. They can give you their best shot, but it is best not assume that this is the final word. What they say is fixed and will not change is likely to. It will also change at the worst possible moment and have to be done in the fastest way possible. It is wise to design the internals for reports and data presentation with flexible tools and small utility routines that can be coupled together in many different ways.

Terminology and semantic units will also vary over time. Decouple the user's terms from internal semantic elements and design from the beginning for hierarchical relationships, sequential coupling of units into a larger element, and meta-units that act like onion skins with thin wrappers for each level of functionality and meaning. Use tables and mappings for output so that you can go multi-language if desired.

Via experience, the users will want meta-units disaggregated, and other atomic elements and actions aggregated. Then they will want the same items re-aggregated and disaggregated. This is because things change over time and if the system is in use for many years (e.g., ten), there will be cycles where you will do something, un-do it, and then re-do it again. This will happen and you cannot avoid it.

Over the life of the system, one task initially done by one person might be done by two people at the same time or in sequence. Two tasks might also be combined and the need for a second person avoided (perhaps routinely, or only on the weekends).

For example, I once had to build one system for a dispatcher and one for a scheduler. The dispatcher had the next forty-eight hours and the planner had the time after forty-eight hours for sequencing and lining up work. I built something that looked like two tools. The two tools were fine during the week. On the weekends, one person had to do both jobs. This meant three tools. This was easy to do since the same program did all three tasks — the dispatcher, the planner, and the master mode of both. There were different task menus, dialog windows, functions, reports, and GUI for each, but there was an overlap of say 60%–70% between scheduling and dispatching when it came down to background data and basic screens and such. The users were able to easily set up and run the various versions because the recognition was based on the executable's file name. The software was able to query its own executable name and based on this, customized the system accordingly. This factory went back and forth — one person, two, then one again — and the system was easily adapted as the functionality requirements evolved. This also made it easier to maintain and make the system consistent between the two functional tasks. I chose to have the system self-recognize and change the menus and functions instead of using a menu and options to pick between the major variants. Occasional users might have difficulty knowing that there is an option to look for and might also have trouble finding it. I also did not want to ask the question every time they started the system imposing an unnecessary interaction 80%–90% of the time. It made more sense to make the system more aware of itself and remove the burden from the user.

My basic rule of thumb is that anything easy to change in the real world better be easy to change in the software and the rest can be rather firm and be reasonably difficult to evolve. I mentioned the search and rescue simulation earlier. The client wanted a tool good for 10–15 years. So, I analysed the problem for the kinds of things that would be reasonably easy to change in the real context and made sure that these features were really easy to play with and extend. The stuff that would cost big bucks or cause the laws of nature and Newton to change, I decided that a relatively fixed internal design would do just fine. For example, what a helicopter can do in terms of travelling and on-site assisting should be easy to change. Making a helicopter act like a boat for an extended period of time is a bit harder. So is making a boat that is capable of breaking ice in the North Atlantic travel across land and participate in a search being conducted in the plains of the interior.

When someone seems fixated on something and keeps repeating "this will never change," I think twice. What are the assumptions being made by the user in this case? If the assumptions are not based on the Zen elements, I do not really go along with the user's belief. I will not necessarily build the stuff, but I will not design it out. Anything built on arbitrary policies and procedures can change in a blink. Anything built on someone's personal preference can change faster. Similarly, if the user is talking about stuff changing and the need for flexibility in a certain piece of the system, I think twice. If the flexibility is easy to do, fine. But, if the flexibility is at the Zen level and is not very likely, then decisions need to be made. If hooks can be left for the most unlikely request without messing up other things, fine. Remember, just because you can do it, does not mean that it should be done. It is also time to repeat one of the main mantras of this book. There is NO one way to do something or one right decision that covers all possible situations. Throughout this book I will appear to contradict myself, even be hypocritical, but the guidance and decisions are context sensitive. And, the guidance changes as a project evolves: what I do in the early phases of a project is not what the final form might look like as the project evolves and is refactored. The goal is always a useful, functional, and appropriate system for the users while taking into account maintenance, support, ongoing evolution, costs, and time constraints. The goal is not making pretty or cute code at the cost of everything else. Some things in a system will be flexible. Some will be rigid. Some will be loosely coupled; some will be tightly coupled. Some things will be designed to evolve easily; some other things will not. I have yet to be involved in a project where you could do everything via the textbook and still deliver the project on-time, on-budget. System development is full of compromises and decisions.

Taking evolution into account during the analysis is important because the future existence of evolution not accounted for may be costly and problematic for the users. You want to be able to respond quickly without destabilizing the world. You need to have techniques for doing oopsy and rolling back to a previous version (and you better give the user the ability to do this at 5am in the morning if that is when the software is used). You might be able to build in self-detection logic and automatically control evolutions and the possible impacts. If you design for evolution, you might be able to refactor the initial instance into a toolkit and use it elsewhere — a bonus. However, you have to be careful not to overdo it and there are reasonable things to do and foolish things to do. You need to assess and make reasonable trade-offs and understand what to build in, what to design for but not build in, and what to ignore.

Unfortunately, there are no real checklists for this and it is hard to know what compromises make sense or do not make sense until you have a number of long term projects under your belt. A senior designer or architect might suggest something and a junior developer might not understand, see any value in it, or agree with it. Interpreting requirements and understanding a problem space takes a level of cognitive skill that is acquired only through experience and reflection. You cannot learn this from a book or from a course or from doing a few projects.

Chapter 12
Pulling It All Together

By this point the eyes might have glazed over. Lots of abstract ideas not normally found in software engineering types of books. Ethnographic methods, fishbone diagrams, bizarre pillars of something called *ZenTai*. Lots of weird stuff. In practice, you do not do all of this stuff on all projects all of the time. You learn when and where to use the ideas. In some cases, you will emphasize certain aspects and downplay others. By learning how to use the tools, you will learn when and where to do what.

In some instances, you might spend minutes or hours on something, while in another case it might be months. The closer the project is to a Class VI rating, the more effort is required to make sure that certain things are done right. This is commonsense. It is not mystical or something that comes to you under a tree mediating about the meaning of life. You are not levitating and an aura is not to be found around your body. It is about using your brain and not just doing the same thing each and every time, nor just doing the first thing that comes to you. You need to use different methods and approaches to solve different problems and this is one of the things that make someone an expert. If you are always building the same type of software, then you might be able to use the same methods every time like a robot. However, it is a pretty good bet that different problems will require different solutions and if you constantly use the same method, why would you think that the results would differ?

You also have to practice this stuff until it becomes second nature. When you are doing it all of the time, you will sometimes be able to quickly contemplate a situation and get a feel for the essence of the problem and what the problem and solution might mean to the user group(s). You will be able to identify major groups in the user community and create profiles that will help you understand how the user group will use your solution. As you do your analysis, you will pick up on issues that will cause comfort problems, and you will observe aspects that relate to experience and evolution. With time and practice you can do it faster and faster, and with greater accuracy and confidence. Do not expect your first attempts to be great, roaring successes either. And, do not feel bad if you cannot do it. Not everyone has the same brain wiring and I believe that certain skills and abilities are needed for someone to be a good analyst and functional architect. It is wrong to assume that just because someone can test code that they can write code, that just because they can write code that they can design programs, that just because they can design a program that they can design a system, that just because they know the technical aspects of a language or system that they know what to do with that knowledge, and that just because they can do a technical solution that they can derive a functional, user centric solution. Figure out what you are good at and develop expertise at that level!

Assuming that you have some ability to do analysis and are sympathetic to user centric designs, here are some summary thoughts for how to do the requirements analysis:

- First, immerse yourself in the situation if possible. Spend time with the users before thinking of solutions. Try to focus on what they do, why they do it, and what their challenges and opportunities are. This will help you figure out what the problem is.
- Second, isolate the essence of the problem. What makes it what it is? Identify the relationships between the pieces. Try to make a simple, process model of the problem, only involving the essence.
- Third, using the model, try to document the most elementary value chains that exist in the situation. What are the people trying to do, what are the people trying to get out of the experience? How do the elements that form the essence contribute to the value process? What are the basic inputs and what are the outputs? There might be many smaller value chains, but your first challenge is to understand the basic challenges or problems.
- Fourth, develop some simple user profiles that describe the characteristics and assumptions you are making about who is involved, when they are involved, what their existing skill and knowledge base is, how frequent their involvement will be, and what kind of skill is needed to obtain value out of the situation.
- Fifth, extend the simple model to include more of the situation, indicating user interactions. Use this extended model and perform an information audit on the problem situation; after all, you are contemplating an IT solution.
- Sixth, use the extended model and review the basic value chains from the third step in light of the user profiles generated in the fourth step. In doing so, use the concepts from the *ZenTai* value pillar. Think about the life cycles, types of interaction, and the value on each interaction. This will give you the most important functions or tasks that must be done by the system; not what they will look like or how they will function. This is at the semantic level.
- Seventh, extend the user profiles for the comfort, experience, and evolution bits. Imagine yourself as being each of the user types and take on their physical, mental, knowledge, skill and experience profiles. Think through what they have to do, and how often.

These seven steps should work in most instances and give you a starting base from which to work. You should have a good basis for understanding the problem.

Chapter 13
Universal Requirement Factors

There are universal factors that relate to the problem *what* and then there are universal factors that relate to the design *how*. In this chapter, we will explore some of the universal problem *whats* which will also relate to the *ZenTai* concepts.

The discussion about universal *whats* can get murky. There are some truly universal *whats* to consider and some *whats* which are not so universal. For an example of a truly universal *what*, consider something hot which could be dangerous. Boiling water or red hot metal will cause a problem to exposed skin to anyone regardless of age, location of use, etc. There are of course the rare individuals who do not feel pain, but the skin will still be damaged. This is universal. Another universal *what* is the fact that there are various types of color blindness and in any population sample, there are likely to be individuals with some form of color blindness. The percentage might vary, but it is likely to have some people affected with the problem. There are other ones of course. If you are part of a company or organization, do they have a list of what they consider *universal requirements*?

Then there are the universal *whats* specific to the target user groups that the product or service is meant for. These universal *whats* are not universal in the global sense, but universal in the context of the product for all intents and purposes. Thinking back on part of the value discussion will assist with this thought process. If the product or service is intended to be used in a business setting, are there universal *whats* for the majority of products used in a business setting? If the product is meant for leisure, are there universal *whats* that should be addressed by this and all other products? Do you have a list of such common or universal requirements? Some of this thinking used to be called human factors design: where they had demographic data for desk height, arm reach, and such. During the design phase such data was used to help design the product. I am not aware of similar work (but it probably exists somewhere) for the user interface and IT aspects. Developing a persona profile is also important for identifying the more specific universal *whats*. For example, if the product is going to be used almost exclusively by the young, there are universal characteristics about most youth. However, if the product is going to be used by the elderly as well, there will be two sets of universal *whats*. One for the young and one for the old (and one possibly for the crowd in between). In any event, the designing organization could, and should in my opinion, have a document that describes the universal *what* for any product grouping or category, in addition to the universal universal *whats*.

A possible universal *what* about the IT problem space is that there are basically two types of users of any IT system and they may or may not exist at the same time. There might be a human user or a synthetic user made of some physical reality (e.g., hardware and/or software). Each category of users can be viewed as an organic entity. They both have interface specifications for how the two systems will be connected, protocols for interacting, life cycles, assumptions, and expectations. When a problem space is being analysed, these can and should be clearly documented.

13.1   The Human Element

Human users have a set of tools available to connect with a system. While there are specialized systems and tools such as taste and smell, the more general set of tools applicable to an IT system are:

- Voice for speech input.
- Ears for receiving audio output.
- Eyes for receiving text and graphical output.
- Fingers for entering text, graphics, functional commands, and selections.

Each of these is possibly connected in one way or another to the IT system. One or more of the tools may be the way that the user obtains the desired value from the system. They may also be the way that the user controls what the value might be. The tools are subject to the comfort, experience, and evolution dimensions.

The universal *what* relates to the possible pervasive characteristics of the human connection. For example, what color is the best for the perceived average user, taking into account different forms of color blindness? This is a universal *what*. The eventual *how* that addresses the *what* is the use of this color on the appropriate button. Each of the human tools has an operating range and the range may vary with race, age, training, and physical condition. A standard checklist or profile can be considered for each of the tools: what size of text can be read by the average user, what color is the best to use, what icon shape is the easiest to recognize, etc.

There is a fifth tool that must also be considered as part of the connection and that is the user's brain. This incorporates the cognitive skill, assumptions, and semantic recognitions that the user brings to the game. The physical eye might process a certain color better than another, but the brain and thinking process must then process this information. Profiles can also be prepared for these areas taking in account any related cognitive skills and semantic processing associated with value generation or value direction. This is the stuff that relates to experience, comfort, and evolution.

A systematic approach can be taken when considering the universal connection *whats*. For example, you can identify populations and sub-populations that exist within the user world. This might include age, skill, training, and physical abilities and also include organizational groupings; management types versus clerical versus hands on factory workers. You can consider each one and ask if the ranges are relatively stable or if there is high variance in the connectors. For example, if an IT system is specifically designed and marketed to the 20-year-old crowd, you will have a tighter range in each of the categories in terms of eyesight, physical mobility, learning speed, recognition ability, etc. compared to the situation if you also have to include people over the age of 60. You may choose to ignore the range, but you should at a minimum be aware of the range. If you choose to ignore the issues, you should not be surprised later when one major user segment has difficulty using your product. A problem with low variance is easier to deal with than one with high variance.

The *ZenTai* aspects help with thinking through the factors. What about the problem or user profile that potentially creates stress? If the user group includes people middle-aged or older, the ability to read small fonts can be a problem. The ability to perform fine motor actions might also be an issue if the user group involves the elderly. If the user group includes two major categories such as power users and simple users, a simple interface can create discomfort for the power users while a powerful and rich interface can create discomfort for the lite users. If the user group includes one set of frequent users and another of infrequent users, the experience factor needs to be understood; what will be remembered and retained or not? What are the previous experiences that the users are assumed to have had? The evolution aspect is also considered. If the user group includes people accustomed to change and another subgroup that rarely experiences change, how will system evolution and change be addressed and accepted?

For a category of products or problems, it is possible that a consolidated profile of universal requirements could be built that would guide any *how* as a solution. I consider it universal because the specific application does not really matter. It is a matter of accessibility and connection requirements for the subgroup: payroll, video game, whatever. If the elderly are part of the equation, you have to deal with the elderly effect. If young people are part of the user group, you have to deal with their basic abilities. If you do not, the user subgroup might feel ignored or become frustrated when trying to use the system.

The following is a partial view of what might be considered universal requirements to consider for each user subgroup while being aware of how they may change or introduce variance (e.g., difference between a young person and an older one, partial or full impairment):

- Speech.
  - o Vocabulary — the set of words to recognize.
  - o Language and dialect — basic language and variances within language.
  - o Pronunciation — how individuals may be different or how an individual may differ if tired or excited.
  - o Possible speech defects such as stuttering or a lisp.
- Eyesight.
  - o Colors — relating to various types of color blindness.
  - o Contrast — ability to distinguish between two adjoining colors or tones.
  - o Brightness — the conditions in which the problem space exists and how the human eye processes differences in brightness.
  - o Focus — the degree or angle of view that an individual takes in when looking at the

screen or system — what might be seen or not seen.

o Near or far sighted issues (or both as you get older).

o Monocular or binocular vision.

o Cataract or other vision issues that cause deterioration.

● Hearing.

o Tone — frequency and pitch of any output audio.

o Loudness.

o Background conditions — what may make the hearing better or worse.

o Hearing issues such as deafness.

● Finger and Hand Motion.

o Dimensions in space — virtual reality in 3D, mouse movement on a desk, pen input on a vertical screen, typing up/down on a keyboard.

o Concurrency — what can be done at the same time and what must be distinguished between close inputs.

o Speed, motion, and pressure requirements.

o Accuracy and placement — thresholds for fine motor control.

o Restrictions and constraints on finger and hand motion — such as arthritis, carpal tunnel syndrome, loss of a finger or limb, sprains and broken bones.

● Mental Capabilities.

o Skill and expertise levels.

o Spatial transformation ability.

o Memory — muscle, short and long term — frequent versus infrequent use.

o Prior experiences — skill and knowledge.

o Recognition — whatever the seeing, hearing, and feeling are providing.

o Logical sequencing — learning long or short patterns of input and output actions.

o General learning — time and effort it takes to learn minor or major functions.

o Impairments that restrict any of the above.

These can be considered a starting point. You might explicitly or implicitly consider them in the user profiles or problem space description. Depending on the system they may be relevant or not and it is not possible to have a general rule. However, each of the universal requirements will have a range of expected values; the values within one or two standard deviations and the extreme cases. You have to consider what has to be a requirement and what is not. You can also include the environment, work space, usage space, time of day, day of week, week of month, month of year, year of life, lighting effects, and other such nice topics in your analysis as some may be relevant and some may not be.

13.2 The Synthetic Element

The above focused on the human connection. A similar story can be written for system interconnections. For example, if the problem space includes a legacy system, this can imply certain issues that the *how* has to address. Or, if the problem space includes certain internet or telecommunication characteristics and conventions, these become part of the problem space. Some of the universal aspects for a system connection in such a legacy situation might be a combination of hardware and software:

- Physical media interchange (e.g., moving a physical magnetic or solid state media from A to B) in a batch fashion.
- Or telecommunication interchange (e.g., dedicated lines, WAN, LAN).
- Web service.
- Or dedicated, unique protocol and interchange.
- Or import/export for batch interchanges (solicited, unsolicited, manual trigger).

In addition to these types of core concepts (i.e., ideas like interactive interchanges versus batch), there are additional universal design requirements for the non-human level:

- Initiation — how are things started up and synchronized.
- Shutdown — how are things taken down.
- Restart — how are systems.
- System level functions — e.g., health checks, heart beats.
- Application level functions — the actual services or functions.
- Updates, patches.
- Support for solicited and unsolicited interaction.

The *ZenTai* factors also apply, but must be interpreted differently. For example, software can be considered to be stressed and uncomfortable if there are unusual work arounds or hacking to get something to work. In most systems and languages, there are obvious and easy ways to do most actions and hard ways to get to the same place. If the situation requires extreme programming to make something work, then this is noteworthy as part of the analysis as it stresses the software.

The value aspect is relatively straightforward: what characteristics of the system connection return added-value results. For example, a dynamic web service link that exists in a current system provides a different type of value compared to a daily batch file transfer that exists.

The existing system environment can also be reviewed with respect to evolutionary stability in the past, present, and expected future. For example, does the problem space, the one in which the new solution has to play, have well established and long set standards or are they evolving yearly? Or, does the problem space imply the certain use of new or relatively new technology where new libraries and features are rapidly created? These are evolutionary parts of the problem.

The experience part of *ZenTai* is a bit more abstract when considering a non-human element of the problem space. The experience pillar relates to the experiences and knowledge that exists before using the system, the experiences that result from using the system, and the experiences that might be gained in parallel while using other systems. In this case, the universal requirement factor for experience is subtle. Do the existing or anticipated systems have adaptive learning elements that will adjust behavior (short term or long term) based your solution's behavior? If so, how might this affect the requirements for the solution? This is the external experience factor: how will the other systems use their experience with you? There is also an internal experience factor for how the solution being considered. First, you can use past experiences with similar systems in the requirements. This is obvious. Second, you can contemplate if there are patterns of connection that are repeated over time which can be used by the solution to improve quality or speed of the added values.

Chapter 14
ZenTai Summary

It is possible that you might read or use the first half of this book without delving into the second half. The first half of the book is not dependent upon Agile or any other actual coding method. It will help you when trying to do Agile in extreme Class IV–VI situations, but you need not code yourself; you are worrying about the value and what the system will do!

I am writing this summary two years after writing the first draft. This has given me time to reflect and think hard about some of my thoughts and suggestions. I have been working on a reasonably large project for the last three years and when doing the final edits on Parts I and II, I have tested myself; am I reasonably true to myself and my suggestions? Can I be honest and say that I generally do these things, think about these things, and try to design user functionality according to the principles I have described? It is impossible to be unbiased in such self-assessments. However, I think I come very close to walking the talk. As I was reading, the first dozen or so chapters, I was mentally checking off features and concepts that I used during the analysis or built into the user experience. Not textbook perfect and not 100%, but I would give myself more than just a passing grade.

During my career, this approach has not always been understood or appreciated by other developers or by some of the actual users. In some cases, the acceptance only came with usage and the passage of time. It is possible to speculate about the future, but history will usually expose what has actually happened. Thinking this way has helped me for over three decades to craft software that the users have valued and have fought to keep using. This is true for software I wrote in the 1970s, 1980s, 1990s, and in the first decade of the 2000s. I cannot guarantee that these methods will, or can work for anyone else. I suspect that most analysts will find some value in parts of the *ZenTai* approach though.

What would my final words be on this topic? Difficult to choose or craft a simple, pre-digested thought for this purpose. Perhaps it is the simple realization that doing a good requirements analysis for a complex situation is hard work, is not to be taken lightly, and cannot be relegated to an ad hoc process that some wrongly interpret Agile as being.

At the end of good development, you want to be able to say: *quod erat faciendum*. This is Latin and basically means that what you have created is "which was to have been done," or what was required.

Part III

Architecture & Design

Overview

Parts I and II dwelled on the **what**. The *what* is very important because if you do not get the *what* reasonably right, it does not really matter how you proceed. You can have textbook pretty code and really cool software components, but the project will still fail. Parts III and IV assume that you have got it reasonably right for the type of situation you are faced with. A Class I situation at one extreme or a Class VI at the other. Just as you would need to do a different style of requirements study at each extreme, the design process and ultimate implementation is also different. If you do not think that you have got the requirements reasonably right, please re-read Parts I and II and give it another shot (or two).

Part III is at the high level design level. I am not going to discuss coding conventions, UML, ERDs, or where to put comments in the code. This chapter is how to think about design abstractions, layering, stability, robustness, and long living systems. It is about how to match your technical marvel with the problem identified in Part II.

Chapter 15
Universal Designs

This chapter follows logically from Chapter 13: universal requirements and the *whats*. A universal design is something considered possible to provide in many solutions (not all — there are always exceptions) to address a common, universal requirement. You cannot do a universal design concept without first identifying a universal requirement. For example, you identify the need to deal with color blindness and through research identify what the appropriate colors might be that satisfy the greatest part of the population. This is the requirement. How you incorporate or provide the solution to the requirement is part of the design *what*; that is *how* the requirement is basically addressed in the final product or system. This might be the physical color of a button or the color of the same functional button on the screen.

In any given system with IT in it, you have a relatively closed set of options for input and output. There are extreme options available if you are able to include virtual reality or motion detection, but the run of the mill, day to day system has a smaller set of options. These are your weapons or tools with which to address the universal, and other, requirements.

For input, you have:

- Physical keys you can press.
- Physical keys can have color, shape, and texture and pressure or response profiles as they are pressed.
- Keys that have a physical or sticky up/down status.
- Possibly dedicated buttons for specific purposes.
- Things that you can click — on the keyboard or attached to a pointing device.
- Pointing devices — i.e., ways to guide a cursor on the screen — with speed, motion, acceleration, coordinates, and possibly pressure or force detection — there are many options for how this can be implemented: touchpads, mice, stylus, joysticks, etc.
- An area in which the pointing device works which can be single purpose or have special areas (e.g., the edge features on some touch pads).
- Touch sensitive screens — where your fingers become the pointing device and virtual buttons can be pressed or pointing/motion activities performed.
- Vision — cameras.
- Speech — microphones.

- Single click, double click, click and hold options.
- Special keyboard sequences that can multiply functions (e.g., shift-click).

For output, you have:

- Simple, single line, limited output displays.
- Full screen displays.
- LED or LCD types of section displays versus full character resolution.
- Active versus passive displays.
- Backlit or lit displays.
- Displays with brightness control.
- Displays with contrast control.
- Partial normal screen resolutions (e.g., VGA vs the latest and the greatest).
- Full color spectrum, limited colors, or monochrome.
- Icons and shapes — single or in combination.
- Fonts, font size, color, blinking (ugh), highlighting.
- Pictures, graphics, and video.
- Speakers.
- Pop-ups, dedicated areas on a screen.
- Possibly separate monitors or output devices dedicated for certain information.

It is relatively rare if you have all of these in one system. A class of devices might have a common platform or set of options. For example, remote controls for televisions and audio rarely have displays. They have many buttons. Some might have pointing or selection mechanisms. How a remote control can address the various universal requirements will be limited by the input and output options available to the hardware designer. The fewer choices you have, the more opportunity there is for compromise or absolute lack of satisfaction. For example, a remote control device might be used by the elderly and with only buttons, the control device might be unable to adequately address vision, limited mobility, and other factors that this user subgroup implies. The remote control might be able to use color for the buttons, group the buttons in a logical fashion, and use shapes and textures to help with some of the requirements, but it is unlikely that many of the universal requirements will be satisfied unless the user group is very homogeneous and has low variance that can be addressed by a button-only device. Similarly, a simple single line display on some audio equipment can display one size of text and while this is OK for those with normal or good eyesight, it is not necessarily OK for other user subsets.

If you are a designer of systems, you can possibly create a table of the tools you can use (e.g., types of displays, buttons, pointing devices) for a class of devices on the vertical access and the universal requirements implied by the environment and user groups on the horizontal. This tool usage and application document should be created in all organizations and it should describe the truly universal and the slightly restricted universal *whats*. You can then systematically identify what requirements will or will not be satisfied in whole or in part and how (the design how). If there are key requirements not addressed by a certain toolset, this can identify areas for development or research. For a larger organization with many classes of products, a common approach can be used when the input and output choices are the same: the same button shape, the same color for the same purpose, etc. This can give a common touch and feel across products.

You will notice that the discussion thus far has focused on the human aspect of universal design requirements. It is possible to do the same for the system or synthetic connection level. You can catalogue and think about what the standard tools are for building the system level connections. For example, you have interrupts, ports, and so forth. You have protocols and mechanisms for data interchange. I will not expand this topic, but leave the idea with you to explore.

As a designer, you have choices and one choice is about how you go about designing something. You do not want to stifle the pure innovative or genius moment when something hits the WOW! factor. Not all aspects of a product or system are of this type, and I think it makes sense after the initial brainstorming and creative moment to systematically think about what has been designed. That is, you do not use a rigid structure or checklist in the initial phase unless you are stuck for ideas. You use the lists to make sure that you have not overlooked the obvious, forgotten something, or were blinded by your initial, emotional thoughts. You use the lists to discuss alternatives, assumptions, and to make sure that you have thought things through. You also use the lists to compare one product or system to another and wherever reasonable and possible, a *universal* solution is chosen!

Chapter 16

The Big Picture

16.1   What Is Meant by the Phrase: Big Picture?

The first thing to do is to think through

what the analysis means and

what the actual solution will be at a high level.

You should have a conceptual, high level solution suggested by the requirements created by the Part II methods and processes, but that is still removed from anything resembling the implementation. The solution at this level is more of a wish or want list of capabilities and functions. The user needs to do blah-blah and blah. Perhaps hundreds of pages of blah and blah. The output of Part II is not 100% oriented about computers and IT; it is a holistic view of their information situation and their life, their tasks and their environment of which IT or the potential use of IT is only a part.

The question at the top of my list is: will a computer be of any use at all? In many situations, there are many things to do manually and organizationally to clean things up, creating standard processes, and sorting out the spaghetti mess before you get to the sanity level that computers require. Totally irrational and screwed up processes should be sorted out manually before you waste everyone's time trying to computerize chaos. This is the type of stuff that I reflect upon when I start thinking about what an analysis might mean. I do not unilaterally assume that a computer solution is what is needed.

I did not keep accurate count, but in about half of my consulting or advisory situations, I have probably recommended non-computer solutions as the first step. Admittedly, this caused some raised eyebrows when certain folks wanted me to recommend expensive computer solutions. In one case, I recommended hiring summer or intern students instead of writing the software system desired by the VP. The students would have had more common sense and be better at the task than a computer would have been. Also cheaper. In this specific case, this recommendation came a year after the company started to build the system. They were trying to build a system that did not make sense to build.

Remember, while you can use Part II to look at the problem strictly from an IT perspective, you can also look at the total user situation or problem using the *ZenTai* viewing glasses. Did you think about this while you were reading the sections? This gives you the big picture from the user side and a holistic view of what will help the user. The holistic solution has to consider the holistic problem and there may be portions of the solution which will be computerized and some portions not. Remember the way David Tse did the hospital problem analysis? His process model was the holistic view and was not restricted to the computerized process. He then overlaid and considered the IT in the context of the larger problem.

In some cases, the model or big picture will be very close to the model(s) you might have created while figuring out the *what*. This would be very true in certain types of software situations, such as simulations, or data driven applications where a data model would have been a logical part of the *what* analysis. The big picture I am trying to describe is the proverbial diagram showing the major bits of s/w or h/w and the links between them. The first big picture does not necessarily worry about what is physically separate or not. It does not worry about what might be on the same computer or not, what is in the same room or not. It is still conceptual and deals with the solution in an agnostic fashion. For example, a big picture might just identify web servers and data servers without getting into specifics about Apache, Linux, Windows, IIS, SQL, DB2, Oracle, etc. And, unless there is a good reason to do it, I would suggest that the first big pictures of the solution should always be at the specie level and never specific to a vendor or special piece of technology. By this I mean that operating systems are generally alike, as are databases, as are email servers, etc. There are times that you will base a solution on a specific piece of bleeding edge technology, but in your career, these occurrences will be on the infrequent side.

I have several reasons for this recommendation. Perhaps the most important is that I want my software solutions to last for a long time (if they continue to provide value). I have found through experience that designs that are based on generic capabilities of a class of software are the strongest and easiest to keep alive. They are easily ported from one technology base to another and there is no reliance on a vendor who might change things, no longer support a certain capability, or go bankrupt. To keep my head straight on this, my first designs do not care about languages, operating systems, or specific libraries. I take off my glasses, get a good blurry image of the class of software and force myself to think in terms of basic capabilities. Once my initial design is done, then I consider what is in my toolkit, what I can shop for, what to compare between, and so forth. If I start off with a specific language or system in mind, this can force my design into places I do not like; and I really do not like it when the solution drives the problem. When possible, the technologies should be flexible enough to fit the problem and the problem should not be changed because of bias of thinking (we are a Unix shop, we are a Windows shop, we are blah-blah).

This big picture approach is not just for big architectures or big projects. Whenever I am asked to do a simulation design or smaller bit of s/w, I take this big picture approach as well. I believe that as a craftsman, you can usually bend, fold, and mutilate the lower level technologies (such as languages, db systems, OSs). I get a good picture in my head about what are the key capabilities I need to provide the value (and other *ZenTai* pillars) and then I go shopping. If I have to, I will scan thousands of pages of technical material after I know what I want to do from the user side, and I try to avoid doing this before. This is of course when I am going into new areas of development, but I still try to avoid language, OS etc. specific pictures even if I know quite a bit about the languages and systems! It is sometimes not easy to turn off specific thinking and think capability-wise; but it is necessary. It is like a grounded theory approach to software design (homework assignment: look up grounded theory).

Another reason I avoid specifics is that I like to design in horizontal layers of equal granularity and functional power/capability. I do not like having to think about the color of the knob on the stove while debating the need for a stove. If I can avoid thinking about the actual supporting technologies, I can keep my head clear and not even know if the stove has knobs. I assume that there are ways to start things, stop things, obtain services, and so forth.

A third reason for avoiding specifics in the first big picture is to avoid the holy wars that erupt between developers. The developers I have worked with in the past who, in hindsight, were the most disruptive, and mostly likely to never get a reference from me are the ones who have an emotional attachment to a movement, method, concept, language, OS, CPU, or vendor. A nice generic big picture design can avoid some of these religious debates in the early phases. The big design can also flush the little critters out of the woods too. You will get the initial push-back and get a reading for future fun and games by using a generic approach and sticking to it till the ideas are crisp. People who jump to solutions immediately are not likely going to be your valued decision makers.

While I am ignoring specifics, I am also ignoring what people might consider traditional problems and traditional solutions. Until sufficient analysis is done, you do not know if your problem is traditional or not. There is great danger in assuming that all problems have now been solved, that there are known patterns and designs for all known problems, and that everything has been done hundreds of times already. There are lots of assumptions in this thought process and they can be problematic. It might be true in many or most cases that the problem is indeed traditional (e.g., Class I–IV rapids), but until you know if the problem is traditional or not, it is best to keep an open mind. You should never casually and carelessly undertake non-traditional solutions or problems without thinking through the costs, risks, and possible compromises. It might be possible to compromise on a seemingly non-traditional requirement and turn a Class V into a Class III or IV situation. However, it might not be possible and in those cases you need to be ready to develop a non-traditional solution to a non-traditional problem.

There is of course the usual caveat. You cannot do a big picture or generic design until you know enough about the classes of technology and know what is common versus specific. You also have to have enough experience to know what are risky assumptions and which ones are not. You need to know what to expect from any OS, not just the one you have grown up with. This goes back to the apprenticeship model. To be a good, nay expert architect qualified to direct a big $ project with significant risks, it will take about a decade of being a developer, junior designer, and so forth learning about architecture. Although this might sound or appear daunting, you have to start sometime, somewhere and just do it, do it often, and reflect upon your process and how things are going.

You will need to learn about your judgment making capability and if your choices for design variables, decomposition, interface design are actually good and will last the test of time. If you have had multiple software experiences lasting over many years (and I mean at least five to ten years), then you would have had lots of chances to do things, and to see if you made the right decision or not. If you have learned from this and reflected upon your experiences, then I would consider you a candidate to be an architect. But, if you have not gone through multiple releases, legacy conversions, technology base changes, and many other things that can happen, you might not have a good base to work from. This does not mean that you had to have been the head person on all projects, but it does mean that you had to have a chance to make decisions, see them implemented, and then see how they live. I might trust you to do some architecture work on a subsystem or minor bit, but I would probably be a bad manager if I let you do a major system without a track record of creating multiple systems that were in use for extended periods of time. This does not mean that you are not smart or that you are not a wizard with code. It is a simple matter of experience and what experience versus text book knowledge implies.

The big picture should have the main, Zen level, building blocks that will deliver the value to the user. The blocks should be at the same level of granularity and if you remove one, the system will not really make sense or deliver on the required functionality. So, one way to look at the big picture is to critique a starting point and debate the existence and level of each component. It is rare to create a big picture and have the first instance or idea be good enough or close enough to be the final form. If I consider my own work, I am not that good and I have been doing big pictures for a long time.

The big picture model should have many of the same characteristics noted in the chapter on abstract models. It should be easy to use and explain to people who are not technology savvy. You should be able to map functionality groups or concepts onto the diagram so that people will know what is providing what. You should be able to use the big picture for multiple purposes and use it for meetings and presentations; just enough info and not too much. The big picture is perhaps the main communication vehicle between the architect and the user; using the big picture to explain to the user what the problem is that the picture will solve. It is an excellent vehicle for validating your understanding of the problem and for creating credibility with the user groups. Thinking about this use might help you consider what goes in the picture and what does not. Would your user understand it? Can the user picture how the system will work and deliver on the lofty promises you are making (or selling)?

A big picture might not actually be visual. It is a concept after all. However, if you are lucky enough to be able to craft a visual of the big picture, the first big picture should be similar to a quick diagram of a housing development: where the basic roads are, where the houses will be, where a park or green space might be. It should probably not include the type of deck on a house as this would not be at the same level of granularity. Nor, should it talk about brick color, or type of appliance to be found in the kitchen. Once you have the first, congealed view of the solution, you can then create other pictures, do your shopping for technology, engage in debates, etc. People love details though and it will be easy for the discussions to wander down to the specifics. As counter-intuitive as it might be, it seems that people have an easier time with the specifics and details than the big picture. The old saying: you cannot see the forest for the trees, comes to mind.

I am often asked where do you start and how do you *see* the solution; that is, where does the big picture or grand scheme come from. What I find easiest, is to think about the system as an organic system, something with a life cycle, something that moans and groans, something with human traits (i.e., anthropomorphic projection of uniquely human traits onto an animate object). As I am doing the stuff in Part II, I usually start playing scenarios in my head. These scenarios are like a day-in-the-life of blah, or a week in the life of, or the full life of something from conception to cremation. This helps me think about the main bits that are needed to deliver the value. I need to sit back, close my eyes and project myself into the situation. I become the tree; I become the system; I **AM** the system. I try to get a feeling for how things are connected and how things will flow through the system and what needs to happen when to what. I follow the user's meta objects and perform the user's meta functions as I visualize how the system will behave. The goal is to discover and understand the key, critical bits. The bits which are vital, are used constantly or frequently and which give high value. You can live without a finger, but you will need special assistance to live without a heart or at least one working lung.

Sounds corny, doesn't it. Drugs are not needed, just a very active imagination. And, please avoid doing this when handling sharp objects or operating a motorized vehicle. Once I feel comfortable enough with the problem space, this projection and playing computer seems to come naturally. If I force it too early, *taurus excretus* is generated. And, don't ask me when I know when to do it and when not to do it; it just feels right. If I get confused and my head hurts, the ideas are not ripe and mature yet in my head. If it comes quick without really thinking about it, the solution is ready to pop out of the toaster.

As I play computer, having a really good imagination helps with the various *ZenTai* elements. You want to imagine yourself being the various users, in the various situations in which they will find themselves, trying to get the job done. You need to take on multiple personalities and not just play the scene from one view. You need to drop the "I think" or "I do" thoughts and really try hard to think like the stakeholder. This is not easy. These schizophrenic personalities and consciously being aware of value, comfort, experience, and evolution using the different perspectives will help identify the natural (or more natural) components for the big picture. The bits are probably the most powerful and natural when they match the system without forcing. If it is Zen, then the bits will just feel right and be obvious (in hindsight) and everyone will comment on how simple the design is. The design will be easy to explain and not require extensive commentary or discussion. They will wonder why it took you so long to come up with something so obvious and simple. Often the final system is childlike in its overall structure and simplicity. Fancy stuff might go on within the blocks, but the main design and decomposition should be simple. When doing design sessions and the client comments on the simple solution in the context of the time and cost it took to derive it, my inner comment, not verbalized, might sound like: if it is so obvious and simple, then why isn't your current system so obvious and simple, and why isn't the competition doing it either? It takes skill and practice to think simple. You have to control urges to make things complex and fancy. I know some programmers that think skill and expertise is demonstrated by the complexity and technical geekness of the solution, using exotic options, the latest techniques, and super libraries and functions. Hey, look at what I know! My own opinion is that skill and expertise is demonstrated by simple solutions that are robust and deliver high value to the client. You might know the technology, but you also know when to use it and when not to use it.

Remember, you want systems built and designed so that if it is easy to change in the real world, it should be easy to change in the software. If you keep this in mind, the architecture will have a form of inherent ability to take the knocks and blows that will inevitably come along.

Creating a big picture is right brain thinking.

16.2   Good Architecture

The title of this sub-section is a bit off actually. It is not so much about good architecture as about weak or substandard architecture.

The big picture is the initial architecture or design, and this will lead to detailed designs. As the design develops, there are many decision points and it is possible to have many designs for the same problem. So, what makes one architecture better than another? Hard question. Unfortunately, there is no mathematical formulation or algorithm that I know of that you can blindly use. It is easier to say what makes an architecture weak or risky. For example, my top dozen warning signs of a problem design are:

1. An architecture that is dependent upon one proprietary technology is a risk in my view. Sometimes you cannot avoid this, but you should consciously be aware of what you are doing and the risk you are taking.

2. An architecture that is based on a developing standard that can be expected to rapidly change is a risk.

3. An architecture that does not allow the easy and rapid replacement or re-development of the building blocks is risky.

4. An architecture that does not allow logical extensions of a building block without mucking with users of the block or other building blocks is not nice.

5. An architecture that does not allow for isolation and decomposition for analysis and testing is also a bit ugly. Bad things will happen and you want a system that can be quickly analysed and studied.

6. An architecture that crumbles like a stale cookie when something goes awry is not a great design. Overly sensitive designs with many assumptions and tight constraints have thin skins easily punctured.

7. An architecture that has many single points of failure is not acceptable if the system is supposed to have high availability.

8. An architecture that inherently creates instability and variability is playing with fire. Consistency and predictability are good things. You do not want systems that are like the wagon trains of the old west: you do not know where they are, not sure where they are going, and have no clue when they will get there.

9. An architecture that must deal with high peaks and valleys of demand, but is not scalable, or that has a clear bottleneck or funnel that everyone must go through.

10. An architecture that makes a geek's eyes gleam with joy and anticipation, and makes the geek's adrenalin flow like Niagara Falls, makes my blood run cold.

11. An architecture that does not gracefully degrade or lower performance. I prefer systems that allow some of the system to work while otherwise limping along. It is a form of s/w leprosy if you think about it. A good system can lose a finger or a hand and still function.

12. An architecture that has many vendors involved with tight coupling that implies coordination between many vendors for upgrades, compatibility, and migration issues is not good.

Can I measure the above? Can I run a software profiler on the system and give a rating? Can I give a more precise checklist? Can I tell you how to view systems that have some good and some bad, and know when it is good enough?

At this point, the answer is no to all of these questions. This is not a good answer. Would you prefer "I know it when I see it," or "The hair on the back of my neck stands up," or "My gut tightens," or "My head hurts"? Note: your head should hurt when learning, but not when understanding a design if you are already supposed to be a designer. Sometimes it is subtle and it might not make a difference in the immediate future, but you JUST know it will matter in ten years. When you are an expert with a demonstrated record of success, you can probably trust your gut feel. When you are a novice, just starting out, you can listen to your gut feel, but you should use reasonable judgment if you are going to trust it as a guiding light. I was able to generate my dirty dozen list because I have encountered quite a few systems with these flaws and have had to deal with them. I have even created systems with some of these flaws at one point in time or another. However, I have tried to avoid repeating the same architectural mistake in subsequent, similar, situations. I have not always been successful in avoiding the mistakes, but at least I try to be aware of the issues and make the compromises in a conscious fashion, not by accident.

If you are encountering an architecture that you are being forced to use, or asked to critique, you should also take into account your own skill and expertise. You want to avoid being labelled an arm-chair quarterback or art critic. You might be quick witted and have a sharp tongue with lots of one-liners to put down, or praise something, or someone. You have opinions. Everyone has opinions. Most opinions should be kept to oneself though and not heard by others. Another old saying: better to be thought a fool than to prove it. Until you have enough experience and knowledge demonstrated by your own sound judgment, you might wish to keep quiet, observe, and learn. It is dangerous to say that you do not think something will work, something is wrong, something should not be done if you are speaking from opinion only. It is also very dangerous to regurgitate or pontificate some distant guru's one-liner or pre-digested thought out of context. Citing best practices or what so-and-so is doing is also dangerous if you are a faux expert. You might be in a room with a real expert and either they will take you down, or will quietly write you off and discount your thoughts, which is not good because you might actually have a good thought or two and no one will notice.

At a very detailed level, there is science that looks at module complexity etc. as a proxy for architecture quality. And, there is science that looks at the amount of functionality that an architecture satisfies as a proxy for quality. While these and other formal computer science and software engineering concepts give some insights about the design, they have difficulty providing insights about my top dozen concerns.

16.3   Layered Analysis

Layered analysis and design implies that you iteratively work from the highest level of concept down to the details. This is not the same as top-down or bottom-up design. Normally, the terms top-down or bottom-up have been used to describe the order in which the layers of functionality get defined and designed. For example, people would design the top layer in great detail (and possibly do some implementation), and then design the next layer etc. This is *not* what I mean by layered analysis.

Each conceptual layer deals with the system in a holistic fashion. From top to bottom, from left to right. A layer has the same level of assumptions, granularity and specificity. It is different from the traditional top-down, bottom-up, inside-out, whatever approaches because you make the system coherent and consistent with itself at each layer of abstraction. Got that? You do not do myopic optimization or design decisions subsystem by subsystem, but you make the architecture work for the whole thing. This is similar to a symphony or other musical pieces. The whole feeling is needed and then the details are fleshed in. You do two or more layers and then you go back to the top layers and re-visit them, perhaps looking at different aspects (but again, at the same level of granularity and purpose). It is iterative, up and down like a yo-yo.

Systems layering works by drilling from the highest level Zen down to the more specific levels of exactly what software is used or crafted. At each layer, you stop and think about what you have just crafted. You make sure that your design satisfies each of the relevant levels of abstraction implied by the *ZenTai* requirements analysis. This might be three or four layers for a larger system and each design activity might take days or weeks (or months) to complete. At some point you stop or you re-visit higher levels to clarify thoughts and concepts. I usually stop thinking about architecture once I have gone down several layers, and I have done my initial shopping and matching of possible technology solutions suitable for the problem. Remember, that I will delay the shopping and technology investigations until the base, functional architecture feels right. I am then ready to discuss the technological solution, and then identify what bits will be in the initial toothpick software build.

Each new layer is not developed top-down or bottom-up either. I do not usually take a systematic approach to it (in the normal sense of systematic). It is more likely to develop through my mental simulations as I follow the bouncing ball through the system and I will expand and further refine bits as I encounter them. I will take a high level function and rapidly develop the various pieces top to bottom that need to be there. I suspect that if you attached a focused light to the process, it would look like random traces.

I have two observations to make about layering. First, I think it is a very efficient and effective way to think and design. It seems to avoid going back and making things consistent and in harmony after the fact. I do not create something rigid while something is fluid; the various subsystems all start at the same level of detail at the same time and keep in synch. You will always have some backtracking but the conceptual layering seems to minimize the re-designing. Second, this is not the way novice thinkers typically go about thinking. Novices do the top-down and bottom-up, detailing as they go — painting from the corners, hoping that they do not have to go back over the canvas and scrape of the paint and start over again. They are taught and think that design is linear, recipe like, and sequential. I may be doing it all wrong, but what I do is certainly not linear.

This is not picking on novices or making fun of them; it is just a natural way of thinking when you are starting out. You do not know enough about the lower levels to stay at the high level and have the confidence (supported by evidence) in your ability to make the necessary assumptions. You need to know general capabilities and what can be done at all lower levels so that you do not make a dumb decision at the highest conceptual level. Hence, the novice gets an idea and drives it down the pipe to see how it can be implemented. Then another idea is driven down, and then a third. It is possible, and even likely, that as you drive down the second, you do a little oopsy and have to go adjust the first stovepipe you created. Then you finish the second. Then you start the third and part way along that pipe, another oopsy. Then you back up, and perhaps change the first pipe which then forces the second to change. If you are lucky, these changes do not result in a change to the third which started the ripple effect like an Escher painting. I call this top-down or bottom-up piping *single-threading*.

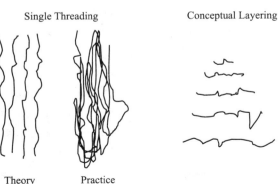

Single Threading   Conceptual Layering

Theory   Practice

You need to do this single threading when you start out, but you should try to learn the species and general characteristics as fast as you can to change from the single-threading to conceptual layering for design. If you work on this and think about it, you will find that over time you will find conceptual layering easy. But, be warned, it does not come to everyone and many decades of being a s/w geek does not mean that the geek thinks in layers. I once worked with someone who had been developing the same number of years as I had, getting the same pay, and could only single-thread. This was frustrating since everything for him had to be researched from scratch, from first principles. As a senior developer, he was being paid handsomely and in theory for knowing solutions, or for knowing how to make decisions. Senior folks are supposed to be more efficient and effective than the young folks. Else, why get more money and more decision making power? He was senior, but his thinking was junior.

Anyway, I think you get the picture. You inevitably single-thread when you start out, so do not worry about that. But in my opinion, if you want to be a true hotshot later in life, you will have to develop the skill to do layered thinking. The natural tendency is to jump to the details, the specifics, and use stereotypes: they are nice and they are friendly. They are not vague and you are not making assumptions and flying by the seat of your pants. Remember, this is understandable and totally expected when you are a beginner and it is in your comfort zone. Over time though try to wean yourself off this habit! You should also be careful about trying to do layered thinking when you are full of confidence based on limited, unproven experience. Just because you think you know, does not mean that you do know.

What about coding then? Do you single-thread or layer when you get past the design? Isn't the toothpick a single-thread process? Are you then a junior programmer if you single-thread coding? Agile/Extreme implies that you will single-thread functions as you work with the client. That is by definition and implication. In a Class I–III situation, you should have enough tools and infrastructure at hand that the single threading is not deep. That is, you have a reasonably shallow code structure that delivers many functions and little code will be from first principles. In a Class IV–VI situation, it is likely that you have to build the infrastructure as you go, or extend it. In this case, the single-threading you have to do is deep and not for novices. You have to be prepared to refactor a lot and you need to be able to refactor fast and accurately. In a Class V or VI development, you will likely have to do some infrastructure creation first, and then start single-threading functions and then go back and work on the infrastructure.

A similar situation arises when thinking about standards and traditional problem situations. If you think from the beginning that the situation is standard, or must be a traditional problem, you might not see the non-traditional bits, or you might dismiss them. A small non-traditional aspect might be reasonably and rightfully dismissed, but once you dismiss too many of the non-traditional requirements just because they are non-traditional, you fall prey to the old axiom that everything looks like a nail to my hammer. You are changing the user's requirements and problem to fit your traditional solutions!

I recently did a reasonably large architecture and design. This followed from an extensive analysis on the problem, using many of the techniques described in Parts I and II. My initial big picture had generic blocks of functionality and services: login and configuration, web server, database servers, search engine, logging and analytics — those sorts of things. There were probably a dozen big pieces in the picture. But, there were no ideas running around that talked about which web server, what operating system, which database system, what protocol to use, what options would need to be set, and other specifics. The design was based on generic, high-level functionalities of the type of technology envisioned. The system was considered mission critical and I certainly did not want to build in anything fragile or exotic. Nor did I want the system to be dependent upon any single feature or capability from a single vendor. All features and functions needed to be black-boxed and independent from other boxes (if at all possible). The system would start out single vendor (or attempt to be) for the stack as a matter of philosophy, but over time the system is expected to evolve and various components might be replaced with other vendor subsystems. When possible, I prefer systems that minimize the number of vendors, but that is not a religious point either. The user's *ZenTai* comes first in most cases and I will build or craft whatever is needed.

Once the big picture looked good from a basic, generic capability level, I then started to shop. I surfed, talked to people, read blogs and discussions about various tools, and reviewed many manuals and texts for the various functional services; all the while matching functionality to my mental checklist. Oh yes, the checklist. As I do the big picture, I keep aware of key assumptions or capabilities that are necessary to deliver the value and other *ZenTai* elements in this specific instance. Things like keeping the client light, using the server side for persistence, dynamic webpages (i.e., AJAX), etc. While I am reviewing the literature, I am re-affirming my understanding of the basic generic capabilities. I am also looking, or shopping if you like, for the magic pieces that will make my critical functionality come alive. I usually assume that the magic can be found in any of the competing systems, but it might be easier or more integrated in one versus the other. In the system I was crafting, I did the shopping and decided on a specific stack as it was a better match than the other when I considered all of my requirements; development through functionality through deployment. This was actually the system I was least familiar with and it would require me to learn a new language, new set of terms, etc. But, again I am an agnostic and instead of choosing the system based on my own learning curve and comfort zone, I chose what I thought was the best stack for the problem.

The next or second big picture for this particular system had the various technology elements superimposed over the functional design: placing the various pieces. I was working with a large enough team that prototyping work could start on almost all of the services and logic in parallel going to the next layer of design. This was done in a combination of ways involving more diagrams, discussions, documents, and experimental coding. The third level of design was at the toolkit level and the building blocks for each service and subsystem; again keeping the view of tools and generic toolkits for each layer, getting more specific and more detailed at each iteration. Each layer tries to avoid making assumptions that will result in a weakening of the architecture. In keeping with this text, the view was that the architecture may not be the best or the right, but it would avoid many mistakes and weaknesses. The top twelve problems noted in the previous subsection were constantly held up as design objectives (or in this case non-objectives). As the problem space and technology was better understood over time, a natural decision point arose where decisions had to be made about what would remain in the final product. This type of decision usually results in whole layers being abandoned or requiring a major refactor. If the gap between experiment and final is too large, the initial code base must be left behind. This is a major decision, not to be taken lightly, and it can cause some individuals on the team angst. If the initial coding effort has provided sufficient value during its development and exploration, it was a worthwhile effort. Not everything you design and code during a development ends up in the final product and this can be a harsh lesson or reality-check for some. An implied goal of a good design should be the relatively easy replacement of a layer, the ability to replace a limb without losing the patient.

As noted in this section, I like the conceptual layer approach for a number of reasons, and while it might not work for others, it seems to work for me repeatedly.

Another benefit of this layered approach is when the boss or client asks you for the design or the current status. Since you are taking a holistic approach for each layer, you can give a balanced answer and explain things without worrying about a bit of the system you have not got around to yet. It is a bonus that comes with no added cost.

16.4   Interface Definitions, and Protocols

Interface definitions and protocols are at the end of the day standards; either formal, informal, internal, corporate, public, or international. It does not matter. In the late 19th century (i.e., 1880s–1890s) and the early 20th century there were lots of debates about standards and conventions. The world in general was rapidly changing and certain sectors such as manufacturing were expanding into the mass production and consumption era. There have been good standards that have lasted a bit, and there have been standards that have lasted as long as an ice cube at the equator. My personal observations go along with the observations made by the old commentators, and the problems they faced are similar to those we continue to face. There are two big observations: we often try to create standards before we know what the standards should be, and that many standards appear to be crafted by people who should not be crafting them.

The first observation implies that we stand the risk of putting things into standards we do not really need and we leave out things that we do need: major features and options, major attributes and methods. So, we start along the merry path and surprise surprise, we need to add new stuff to the standard. And much to our dismay and sadness, neat things imbedded in the standard (possibly adding overhead and other headaches) are not used. If we launch standards too early, we do not know the Zen elements, and it is my opinion that the best standards and definitions are agnostic, minimalist, and are based on the Zen components. This can be either by design or by accident. Either way, the standard is better. The old people noted the dangers of early standards and these situations still plague us. Yes, we are occasionally forced into early standards, but why not recognize that it is an early standard and design in some flexibility knowing and acknowledging that we really do not know what we are doing?

A standard specified by someone not capable of creating a standard is harder to deal with. The designs are not that great and the standards have problems from the beginning. For example, there might be sneaky little workarounds, hidden hooks and fancy escapes. That is one type of questionable functionality found in some standards, which can lead to wonderful opportunities for the malicious hackers. The other questionable design decisions are when apples and oranges are mixed in the same standard or subset instead of taking the time to separate and decouple. Some of this may come from lack of discipline or awareness. I do not know. But why on earth would you go to all of the bother of a nice layered, protocol encapsulating a black package and then letting stuff within the package to explicitly escape and potentially take control of your system? What is stronger? Meta knowledge about something that allows you to ignore what is in a message or a protocol that forces you to worry about what is in the message, looking for escape sequences, special characters, etc.? I would argue that meta knowledge is better and avoids many problems. Many old protocols used lengths to define the sections within a protocol and did not rely on textual separators. The old protocols delivered the packets of information in a safe and secure fashion independent of the packet content; there were no escape sequences to worry about.

Assuming that the standard and interface is good and clean to begin with, there are three things (at least) that can happen to standards:

1. they can last a bit and then are replaced in their entirety by incompatible new standards;
2. they can last a bit with regular updates as the standard is cleanly extended to accommodate new findings, and eventually after many extensions and beating upon, they are replaced by something new and incompatible, and
3. they can be bastardized and corrupted and are not clean anymore as they are extended, eventually to be replaced by something incompatible.

Notice that there are some apparent truths here. First, standards do not appear to remain standards for long as they do change and they are replaced. This encourages thin wrapping and layered approaches to protect long lasting code from standards which are not long lasting standards. If the standard designers are not skilled, they will actually bastardize and corrupt a standard instead of extending it or replacing it.

Therefore, I would recommend that interfaces and protocols are consciously reviewed and reflected upon to make them as agnostic as possible, based on the Zen elements as much as possible, semantically decomposed and grouped as much as possible (e.g., protocols for managing the web service versus protocols for getting the web service to do actual work for you), and have the protocols layered at ever increasing levels of details and granularity. Else, history would suggest that you will have lots of fun riding the wild donkey (aka jackass).

Chapter 17
Designing for Change

There is an old saying that *the only thing constant is change*. There is another saying that is applicable to software projects:

that was then, this is now…

In an IT deliverable, there are three major domains of change to consider after the software is considered finished with respect to the first version: the technology, the user's problem, and the user. The *ZenTai* analysis helps with understanding the latter. In this chapter, I will discuss how to understand the other two domains, and provide some ideas for how to design for all three. In a later chapter, I will discuss project management's relationship with change.

Dealing with change and the uncertainty of change is the main issue here. There are some generic realizations that I have come to learn the hard way. First, most people do not seem to consider change in either design or project management. For some reason, they just appear to ignore it or blow it off. Perhaps they are novices pretending to be experts, or perhaps they have been living in a void. I am constantly surprised by people's myopic capability regarding change.

There are two aspects of change to consider:

- Change is the way a system evolves and improves (hopefully).
- Change implies something different has happened and the status quo is no longer and this can imply potential benefits and/or risks.

I once spent many months thinking about change and the implications of change. This led to a theory of temporality that I will share later on. Suffice to say that I once again probably had too much time on my hands.

So… the first observation is that not all change is the same. For example, there are some changes that:

1. you can control and either accept, deny, or control the timing of (e.g., a planned change with control),
2. are imposed upon you and that you cannot control, accept, deny, or control the timing of (e.g., a planned change with no control),
3. you will know almost precisely when they will occur,
4. are very likely to happen sometime within a time period (say next month, but you cannot predict the day or week),

5. are slightly random and follow some form of distribution based on frequency and repetition,

6. are truly random changes following no known or documented distribution.

For example, changing or upgrading your development system close to a critical date is like elective surgery and should be carefully considered! If you think for a bit about change, there are a number of other interesting things about it. The following paragraphs describe a few.

From a management perspective, it is no longer management of change, but *management during change*. With the continuous improvement initiatives and the rapid product life cycles, you do not have the luxury of stable periods as in the past. You also do not have the luxury of special change management teams riding in on their white chargers to rescue the damsel and then watch them ride off into the sunset leaving behind the kingdom safe and sound for another decade or two. I find it interesting that many managers are still not aware of the continual nature of change, and the need for having the organization attuned to it. They might be like the classic boiled frog story where supposedly if you pop a frog into a pot of very hot water, the frog will leap out; if you pop a frog into a cold pot of water and slowly heat the water to a boil, you end up with frog soup as the frog is not aware of the changes in temperature. (Note: no animals were harmed in the writing of this text and no frogs were experimented with.) I think that management should work on the whole staff and organization, creating the skills necessary to anticipate, manage, and operate in constantly changing situations.

Often a change has a direct, observable and well documented impact, but that is not always all that happens. Of course, the goal of a planned change is usually a positive impact and most people do not make a change to consciously make it worse (although I have had my suspicions at times). Most people think about and justify a change based on this primary impact. However, there are two further aspects to consider.

First, when a change occurs, there is a period of time that things might be more inefficient and ineffective before they become more effective and efficient. The utility function often looks like a checkmark having a dip before rising. For example, you install a new development system and while you expect gains in the long run, there will be a learning curve and reduction in productivity while you learn the system.

Second, there can be indirect impacts based on obvious or not so obvious relationships. For example, you upgrade the operating system to a new release point and discover that you also need to install more memory or disk. Or, that a magic number was changed (higher or lower) and that this threshold change implied many other changes would be needed. In some of these cases, you can do your analysis and have a handle on the indirect impacts, but in others you will know that a change will happen. You might even know when the change will happen. And, if you are lucky, you will know the direct impact (or not). But, there will be times when all you know is that something is going to go bump in the night on a specific day and what that really means will only be known at the time of change and you better be ready to react.

The last observation I want to share before getting into the three topic areas is that the primary or secondary impacts associated with a change might not be seen for a period of elapsed time. This elapsed time might be associated with when a certain function will be encountered, or when the result of a function will be used or inspected. This delay in impacts means that you do not have a change, deal with the immediate impact and then invoke amnesia. You have to remember recent changes (and some not so recent) and be sensitive to possible problems that might rise out of the tar pit and seize an ankle.

17.1   Technology

In many modern web-based (and non web-based) applications, the system can be viewed as having a number of layers or groupings of software and hardware. One possible way to view the stack is:

Ⅰ. hardware (e.g., physical device level),

Ⅱ. chip level s/w (e.g., imbedded s/w on chips),

Ⅲ. operating systems (e.g., base technology and drivers),

Ⅳ. systems level tools (e.g., servers for communications and database tools),

Ⅴ. supporting subsystems (e.g., email, search engines, reporting systems — using level Ⅳ tools, development environments and other toys),

Ⅵ. application (e.g., the main system — the cause for existence),

Ⅶ. user level logic (e.g., macros, work flow tables).

For systems expected to have life cycles longer than a fruit fly's (i.e., longer than about two weeks), you will likely experience change in all of these levels. There will be evolutionary changes in the form of patches, point releases, and major new releases for all software components. You will also get fixes and new revision levels for hardware, and new product families to deal with. This subsection will concentrate on levels I through V. In these levels, you are mostly at the mercy of the vendors. If you are big enough or strategic enough, you might have a say in the changes, but you should not assume this. Many users have been dated, dined, and loved only to find themselves spurned for a newcomer and dropped like a hot potato. And, not even a Dear John note.

Although I will not dwell on it, there are also times when you want change and you will not get it, or get enough of it, or get it fast enough. This stagnation can occur for a number of reasons. For example, the vendor might be in the midst of a new release or new family of products, and has reduced the effort on the existing product. Or, the vendor might have some revenue problems and cannot respond to the market. Or, the vendor's market has grown so fast (or is trying to grow fast), that the focus is on new installs and not on the product. Or,... you get the idea. There are lots of reasons why products will not change when you want them to. There are also the changes that go bump in the night when the vendor thinks the new or enhanced version is ready, and the software or hardware has a different opinion. Basic point: don't count on a promised change or new version to save your system. And, do not count on something that is promised not to change, not to change.

How I deal with potential change in levels I through V depends on the type of system. The more critical the system is, the more conservative I am, and I place less trust in what a vendor may or not do; and how well they do it. It is nothing personal; it is practical. Of course, anticipating and trying to do something for every possible change and impact does not make sense. For critical systems I do not worry about 100 year storms, but I will try to worry about the 10 year ones. If the system is not a mission critical system, I will take more risks and play the odds. You see, I have the magic touch when it comes to systems. The magic touch is really good if you are in Quality Assurance, but it gets somewhat boring when you are a developer or a user. I think this magic touch must be a form of perverse magnetism: if there is a bug or an early life failure, it is attracted to me like a moth to fire. Send me a product to do a pre-market evaluation and I find bugs and get hard crashes. Hand me your cell phone that you are beta testing for your employer and I will crash it and force a hard reset. If I am one of the first to buy a snazzy new disk drive, I will encounter a design flaw. I will install the new operating system and try the development environment for a few hours and the system will hang — dead as a nail. I tried to create large (1.5 mb) Java applets in the late 1990s and discovered a number of interesting incompatibilities between the supposedly compatible JVMs. The list goes on.

This uncanny ability to encounter problems made me a very good QA analyst (in my own opinion). I was even called pathological by a colleague once. I thought it was a compliment until I looked up the meaning. Given my perverse nature, I still considered it a compliment after I learned the meaning. I think it helps to be pathological, unreasonable, and irrational when testing. This background has made me sensitive to what causes things to crash, self-destruct, and otherwise create unwanted impacts. Learning and knowing how to make mistakes is one of the ways to learn about change and some of the causes of unwanted change, including the ability to recognize a change when you see it. Several years ago I visited a factory and was impressed when I was told that they make bad parts on a regular basis on purpose so that they know what causes them and what they would look like. This factory also had a very long record of zero defects shipped and the *bad part* process was just one of many techniques they used to ensure that an undesired change did not reach the customer. They learned about the parts' design and how this interacted with the manufacturing process, building up knowledge to be used in future designs. As a consumer, I wish that more companies would do this type of thing.

I used to do something similar to the *bad part* idea when I taught mathematically oriented subjects, showing math out of control and math gone bad. I always thought that you should know when the math is working and when it is not and be able to tell the difference instead of blindly punching a button and believing the answer. I guess I am old fashioned in that regard.

I have learned a great deal from these and other experiences over my career. I am certainly not the first to comment upon change and the possible implications. For example, Sweet in 1885 noted: don't be surprised by the unexpected when dealing with new inventions (Sweet 1885). My matching hypothesis to this observation is that **any** change in the status quo implies a potential risk. I have incorporated these and other thoughts into my design philosophy:

1. Design the system using generic building blocks based on generic capabilities: vendors' and your own. Over time the blocks will evolve at different speeds and you want to be able to

isolate and control each block independently of other blocks.

2. The design blocks should not (when possible) know the technology or innards of other blocks. Make them agnostic and when necessary thin wrapper, encapsulate, etc. These techniques will create a form of elastic shock absorber between the system pieces. Remember those nice metal balls hanging on strings, touching each other illustrating kinetic energy: you lift the one on the left, let it drop and the ball on the right is the one that jumps? You do not want kinetic software that reacts the same way. You want each bit of software to have sea legs and not be rigid. You want to avoid software that preserves kinetic energy throughout the system.

3. The components should only use the generic assumptions for the basic class time of functionality. For example, modern day communication servers do blah-blah and NOT assume that a specific vendor's solution that offers a unique, nifty capability will be used for the complete life of the product.

4. For mission critical systems, avoid the first version of anything and everything if at all possible. Try to use the second version of software and hardware at the toolkit level, and at the supporting infrastructure level. You can be a pioneer and do it knowingly, but it is awful hard to do this for 7×24 systems unless you have a very large bankroll and many super-geeks.

5. Always remember to design in automated testing for regression and stability tests. This implies regular system wide tests with life cycle concepts, not just isolated tests on the changed component.

6. Build in layers or services with suitable logging and analysis between the components so that you can trace and analyse. There are few things worse than having to hunt down that random crash with no tools at hand.

7. Use the basic components that have been around for a while (e.g., in Java's AWT, the component class) and avoid the latest and greatest extensions and add-ons. Such goodies are more likely to change and be unstable than the basic components.

8. I did not invent this old saying, but I agree with it — *if the user can do it, they will*. No matter what they are told, or what is in the manual. This does not always impact the technology level, but it can. Design accordingly.

9. Design the components independent of activation and use a two layer design (if not more). One level receives the commands and requests and if possible normalizes the inputs. The next level does the work and does not care about who requested the function, what was the mechanism for invocation, and so forth. The return functions are also so isolated. This makes it easy to drive logic from virtual reality, disk files, web services, etc.

10. One final and very important suggestion for design is to delay binding till the last possible moment in the design and keep as much of the system agnostic as possible. Delaying specifics is one of the most powerful ways to introduce flexibility and responsiveness in a system. It is also a way to reduce variability and the added logic and costs throughout the

system. You should delay processing variability till you actually have to deal with it. Variability and variance are killers when it comes to efficient and effective operation and the more you can do in the design to reduce and delay variability and variance, the better your design will be.

If you take these ten suggestions about change and the previous observations about weak architecture, you get my top twenty-odd design criteria about s/w in general. These observations are at the basic girder and foundation level. I have other observations about how to design for change in the problem space.

17.2 The Problem

In section 17.1 we discussed how to design for change at the technology level. While you can do some stuff at this level, the fun work is usually in the problem space. And, the basic technology levels are not likely to change as often as the problem space. Of course, the more types of technology bits you have, the more likely you are to have something changing all of the time.

Why do I consider the problem space fun? I am not sure. Probably some form of sick masochist thing. It is like a cat and mouse game; the user will try to tell you what is firm and what should be flexible, and sometimes it will be this way. But, they are often wrong or are giving you a partial picture because they are using their own personal perspective, not using abstract models, and they might be partially blinded by temporal locality (big words for simple concept: what has happened recently). When a user says something will not change, you have to probe the assumptions and think about the stability factors. When a user says something will change and must be flexible, you have to consider the trigger's feasibility: what are the barriers that might prevent or slow down the trigger, or otherwise lower the probability of the trigger actually happening. You also have to consider what the scope and impact is of the potential flexibility; how many people will be affected, and so forth, and weigh off the cost and impact of flexibility and inflexibility. Some people also love to dwell on the extreme cases or the exceptions and this relates to changes in the status quo. You can never account for everything and there is the law of diminishing returns when you try to catch or provide for EVERY extreme case. I have encountered two forms of the extreme case users. The first form gives you the extreme cases and is reasonable in compromise and moves forward. Some cases will be covered and some will not. The other form of extreme case user ends up in paralysis and cannot seem to move on, or easily accepts the fact that not all cases will be covered. The second form of user will be a liability and not an asset in getting the project done and should be avoided if possible.

This is where the Zen and really getting into the problem is the key. You cannot design everything flexible and changeable in every which way, and you have to make compromises and trade-offs. You have to project into the situation and think about what will be easy to change (regardless of what the user might say) and what will be hard to change. This is usually my guide for compromises, and where I will invest my time and energy for flexible design options.

The flexible portions are usually table driven designs, as much as possible in external tables or definitions. The most extreme case I have done was a prototyping system for factory planning systems where almost everything was defined via tables and the tables were all temporal with effective dates. One set of tables controlled the GUI, reports and logic at the technology base for the stuff I knew or expected or suspected might be flexible. Another set of tables defined the user's problem space in terms of how resources and work were described, including crews, shifts, calendars, etc. The goal was to go into a factory and within two weeks create a custom system for dispatching, scheduling, or planning. Actually, what I built was a prototype of the prototyping system to see if it could be built, if the temporal modelling would work (everything was sensitive to time), and if a simple learning mechanism could be built to learn about the temporal reasoning. The various functional components were also listed in tables and the logic was dynamically configured based on time (when the system actually executed and when on the scheduling horizon decisions were being considered). This extreme table design allowed for almost everything to be mixed and matched as needed. For this to work, consistency was important and all of the logic and pieces had to play by the same rules and not take short cuts or make assumptions. This is part of a toolkit approach for the problem space: be very consistent, anally so, and do not yield to the temptations of "I know this, so I can cheat a bit at ...." Bad.

I built this prototype of a prototyping system after twenty years of research, (theoretical and practical), s/w development, and consulting in the domain being prototyped. This experience is what helped me understand what needed to be flexible and what could be rock solid. I think I spent thirteen years thinking about this off and on, from my first general design in the late 1980s to the prototype in the mid 2000s. Rarely do you have to do this, or have the time to do it. As noted, this was my most flexible system. I have crafted quite a few flexible systems, and usually a nice data model of the user's problem is needed, as is a corresponding task model or process view of what the user has to do. On that note, here are my top ideas for dealing with flexibility in the user problem space:

1. Do the Zen thing and understand the essence. This is the stuff that will likely be more firm while other aspects will be fluid.

2. Drive the system with data tables containing directing logic and features, moving the specific choices from within the code to levels of re-direction and isolation. Look for patterns of patterns, patterns of similarity, and patterns of differences and let these patterns guide you.

3. Consider what is easy to change in the real situation and what is expensive or hard to change. This can help you make your own decisions about where to spend your time.

4. Make the meta-objects and functions match those of the user addressing the user's verbs, nouns, and tasks.

5. Break tasks and processes into independent and decoupled little steps with clean, unassuming interfaces between each step.

6. Not just break the user's tasks into small and decoupled pieces, do the same inside at the

functional level and avoid long and drawn out software threads. You might end up with a few monster routines and modules, but the majority can be reduced and refactored into bite-sized pieces.

7. Create toolkits of the most basic building blocks and support these in libraries so that the functional units can call them; thin wrapping language and system functions when appropriate.

8. Provide a mechanism by which you can create macros or internal routines that basically use other internal functions to get the job done. It is a form of incestuous or narcissistic behavior. The caller sets up the necessary objects that will be worked upon, and then drives them through the contrived logic in the callee instead of replicating or creating new code.

9. Above all else, ensure that you separate invocation from function as it is very likely that user interfaces and triggers will change: why, when, how, and under what conditions.

10. Use some of the concepts from Michael Jackson's design methods. You can write exploratory modules that are self-healing if they do not like what they are doing and then pass the repaired or unchanged world back to the caller to deal with.

11. Think through and play computer again. Think about what can change and how, and then consider it. What would be needed to do in the design if it did indeed change? You do not want to build in all of the stuff, but you do not design it out.

17.3 Users

Once again we encounter the user. Over time the user will evolve and learn as they have more and more experiences. That was addressed in the *ZenTai* analysis. There is another form of user change that you need to consider in the design. In one of my systems, there have been a hand-full of different users actually thumping on the keys over the years. There have also been several different first level managers, a couple of second level, more than a half-a-dozen senior types and about that many presidents. If the system is relatively low level and used at the personal level by an individual, then many of these personnel changes will not matter. You will only have to deal with the immediate users, the learning curves, personal preferences, and possibly different tasks and such. But, if the output of the system is directly or indirectly used by management and others, there will be other changes you will encounter (McKay and Black 2007).

The big people can be internal, corporate, or major customers and they may ask for a wide variety of changes to be made. For example, at one client the following changes occurred:

- Different things to be emphasized.
  - o It does not matter if it is an aggregate status report for complete product lines, I want to see this specific part's info.
- Different times for the system to be used for specific functions.
  - o I do not want it at 7am, I want it done at 6am.
- Different styles of the same report.

- ○ This is what it looked like at my last factory.
- Combinations of things.
  - ○ Take these bits from these reports, from these different systems and create a super, single page report with some new analysis.
- New functionality of a minor nature.
  - ○ This is the report I used at my last factory.
- New functionality of a major sort.
  - ○ Create a complete planning system for the plant including HR, industrial engineering, and finance.
- Resurrection of old functionality.
  - ○ I want that report and function to be used again — the special one we did three years ago.

Designing things in layers and in smaller bits helps here. It also helps to write the actual internal logic in more layers than you initially think are needed and if in doubt you can thin wrapper it and isolate anything with a verb in it. Of course, the big people want everything done yesterday which is why it is important to have all of these layers and obvious places where you can add, separate, decouple, re-use, and abuse the code.

The tables and flexibility outside of the actual code is important. This approach dramatically reduces the actual coding affected. Sorry, I know that as a geek you want to do the code level changes and not just muck with tables, but remember it is not about you; it is about what is good for the client and the system as a whole. It used to be the general rule of thumb that if you had to muck with a line of code, there was probably a 15% chance of messing it up and introducing another bug. I suspect that times have not changed. Hence, you have to be careful at all levels of the design to minimize the number of things you have to modify when one of the usual changes comes along. They will want new; they will want combined; they will want old revamped. If you break things down to the line of code level to ensure that you do not do two things on the same line, you will have a better chance of maintaining and changing the code in an effective and efficient fashion.

Part of the fun when dealing with new users is déjà vu. They will ask you to do things that have been done in the past, or ask you to turn things off and then ask you to turn things back on again. You have to be patient and if possible challenge the situation to see what assumptions have changed since the last time the idea was done or used. It might actually make sense to do this again. In the design, you can and should always keep the variants around and have easy ways to invoke them: switching stuff on and off, switching between variants.

Being big people, they will also want you to change the user interface to match their perception of reality based on their *expertise* (groan). I once built a system and used it for a couple of years. Then, someone else wanted to use it, and then another second person wanted to use it and both of these users had their opinions about how it should look and function. It should look like this. The menus and functions should do this. Lots of opinions. These two new users disagreed with each other and with me. Note, only one of us had actually used any similar software in a similar fashion, let alone this specific software and that person was me. To put the discussions in perspective, the other two users were very important people in the domain that the software was addressing, the leading world experts. And, I do mean world-expert. How do I, as the junior non-expert, try to explain to the big experts that what they had in mind was indeed a nice idea, but it had been tried in the beginning, did not work well, and we had moved onto the fourth or fifth generation of the interface based on real use? I was in no position to debate the issue and certainly not in the position to refuse their requests.

The system had been designed in the layered and decoupled style I have been describing and it was easy to create three views to the user. The user could pick whatever was behind door #1, door #2 or door #3. Namely, the domain junior's version (mine), or expert one's version, or expert two's version. People with the best of intentions will project and think that they know what things should look like and how things will be used. Unfortunately, most of this is based on personal opinion, personal experiences, and most people turn out to be rather poor at projecting reality. Smart in many other ways, but they may lack the skills and background necessary to see and to understand what the situation might be. Ah, can you remember what I noted in the last paragraph? I said that this was the fourth or fifth version of the software. My first version was not any better than what the domain experts suggested, and when doing something for the first time, all you might have to go with is your own opinion! Furthermore, it was not a case of "I am the expert and you are the junior team member" with this one project. They honestly thought that their ideas were appropriate. You can probably guess the result. Two of the versions quickly and quietly died. It was a simple case of easily, and at low cost, being able to put several front-ends together over a weekend and letting nature take its course. Darwin in action.

Lesson: it is sometimes better to create multiple versions and let the users reach their own conclusion than to tell them that they are off the mark and really do not understand. It is easier to just turn something off, or hook a bunch of blocks together than to explain the development cost and effort involved. It is easier to react quickly if you design using a toolkit approach for everything and do the preliminary work to be consistent, avoiding the cute stuff. Try to avoid the fighting and the arguing when it comes to changes that the user wants, especially if the user happens to be the one paying the bill. Protect the old so that you can quickly revert, try to do the changes in a controlled or parallel fashion so that damage is minimized, and avoid peak periods of time when nerves will be frazzled. If you do all of this warm and fuzzy stuff combined with a good s/w base that assumes change, then you will likely have a system that will live for many years. If you create a system that is not responsive to change, do not be surprised when they pitch it out, without as much as a please and thank you.

My expectation for toolkits and rapid development goes like this. I expect to be able to build a screen with simple lists of objects, with attributes, and with the ability to pick something and fire a verb in one to two hours from scratch. I expect to create data screens for an object in one to two hours. These actions include menus, navigation, and the result looking reasonably professional, clean, and robust. I expect to create reasonably complex pages with dependent logic and control in about a day. I expect to be able to wire a different GUI control mechanism to underlying logic in one to two hours. I expect to do most fine tuning and tweaking on the user interface in a half hour or less per change. I expect to be able to wire up a new business logic meta object from the bowels of the system and support its life cycle from the GUI in less than a day. I expect to be able to craft most new business logic components in a day. This is not always possible at the start of a Class IV–VI project, but that is my long term goal. If you have this level of response, you can do rapid build cycles, handle change, and be very agile when dealing with the users.

Chapter 18
Stability & Robustness

If you seek excitement, daily fire fighting, living on the edge, and adrenalin rushes, then you should build bad software and hardware. If the system is constructed to be stable and robust, Quality Assurance and other support functions are relatively boring, or at least they should be. This is one way I measure the effectiveness and efficiency of an organization. For example, a few years ago I went with a colleague to visit a factory. The tour guide was the QA manager. The tour lasted two–three hours and not once was the manager distracted by a poor quality problem, or a fire to be put out. Everything was just going along like a hot knife through butter. If this was a manufacturing book, I would explain more about how they pulled this off, but this is a software book. The key message is that in most factories I tour in North America, the QA manager is the least likely person to have the time to give a tour, and is the most likely to be constantly interrupted to do fire-fighting. That is because most of the North American manufacturing systems are not designed to build good products in a sustained fashion. The systems are not stable or robust. They may eventually deliver a good product out the back door, but it is at high cost and effort. Software development is no different. If you do not design in concepts for stability and robustness, the fires will be frequent and the users will be very unhappy campers.

Poor software and a poor process can really annoy users. Sometimes, the developer is not aware of how unhappy the user is with the software and development process. The developer goes and has the usual weekly, bi-weekly or monthly meeting with the user and they discuss the status and any issues. In some cases, the communication will be true and revealing. In other cases I have seen, the users are very careful about what they say to the developers because they are *afraid*. They do not tell the developers how unhappy they are with the software because they do not want to reduce or eliminate any possible chance for getting help in the future, or making any improvement at all. They figure that something is better than nothing and if they lose whatever help they are getting, however bad, they will be worse off. This is a sad state of affairs. I have had the benefit of sitting in a status meeting listening to a development manager carry on about the great relationship they have with the user group; while knowing that the user has voiced the contrary. This was one of those fearful situations where the users did not dare say what they thought. The situation was made worse because the developers actually thought that the customer was happy and that all of the processes were appropriate.

In some cases, the developer might not care too much about when a change is introduced, or about the user's view of the sixty odd items on the to-do list. The developers are in charge and will tell the users what the users should want and when the users will get it! And, the users better be grateful for whatever the developer does. With these types of relationships, the users reserve their criticisms for other meetings and discussions where the developer will not hear them. Yes, this is not the right way to proceed but I have seen it in cases where there is a monopoly, and the developers have attitude and are arrogant. I have been in meetings with senior developers who laugh at the users and comment that the user is always wrong, does not know what they want or what they should do, that the user does not take the training, does not use the fancy features provided to them by the developer, does not read the manuals, and does not do what the developer wants them to do. The user is the problem and is the reason why the system is not considered stable, robust or successful. The developer knows what is best. May be, may be not.

In other cases, the users are just nice people and do not want to impose or complain. While robustness and stability are not the only problems users will have, it has been my experience that users will live with less functionality, and many compromises and workarounds as long as what is provided works, works well, and works reliably.

How well built and how reliable the software should be will depend on the type of system being built; of course. For a mission critical system, the expectations should be high. Whereas, for a minor tool, used infrequently during the year for non-critical tasks, the expectations can be less. In this chapter, I will discuss concepts suitable for mission critical situations and you can dial it down to the appropriate level.

18.1   Levels Ⅰ through Ⅴ — Infrastructure Stuff!

In this section, we will use the groupings introduced in Chapter 17:
Ⅰ. hardware (e.g., physical device level),
Ⅱ. chip level s/w (e.g., imbedded s/w on chips),
Ⅲ. operating systems (e.g., base technology and drivers),
Ⅳ. systems level tools (e.g., servers for communications and database tools),
Ⅴ. supporting subsystems (e.g., email, search engines, reporting systems — using level Ⅳ tools, develo〈plment environments and other toys),
Ⅵ. application (e.g., the main system — the cause for existence),
Ⅶ. user level logic (e.g., macros, work flow tables).

In a mission critical situation, I first of all need to worry about my own problems and my own bugs; the levels Ⅵ and Ⅶ. Hopefully you have vendors and infrastructure support groups for helping with the first five levels. A number of the ideas in the last couple of chapters can help in picking the pieces and designing the system. Many of the ideas are directly applicable to stability and robustness:

- Avoiding fragile situations in the first place and then if a problem does occur, having enough layers, tracing, and elasticity to isolate the problem and to rapidly fix the situation.

There is an actual theory about this type of control situation — Ashby's Law of Requisite Variety (Ashby 1964). For example, if your problem space has three degrees of freedom, your solution space better also have three appropriate degrees of freedom to match the problem, else you will not be able to control the situation. It is a simple concept but often overlooked.

So, what is the bottom line? I think that this control theory approach is true for all levels I through VII: hardware, chip level, operating systems, systems level tools, supporting subsystems, applications, and user logic. You can also help robustness and stability via project management and operational execution and that will be discussed in a later chapter. In this subsection, I will focus on design ideas you can use for layers I – V. The next subsection will extend the ideas into the application and user levels, VI and VII.

If you are using existing components for levels I – V, it is possible and hopefully likely that the layers and components have been designed and implemented using standards and good practices for layering and encapsulating. That is, there are well defined levels, interface specifications, and protocols. A layer should be data ignorant and insensitive to the payload of the message if all the layer is doing is facilitating the movement of the payload from one level of technology to another; e.g., the network layers and so forth in TCP/IP. The layer should not be looking for escape sequences or other workarounds within the payload. If the interfaces are not clean, it will be hard to create a new, next generation standard without the same legacy holes; the existing customer base might rely on the bad features, possibly exploited by some programmer who did not realize that such features should not be used, even if they are provided. Beware though of the problem of too many layers and too much encapsulation. There is a reasonable number of layers for any given problem and it is possible to get carried away and introduce too many layers that only serve to create confusion and problems. You want layers, but the layers need to be obvious and transparent from a design and architectural perspective.

These lower levels (I – V) are usually out of your direct control, and in many cases you rely on the vendor's failure detection and quality control. In a critical situation, you can augment this with additional logic that corresponds to a quick pulse reading and a fuller health work-up. The pulse reading is done by a centralized controller that sends out probes to the different services and components. The simplest or first test level is a "are you there and alive?" message. This is the checking of the pulse, the heart beat, and the dead or alive status.

The second test level is a solicited, "take it around the block" type of check. The conception to cremation process. I call this the *health check*. The service or component is asked to do a complete life cycle check of functionality. For example, a database system is asked to create a database, add a table or two, add one or more rows, add some data, report on the data, delete the rows, delete the tables, and finally delete the database. This takes the pulse taking to a different level and checks actual capability and the basic functions.

The third test level of automated health checking is to have prescribed input and output information that is used for checking the health. This information is matched to make sure that the service is not only functioning, but that it also has some level of sanity. There have been many software packages written that appear to function, but are returning the wrong client's information or doing the wrong thing. You definitely want to check this out in critical systems. You want your tests to be prescriptive and deterministic.

The fourth test level uses time. A standard is set and then compared against for the health check to complete. This can assist you with the detection of a deteriorating system.

In short, these four types of tests help you know if the service is up, if basic functions are executing, if the applications are using the right data in the right way, and if the functions are done within an expected time frame. These help you with the quality of service! If the above concepts are included from the beginning of the project, are launched and automatically included on every build, and the tests grow in functionality with the system, the suite of functions creates a handy, built-in regression test.

That takes care of checking. If the system level items are not happy and are having a bad day, you can try to roll back to a known checkpoint and if necessary you can group services together if they are co-dependent. If the data is separated from the logic, you might be able to activate another logic server to use the same data if the logic server is determined to be out of sorts. Using redundant technology for things like databases (e.g., SQL and ORACLE at the same time) can deal with the reverse case where the logic services are fine but the data level is contaminated.

While it might be desirable to have fine granularity to analyse and study levels $I-V$, it might be necessary to use proxies. For example, if the database server is running on an operating system and the database server is working, we might assume that the operating system is OK.

One idea that can help with isolating problems is to have a minimalist application on each server or system component. This is similar to the health check application and in some cases, it might be the same code. What you want is the various pieces to play together and the health check software might not be sufficient to make sense of the system. The health check for one system component might be good for that one piece, but not good enough to flow something through the stack top to bottom. On one system I helped design in the early 1980s, we wrote the real-time operating system and one built-in capability was the ability to exercise each of the hardware devices in sequence without an application. This made life far easier for isolating and understanding a complex system. This imbedded operating system was also developed on a different system (Unix), and tested with application and hardware test harnesses before being ported to the real hardware platforms. It ran on several CPU bases from two vendors (Intel 8000 class and Motorola 6,8000). This self-testing ability was useful while developing under Unix, and again when integrating with the hardware devices. The application layer was also able to be developed and tested without the operating system or hardware.

I recently encountered one of my old team members and we both considered this project as the highlight of our careers. How often does your team get to write all of the software from top to bottom of a system: firmware, device drivers, operating system, application, application language? Build a scalable architecture? Portable operating systems? We both had learned a great deal from being on this team. I was the technical architect, and was lucky to have a great team to work with. A great functional architect and great technical leads. The product scaled from the desktop to around six or eight meters and in the larger configuration sported two computer systems. There was a blind Intel PC controlled by the Motorola real-time system. The Motorola subsystem mimicked the keyboard and monitor to the PC, and the PC was used for disk and communications. The user's machine could also be used onsite as a development system if needed; flick a switch and voila, instead of a paper processing machine you had a PC ready to go. We probably had built the world's largest (in the physical sense) PC. One nice part of this early 1980s system was that the whole system was softly configured at power up with smart chips on the device boards identifying who was where and who controlled what, supported by a single device protocol and a table driven operating system. There were other architectural gems as well, but the corporation's philosophy at the time was to be quiet and not publicize internal solutions and concepts.

Another idea is to wrap some of the technology with a virtual machine definition and protocol. This enables you to wrap different solutions in a transparent way and it also allows you to trace and analyse higher level instruction streams in and out of a piece of technology.

And remember, avoid being the pioneer whenever possible. Sometimes you have to be, but do so knowingly and be ready to dance. Traditional solutions should be used for traditional problems and there is little to justify a non-traditional solution for a traditional problem. Try to use mature technology at levels I – V and try to avoid bizarre or really neat features. Yes, it might be boring and not leading edge, but it is not about you. I know I am repeating myself, but it is important to keep in mind at all the time that it is about the user and not about you the programmer. It is about creating a high quality system that will work for a long time without a problem!

I used these techniques with the Java library for teaching applets. The library of core technology is about 90,000 lines of code. Not huge, but not small either. The interactive math language, dynamic typesetting, object-oriented charting/animation, and spreadsheet functionality are all supported within a virtual machine environment that uses simple, core Java AWT components as the base. The resulting applets are about 1.5 mb and have something resembling a simple operating system imbedded within, supporting a three-layer application architecture. The core technology was created in 1998 and a decade later, the most sophisticated module still functions with the latest Java technology without any changes being made to the applet since about 2003 and without any changes to the very base technology since 1999. I have not monitored or have had to chase the countless releases and upgrades to Java, the toolkits, or the Java virtual machines. I paid attention when building the initial system, made my decisions, and then played my hunches about what would be solid, what would not be, and what infrastructure I would have to create myself to protect myself from the constantly changing world of Java technology. Maintenance and upgrade cost related to the technology: zero. I stopped building applets with the technology in the mid 2000s and had constructed perhaps fifty to sixty functional applets with no known bugs. However, not all was perfect and there has been one technology glitch. When I last checked, I will have to replace my lite animation routine in most of the applets with my heavy animation method because an interrupt associated with mouse movement has been changed. The lite method worked fine till the vendors changed the interrupt behavior, and luckily my heavy routine with more interrupt handling capability was not affected. This is the only infrastructure hit I have taken with the code in a decade. In 1998 and 1999 when I was developing the system, I was told by others that I was over designing, that I was wrong not to use the fancy features of Java and the libraries, and that I did not know what I was doing. The code tells a different story and illustrates how the ideas above can help create a stable and robust system. The applet virtual machine built on top of the Java virtual machine was the trick. It minimized the risk and the exposure. A minimal technology foot print.

18.2   Levels VI through VII — Your Stuff!

At these levels, you usually have only yourself to blame for any quality issues: stability or robustness. You should do, or at least consider the material described in 18.1 to protect yourself from external sources of instability. Depending on your requirements, your code or subsystems could look like a vendor's and all of the concepts can be applied for heartbeats, health checks, wrapping, layering, isolation, etc.

If the system is mission critical, no surprise, you should avoid the wild and sexy functions and libraries. Dabble with these in small projects and on the side, but do not rely on leading edge libraries for key functionality. Or, make sure that you have dual technologies and a backup plan. I have seen and have worked with developers who chose risky options because of their own agendas, seeking fame and fortune or near death experiences. They dismissed certain concepts and approaches as *old school* and implied that unless the latest methods and ideas gleamed off the web were used, the design and implementation approach was wrong. They said that they would be embarrassed if their peers saw what they were doing or using. True expertise is knowing what to use and when to use it. Just because something is *old* does not mean that it is not appropriate or not best. Dismissing something immediately because it is simply considered *old school* exposes the dismisser's ignorance, hubris and their lack of knowledge, judgment, experience, maturity, and true expertise. The reverse is also true; accepting something blindly because it is old or is common practice is also indicative of a lack of expertise and wisdom. Try to catch yourself if you are inclined to make decisions based on such superficial criteria. You need to understand the problem, and understand the tools and techniques at your disposal. Some will be old and some will be new. Do not inflict pain and torture on the user just because you want to have certain technology turn up on your resume, or just because you think that by using it will make you cool or someone to look up to.

Personally, I have found the Michael Jackson concepts for program design help out a great deal with application software stability and robustness. The structure of the code and the way errors are handled are substantially different if you follow his ideas and if you do not. It is possible that the first version of the code will go up with the same speed and effort (close enough) regardless of the method used, but the difference is seen later as the code is used and maintained. I worked as a member of a small team writing a relational database system in the late 1970s, early 1980s. Every module was consistent and was implemented with the Michael Jackson philosophy. The system grew from about 25k loc to 50k loc painlessly and in two years of systems testing with new versions every term or two (students were using it for a class), one bug was found. While many things probably contributed to this ability to grow and be stable and robust, the Michael Jackson approach was definitely a major contributor.

It is also possible to write the code with Shingo's Poka Yoke concepts in mind. Sometimes, I have imbedded self-checking information in the data structures and these features were then checked upon each routine call. Or, additional checks would be made in the code before anything was done. It might also be called extreme paranoia. If the system is critical, do not trust anyone, even yourself. The code would check and double check, and not assume that the caller was playing with a full deck. Although this sounds painful to do, the end result is usually worth it. The code is rock solid and is quick to pick up any programming or interface violations. I was introduced to this general philosophy in my early 20's when I was writing backup software, and then really experienced the benefits with the relational database development. I have used the ideas ever since.

Remember the humility assumptions: the third one is that the code is buggy and you are not as great as you think you are. Assume that you do make mistakes and try to detect them. Assume that major consumers of your software might not code requests correctly or in the right sequence. Assume that you might get stuff meant for someone else. In some cases, you add extra stuff to linked lists, or put headers and trailers on things (like matching parentheses), but in general all you have to do is practice safe coding.

There are limits to this of course. You should always protect yourself from external, incoming functions and calls. If you are connected to another system that is under your own control, as part of the same development, you can sometimes make assumptions that the edge interfaces will do some checking and triage for you. If you have standards and conventions within your own team, you should be able to assume that violations of the standards or protocol will be caught during unit test, and that you will not have to build in permanent checking against any and all illegal calls to your software from your own team members.

Silly checks. That is what I call them. I do not build them into every system, but if the system is mission critical they are likely to be included. The first I did it was back in the mid 1970s as a student, when I was building the backup system for the University of Waterloo's VM/370 system. The idea is very simple. You make some assumptions about what a healthy system should look like and occasionally check it. If the system is out of whack, then you do something creative like go back to the last good checkpoint and have amnesia about what has happened since. You can look for things that go too long, too short, run out of something that should not be run out of, more than one of something when only one should exist, etc. The silly checks catch all kinds of things you cannot predict or anticipate.

You do not know the causes or what actually happened, but you monitor the vital signs and outward behavior. You can think of it as a kind of instrumentation of the system, setting gauges and limit switches that will squeal when things are not what they should be. It is a form of anti-pattern match. If you are looking for a specific situation to develop, you might miss others. Worse still, it might take too long to pick up the odd pattern you are looking so carefully for. I prefer to look for things that break patterns. This should sound familiar at this point if you remember the management of change concepts. The 'any change in the status quo' point about potential risk. Same idea. I have not thought of a better name for these thingies. They have caught operator action that was of the "we would NEVER EVER do something stupid like that!" and other silliness.

Hmm. Anything else? You can do nice things if you know that the system is meant to last a long time. For example, you can bury in flags or settings that indicate what versions of software can run with various fields and what version that the data definition is at. This will allow software to know that the software is running in a degraded mode, and this might mean that certain features might not want to be executed. You can have logic that crosses patches, point-releases, and releases and have this linked to data record design and logic. This does not work too well on large databases, but if the databases are only a gig or two, give it a shot. A gig or two database is child's play these days, so what interesting things can you do if you add meta data about the data in the database?

If your system is lucky enough to be used for an extended period of time, you can assume that there will be logic added that is 100% dependent on older data definitions, and some logic that is based on 90% old and 10% new data added for this specific release. In the latter case, there might be a degraded state that will work (and make sense) with the 90% data and ignore the 10% new data. If you have logic that is 100% dependent upon new data definitions, not much you can do.

In some cases, you can build in spare fields for all of your common data types (e.g., ints, longs, doubles, strings, whatever) and use these for quick patches without mucking up the whole record structure. You can add unique identifiers to transactions and know what was the last transaction that played with what. You can add a version code and an extender id to another record for point releases. The software application should know what versions and data records are valid for itself. Newer versions could back-fill one level to ensure that the n-1 version of code could still function in a limited fashion.

Imagine what you could do if you actually thought about what might happen to the system over a ten year period? You can guarantee patches. You can guarantee point releases, and you can guarantee full releases. If you recognise this, why not build in some logic that does something with this knowledge. You can also guarantee that not all patches will work, nor point releases, and that you will have to roll-back after the oopsy. Why not anticipate this and build in some mechanism for roll-back? Code will go through the throws of crawling on its belly, the terrible twos, trying threes, eventually become an adolescent, young and wild, and if you are lucky become grey haired and senile. The code may also become annoying and you wish it would leave home. Do you code for the future?

You can also consider what happens to a system while it is running. Are there simple and routine events or changes that should be, or could be, accommodated with the system running live without the system being taken down, re-built, re-published, and re-started? Is it possible to introduce logic and code into the running system in real time? This can be dangerous and should only be done in extreme cases, but if you can introduce fixes and new features while the service is up, it might be a useful feature. Infrastructure levels often allow you to do this type of thing (consider operating systems, database systems). Why should applications not do this as well?

When doing a high velocity project in the Class V or VI range, it is also possibly wise to consider what assumptions are being made in the code itself about fields, field sizes, and other similar artifacts. For example, how many characters should be in a field in the database? If you pick a number, what are your assumptions? I think it is better to allow for lots of slack in any string field — these things always seem to sneak up and bite you. How many blah-blah's will someone have assigned? You think that two blah-blah's is reasonable and all that is needed? Best to program for a safety margin, or better yet, make it variable. The client or user will almost always come back in the near future or distant future and increase almost every number that is used as a constraint. For example, if the system has a number of user preferences, do you want to set hard fields for each of the preferences? Or, do you want to create a dynamic list of preferences with self-identifying preference names in the data to allow for easy addition of new preferences as they are identified? Ideally, you might even want to add new preferences without requiring a compilation change that breaks multiple projects or solutions.

What do these ideas have to do with robustness and stability? Lots. Remember, that by using the Agile/Extreme type of development, you will be using the scrums for refining the functionality. During the rapid prototyping, you will almost certainly bump into a variety of enumerated things that are not what you were told or were expecting, and it is better to have built in absorption capability than to demonstrate what rigid code looks like, or to have to show what a long or big refactor feels like. If you cannot quickly extend or refactor while maintaining robustness and stability, the users will become very frustrated with you and your development team. How quick? In a tight schedule, the users might accept a day or two, but little more. If you have a system designed to be extended and with built in elasticity principles, you will likely have a system that has the potential to be robust and stable under a variety of situations.

If you are having a hard time thinking about this decoupling, think about earthquakes and what happens when everything is tight and there are many points of coupling. Now, think about what happens when things are separated, and there are few points of coupling, and the coupling is designed to give some movement. Between large software subsystems, you want the same type of decoupling.

Chapter 19

Tempus/Temporis

This chapter could be considered a small tangent. It is about *tempus, temporis* (aka time). I am going to lead you through a process of thought that developed a taxonomy and approach for thinking about temporal situations in decision making. Consider it a mini-case study.

Since the late 1980s I have been occasionally possessed by the idea of having software that is smarter than a fence post or tree stump. In my research of planning and scheduling, it became very obvious that software systems that pretended that context did not matter were going to have a difficult time dealing with any planning and scheduling problem in a real factory. Of course, there are many contexts to consider when thinking about semantics with a big S. For example, the geographic context. Or, the who or why. These and other contexts can impact the design of the software if you actually recognize them and consider them. This is some of the reasoning behind Chapter 8 — Value. But, for planning and scheduling there was a context much more impressive. This was the temporal context; the time dimension. It is a key context if you are doing anything with work flow or tasks.

Time can be interesting. In 1991 I spent a few months researching the way time was referenced or used in research. I looked at mathematical, engineering, artificial intelligence, computer science, decision sciences, psychology, sociology, philosophy, and logic journals. While I had a hard time staying awake, the browsing was educational. I might have missed something, but most of the publications I looked at had time as an abstract time series: $t_0 \cdots t_i \cdots t_n$. All of the time references were context free resulting in each time unit being isolated and independent of any real world context. Some of the research was trying to figure out ways to represent points of time and time durations (some of the AI and CS work at the time), but usually time was not itself the focus.

Why did I bother to do this research? At the time I had just finished designing a scheduling system with some temporal knowledge in it, and had some programmers build a prototype of my design. The scheduling system knew about the time horizon and various bits of the system would change with the decision logic. It would use different logic and data structures for decisions in the near future, other logic in the mid-future, and yet another set of logic and data in the far future. This enabled the system to emulate in one system, dispatching, scheduling and planning. This was done with one scheduling engine and one database. Most of the existing literature at the time was based on an assumption that you needed a hierarchy of separate logic engines and databases in order to do the three tasks. Since we had built a temporal system from an Industrial Engineering perspective, I was curious to see what other research disciplines had done with time, and to have some basis for comparison. Hence the literature review.

But I still have not answered why I found the temporal context so interesting. In my scheduling research, I had noted that schedules were crafted differently depending on the day of the week, shift, month, and season (just to name a few reasons). Not only were different pieces of data included in the reasoning, but different heuristics were used and what would be considered a good schedule on Monday would not be considered a good schedule on Tuesday, all else being the same except for the day of the week. I realized that the scheduling system had to know when it was executing (e.g., it is the first week of January and I am about to create a schedule), and where the decision was on the planning horizon (e.g., two weeks in the future). The planning was different if the situation was July, and a schedule was being constructed for the next two months. All of this pondering led me to the point of thinking about time and the meaning of time: the semantics of time if you want to think about it from a semantic point of view.

There were some aspects of time incorporated in the various research areas I looked at. There is the area of forecasting and that is all about time. The seasonality concept is sensitive to the meaning of cycles. Learning curve research also includes the concept of time and the meaning of elapsed time, investigating various ramp-ups when learning a task. You can also find specific research for crews and shifts, and when the crews are allowed to pull double shifts or need rest periods. But, in general, time has been viewed as an abstract time series, and most work flow, scheduling or planning software does not know if it is making a decision for a hot humid day or a snowy one.

In a nutshell, I did not find a good model or way of thinking about the meaning of time, nor a way in which I could incorporate the meaning of time into my software systems. The following subsection gives an example of what I mean by temporal context and how I started to think about time. Work flow is used as the domain space since it implies activities and outcomes.

Work Flow — Taxonomy/Design Example

Work flow is a subset of the general planning and scheduling problem space. Specifically,
- Activities have a start time, end time, and a duration.
- Activities have effort expended during the full or partial duration.

The trick is to figure out the sequences, who does what, and when. Before discussing design solutions, let's explore the problem space for a bit. If you did a representation of what might be planned in a work flow or other task sequencing situation, the simple view would be:

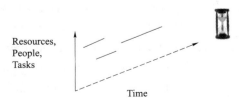

However, this is only a single dimensional view of the decision making. In reality, the diagram is more like:

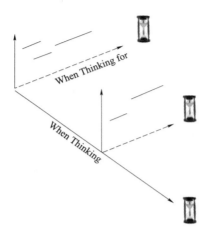

The first dimension *when thinking* might not matter all of the time. But, can you say that you use the same logic, same data, and same approach regardless of time? For example, during construction? Construction in summer? In winter? Do you always have the same way of looking at the situation?

The second dimension *when thinking for* may also matter, or it may not. Are we thinking about a Monday or Friday? Day shift or night shift? The night shift on a weekend versus a day shift during the week? Also important is logic and thinking for relative reasoning; activities two weeks from now, anything one month from now. To pull this together, imagine writing planning and scheduling software, or work flow software that needs to know that it is being executed in February and that the tasks will be performed in April. What the software might have to do with the time between February and April might be different if the software was being executed in June and the tasks will be performed in August. Use your imagination and think about what might affect work flow or tasks, or the planning of work flow and tasks! You might want different objects or meta representations depending on when the software is executing and the time horizon. You might want different logic or heuristics as well.

This leads to a brief note on what you learn from those ethnographic studies described earlier. You learn that time has its own vocabulary:

- Absolute time — September 18th.
- Relative time — 2 weeks from now.
- Business time — fiscal month, qtr, year.
- Environmental time — summer, winter.
- Cultural time — long weekends, holidays.
- Health time — flu season, heat.
- Calendar time — for a week, for a day.
- Situated time — life cycle of equipment, position in treatment.

And, there are probably more. But, you get the drift. Time has points and durations. There are some abstract ideas to note about points and durations. For points,

- A point in time probably has a surrounding zone.
- Possible time duration before point — e.g., the day before a long weekend.
- Possible time duration immediately after point — e.g., days after a fracture clinic and for durations.
- Status quo period — after the beginning we have a quiet time where everything stabilizes, just before ending period.
- Time zone just before the end of the period (the last rush).
- Time period after the end of the period.

If you put these ideas together, you get a representation for thinking about time and events (one of my co-workers has called it the Bunny Hop):

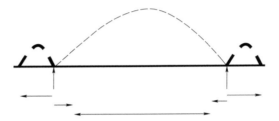

This diagram starts off with the period before the event, the event, a period of time after the event, a longer plateau, then a period of time before the cycle ends, the terminating point for the cycle, and then a period of time after that. This is a seven part decomposition if you are counting and is what I came up with in 1991. This decomposition helps give a meaning to time, and if the meanings of time are important, you should include them.

If you combine the diagrams, a more complete picture of a work flow scheduling problem arises with these patterns of time being active on both axes:

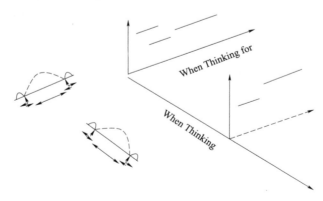

There might be other patterns as well — within a period of time. For example, every day? Every shift? Every Monday during the summer, day shift only? Every second week of the last month in every fiscal quarter?

To summarize,

- If *when* has meaning, it might depend on *when thinking*, and *when thinking for*.
- If *when* has meaning, the meaning might depend on points and zones around points.
- If *when* has meaning, it might not be constant — patterns might exist.

But, what does meaning really mean in this context? In work flow, it probably means that some part of the problem (data or logic) is different from the status quo:

- What might be real or potential inputs.
- What is considered as processes — activities and logical sequences.
- Who/what is to perform or participate in processes — resource allocation.
- What is considered a likely outcome.
- How prior activities are viewed.
- How possible, subsequent activities are viewed.

The types of modification caused by temporal phenomenon are:

- Straight replacement (same, more, less).
- Field or attribute level replacement.
- Modification of base level (*, %, +/-, min and max threshold).
- Nested or hierarchical modification (e.g., long weekend effect in summer vs winter).
- Pro-rated across durations (linear/nonlinear).

Now stand back. A design has to be based on something. Remember what was driving me? I wanted to create a system that would be aware of the temporal context and be slightly smarter than a hammer or screw driver. To design, you need elements and building blocks, real or imaginary. If you have followed this long lead-in, we now have the elements of a taxonomy:

- When — when thinking, when thinking for.
  - o Trigger Points.
  - o Durations.
  - o Zones within and around.
- Patterns.
  - o Replications, frequency.
- Modifications.
  - o What is affected, how is affected.
  - o Base, hierarchies, proration.

In theory, a contextual inference engine built upon such a taxonomy would have a number of benefits:

- Feasibility — more realistic plans.
- Resource utilization — fewer miscues, errors in allocation, expectations.
- Documented processes — accuracy.
- Understanding of variance — why variance occurs.
- Organizational knowledge — capture and use.

In the early 2000s, I built one of these engines in about 10–15k lines of code and it has the following characteristics:

- Virtual machine for database, GUI, logic.
- Temporality extended to GUI, menus, functions, internal logic, data base.
- Pervasive — all records, all fields.
- Learning — detect changes, auto-creation of temporal records.
- Quantitative thresholding — close enough logic.

This taxonomy forms the basis of the design. Everything in the system subscribes to the taxonomy and supports it. All data classes support it, and all GUI and other elements support it. To support the temporal dimensions, it was necessary. Since the taxonomy forms the design of the prototype toolkit, there is a feeling of natural fit between the code and concepts. As the detailed code was developed, there was also a feeling of natural fit. The taxonomy and its pieces provided hints and insights into what software should be written, and how the pieces should interact.

Every record in the database has a half dozen or more fields associated with temporality. There are absolute and relative effective dates, fields for patterns, and guidelines for modification; against the base values, or the last ones instantiated. There are concepts throughout the system for working with the *current* of something. Someone sets up an object or element as the current one, similar to controlling focus, and the rest of the software just works with current. The logic also constantly resolves and looks at the time of execution and for the scheduling logic, when on the horizon. There is some optimization logic to know when to re-resolve a definition as well. Within the logic the modules and methods can either resolve the object and class relative to the time and then call the temporal context engine to retrieve values knowing that a firm object exists, or the methods can call the context engine for a field and let the engine figure it out. Conceptually, the context engine acts as a prolog or wrapper to the data and if the object has not been resolved, first resolves the effective dates, deals with patterns, and then applies the modification rules before returning the data. If you think about it, the logic works the same way a human would figure it out.

I included this discussion on temporal modeling for several reasons. First, if you are going to discuss context sensitive systems and semantic relationships, you should give it long and deep thought. There is more to semantic context than simple mappings of what one considers an address to another's meaning of address. Second, I wanted to lead you through the thinking behind a non-trivial taxonomy and show how it can anchor the design. Third, it was not a passive or static taxonomy. One of the key elements is the modification element: what does time really mean!

Recently, I have looked at another temporal problem. In this case, the problem had lots and lots of time references; many symbolic time concepts relative to patterns and anchor points. For this problem, I decided that the user's work flow engine should use the language of the user and that the system can do the work of translating abstract time references. Again, the design had a natural mapping to the problem and supported the user instead of having the user map to the system.

There is yet another reason for this chapter. It was to introduce this little diagram:

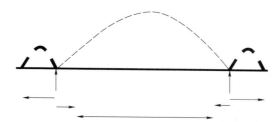

Although it was fun to write software to support this type of taxonomy, the taxonomy has better uses. I, and some others I have shared it with, use this view of time for understanding and decomposing sequences of events. It makes you explicitly sensitive to thinking about the time zone before an event, the time immediately after an event, the plateau or period of time when things are just going along, the time just before the event ends, the terminating point, and the time after the termination. You can even superimpose these diagrams on top of each other and talk about the influence of one pattern of time upon another. I have discovered that thinking this way helps me in many ethnographic studies. That is one good use. Another use comes in project management and somewhat intelligent scheduling and planning. Much of my research in project management and scheduling focuses on adaptive scheduling heuristics that exploit some of the characteristics; such as the time before an event and the time after an event. I will come back to this simple view of time in the project management section.

Chapter 20

Task Oriented Design

This should be a relatively short chapter. If you have got into the spirit of *ZenTai* about understanding what the user really does, you should have a nice model of what the user does, when they do it, and what the information flows and requirements are that match what the user does. Design your code accordingly. Make your semantic classes, objects, and methods match this. End of chapter. You will have a task oriented design in all likelihood. You can stop reading.

You will also probably have a GUI that is optimized for the user's tasks. This is like having high level wizards, not designed for the infrequent novice, but designed for the power user who has to do many dozens of the same thing, one after the other, each day at the same time of day. These wizards are in sharp contrast to the functional menus and interfaces that force the user to pick the pieces out of a maze, click this, open that, drill down here, go back up, pick this, click this, open that, drill down here, go back up, pick this, click this, open that, drill down here, go back up, pick this, click this, open that, ... Does this seem a bit stupid? — typing the same thing four or five times? Think of what a user thinks about your design skills after they do the SAME thing dozens of times in a row each day! It is not just creating a macro that does it all automatically; there might be user intervention on each step and it cannot just run blindly.

The task oriented principles are:

o Create menus or screens that match the major time consumers that a user does each day, week, month, etc. considering repetitive or elapsed time.

o Use larger groupings of steps once the task automation is proven, but allow the individual steps to be revealed and executed one after another as users often like transparency and the ability to step and inspect partial results.

o If the problem changes with time, then have the interface change with time. For example, the task menus or screens can change depending on day of week, week of month: the Friday menu might be different than the Monday thru Thursday menus, and on the first of the month, there is yet another screen or menu.

o Support functional (e.g., the typical document processing or spreadsheet functions) access as well. Let the user brew their own if they wish.

o Identify and support repetitive tasks in high level wizards. Look at reducing keying, information overload, etc. and make it easier and faster to do the task on the computer than by hand.

That is about all there is to it. The users I have built such interfaces for love them and fight to keep the tool alive. It is one thing when they tell me that they like the stuff. They could be lying through their teeth. It is another type of evidence when they defend the tool to corporate auditors who want to get rid of this unique software in favor of a large vendor's tool (many people consider it high risk to use a small firm's software for a mission critical function, especially where the small is a lone professor).

The management also like the tools because they are easy to train people on and even the manager can use the system in a pinch: at 7am do step 1, then step 2, then step 3. It is like having a mini work flow engine running within the application. Inside the code, I separate the GUI elements from the logic (thin wrap it), then have simple methods that do step 1, then step 2, then step 3 where each step is a clean, decoupled functional unit. These steps may then be broken down again into sub-steps. If the decoupling is clean and consistent, then you can quickly build strings and streams of functionality that can be wired into highly specialized GUI elements that minimize the cognitive load on the user, as well as the physical interaction.

The most successful of my designs and code follow this theme. The code focuses on the logical steps done by the person to get the job done. This might imply gathering all of the bits first (in a list), doing some clean up on the list's elements, and then sitting down, driving one after the other through serial or recursive logic nailing each item. I have had users look into the code and be able to understand the architecture and logical flow through the system; the inside of the code and major modules modeled the user's task. It was then very natural and comfortable for the user (with no programming background) to comprehend and be comfortable with the system. My simulations are designed this way and my applications are too. May sound a bit corny and odd, but if you make the software behave in an organic fashion and match the real world, life is better for all. What is hard to change in the real world becomes hard to change in the code (live with it), but what is easy to change in the real world becomes very easy to change in the code. This is especially true if you model the tasks and the sub-tasks used to get the job done.

Enough said. Short chapter.

Chapter 21

Design Sufficiency

This is the last chapter in the design part. When do you stop designing? When is the design good enough? When is it sufficient? These are difficult questions. When can you start coding? A more difficult question.

When is a painting finished? If the painting is of something very concrete and specific that can be blocked out in advance, like a formal portrait or tree or river, and you are painting in the photo-realism fashion, you can say you are done when the image matches the precise image. But, if you are doing an abstract, impressionist or slightly expressionist style, when is a painting, carving, or sculpture done? When I am painting or working on a piece of abstract art, I work till I just stop; something inside says that enough is enough, anything more will not be a good thing to do. It is a gut reaction. If you are doing a project in the *Mushing* style, it is like abstract art. You will know when enough is enough. In software, I have my process or set of heuristics to partially rely on and this helps me understand when to stop.

In the *what* phases, I have been thinking through the problem in an organic or living fashion. If the software is for a real estate board, then I will *be the board*. I will be the house. I will be the buyer and the seller (if they are involved). I will be the buying and selling agents (if they are involved). And, I will be the transaction. This role playing is for their whole life cycle and not just what might be computerized. This gives me an organic, systemic understanding of the problem for which a design must be derived. If I do not do these mind games, I really cannot reliably claim to have created a design based on the ideas from this book. So, this is Step 1.

Step 2 is isolating what might be computerized or not. This involves thinking through the models and ideas from Step 1 and identifying what makes sense to automate, facilitate, or fabricate. You do Step 2 when you think you know enough about the problem. You again take on the roles and persona identified in the analysis and you are once again *the board*. Step 2 gives major process and functions that form big blocks to design. It also gives suggestions for the high level wrappings and interfaces.

Step 3 for me is to play again. I will take the objects and things from Step 1 and flow them through my imaginary system; the KENIAC. The KENIAC is rather slow by today's standard and once got an official piece of mail from an insurance company addressed to ***** Insufficient Memory ***** (yes, an error message ended up in the name field on a label and into my mailbox, and was indeed delivered to the occupant). The insurance company obviously knew something about me that took me a bit longer to figure out. But, the KENIAC system seems sufficient to play virtual computer. Because of its limited memory and speed, the system is forced to work at a relatively high level, else the circuits will overload and fry. If the system feels *right*, then proceed to the next step in the design process. If it does not feel OK, or if I have a worrying thought, I now know from experience that I should stop a bit and re-think my models and ideas. This is my fidelity check. The term hi-fi (an old stereo term before CD's and such) stood for high-fidelity and the fidelity referred to the ability of the sound equipment to match the real concert hall, real instruments, and real voices. Does my high level computer model fit the organic nature of the problem as I know it? This is the fidelity measurement. If I do not do this, I am certainly not ready to code or do the detailed work. As a Neanderthal-type computer I do not know about things like Unix or Windows or Java or C#. This level of abstraction is 100% agnostic. Usually after this step, I can draw a simple picture or schema for the system. For this step, I am not acting the role of the items from Steps 1 and 2. In this step, I take on the persona of the machine. The computer and the items from Steps 1 and 2 become what I process and spit out. They give me orders perhaps and I give responses, but I am the machine. This is different from projecting into the data.

Step 4 takes the picture or schema and iterates. I then think a little bit more crisply about the problem, perhaps itemizing or listing out the task nature, temporal elements, or change dynamics. I might break down a larger task into the individual processes and think about the coupling and interrelationships. I sketch and draw at each iteration so that I can visualize the elements and how they interact. I might draw a circle around a P with lightening rods coming out of the circle. There will be one lightening rod for each type of interface (human or computer) that comes in contact with the product. On each rod I will do a certain amount of design thinking and then move to the next rod. I will not sit at a rod and drill all the way down. I will stop myself from thinking about the details. For the simple Class I white water rapids, Step 4 may appear instantaneous; especially if I have traveled on that path before. You have done the majority of design if you have projected through the system, both as the entities and as the computer, and if you feel the joy that comes from when the two approaches naturally merge into one holistic view of the system.

To summarize, what the above means is that I stop thinking and get ready to code when I see the system at certain levels of granularity, and it feels good from both perspectives. I like to see most of the pieces at the same level (e.g., horizontal progression, not vertical) before I get my hands dirty. This means that I have essentially crafted a conceptual building with floors 3–7 without worrying about the first two floors. But, I know what the first two floors must do and while the top part is levitating, I build the lower levels: the actual code.

I usually do not do detailed pseudo code or precise design notes. If I am passing the project onto someone else, I might do this extra work. However, if I am doing the coding myself, I will not likely do this unless the code is algorithmic and difficult. Usually, I will design down to a level and then rev up the keyboard. I will tape up my fingers (to protect my joints and reduce the pain) and then start. I know I have not answered the main question yet. When do I actually start coding and stop explicit designing and thinking? It depends.

If it is a Class VI rapids, I will likely end up thinking for about 1/2 to 2/3 of the time (if the project is looked at in hindsight) and never really stop designing and thinking. If it is a Class I rapids, the explicit thinking is almost a blink of the eye. After three decades of coding, Class I rapids really do not require much thinking. However, all of it comes down to the same test. I like to go into overdrive and do development in two–four hour chunks, occasionally eight hour pieces. By development, I mean doing the final design, code, debug, test, integrate, and cap. This will be detailed program design on the fly.

Before I start coding, I will know that the module or class has these big, high level, abstract methods and has an organic behavior of blah-blah; but that is all. I keep designing till I have this feeling of what I need to build; the big methods and the organic behavior. I will not usually have detailed mini-methods, data structures, algorithms, or other things thought out. There will not be detailed UML or ERD or the like done in advance; not on a *Mushing* project. I leave the detailed design and final thinking for the coding session. I have found that if I keep my design phases out of the muck of detail, the high level abstractions and design elements are more flexible, agnostic, and decoupled from internal assumptions. I will make the code inside the module do whatever is expected of it, and I do not bend the design to fit what might be done in the code. By leaving the binding of the final design to the final session, I do not waste design time; designing something that as the coding development evolves becomes outdated. Everything is focused and intense, and as I code, the design is also fresh and dynamic. If I have done the higher abstractions well enough, then there is low risk of re-designing things. Everything is on the horizontal layers and is consistent. If I designed and developed in vertical slices, this would NOT work!!!! I repeat, vertical threading is not what I do. I typically code following the organic life patterns of what I am creating; following the life cycle and the basic functions. I then go back and add the warts and eye shadow as needed. Remember the toothpick? This is how to create the toothpick.

This approach is based on four additional assumptions. First, you know how to touch-type. I could be wrong, but I think it used to be the requirement for secretaries to type at least 120 words per minute. What is your typing speed? I am slowing down, but I used to be able to do the 120. If you can type at 120, that is a lot of code you can pump out! Especially with all of these new-fangled development tools that complete your typing for you. Imagine what the 120 translates into? In the 1980s and 1990s, there was lots of effort trying to design development environments, debugging tools, new languages and some other neat gizmos; all to improve programmer productivity by perhaps 5–10 percent. I always thought that a far better solution would have been to send all of the programmers to typing class; save the money and effort on a fancy tool. But, taking typing class is not glamorous and not very scientific, so people keep trying to improve programmer productivity the hard way. If you cannot type fast, you might have to translate my four and eight hours into days. After a decade or two, you can probably hit and sustain about 800 lines of code per eight hour shift; the design, code, debug, test, integrate, and cap process. I regularly do 10,000 lines of non-bloated code in 10 working days. It is not that I am a hotshot programmer. I attribute a great deal to my typing and the way I go about designing. The layering, wrapping, and decoupling at multiple levels of abstraction reduces the need for refactoring and re-developing code. When I finish the code in a session, the majority of code is not touched again. I suspect that many people actually do close to 800 lines of code in a day; the problem is that only about 100–200 of it remains at the end of the day!

The second factor and trick of higher productivity that matches this design sufficiency concept is the cone of silence. By this I mean turning off the email, phone etc. I have heard that many people get an interrupt every few minutes, at least 20 emails a day, and this is not good if you are actually trying to get something done, not lose concentration, and do a decent job. Social networking sites and addictions make this even worse. Batch your communications up, and do them at regularly scheduled times and do not consider it "I am important because I get emails," or "I am so important that I must respond" either. I have been asked by some programmers for a second monitor and when I look, they have their on-air persona active on it displaying their *important* emails and messaging links. In some cases, the second screen is actually being used for development, but not always. I am not, and you are not likely to be that important either.

When I am writing or coding, things can wait till I am done. Might be a tad rude to some but unless someone is REALLY stuck on a response, I will not likely fire back immediately. It takes time for most people to switch cognitive processes unless you have an extreme Attention Deficit Disorder (ADD) and are multiplexing. Even with extreme ADD, you lose something on the switching and focus, and this is not good. Count the number of times you get a communication and respond. Count the number of times you bother other people with trivial communications (email, social networking, phone, or in person). Did you REALLY REALLY have to send that communication now? Did you REALLY REALLY have to talk to so-and-so now? Did you REALLY REALLY have to ask that question?

How much time and focus did you just mess up? How often each day? How much for the week, month, year? Did the other person really need to hear it? Why not wait and batch up a few of these points? Don't flatter yourself. Many of our thoughts are drivel (including mine) and should never leave the mouth or keyboard. If you reduce and slow down your interrupts, your productivity will increase and perhaps you will not have to do the hero thing and work those long hours. I have heard that some companies are doing email-less Fridays. These are BIG firms. Makes sense to me. There are times that I need to be on air and responsive when I am indeed the orchestra leader. There are other times I should shut up and I wished others would do so as well. This is not to say that some social chit-chat is bad or should not happen. But think about it first and see if you are being disruptive and remember that just because you do not have something to do, does not mean that the person you are bugging is so fortunate! If you cannot control the urge to disseminate possibly loose thoughts in an ad hoc fashion, you will have to re-adjust your estimates for output and what you will accomplish in any stretch of time. I do not use my fancy phone for email; I have one office without a phone, and I try to control my anxiety about opening and responding to emails asap. In *Mushing*, design sufficiency is tied to productivity and what you can get done. People ask me how I get so much work done, and one part of the answer is that I try to focus on work, keep my mouth shut, and do less chatting and socializing.

The third assumption is actually noted above: use virtual machines, decouple, layer, thin wrap, and do this in horizontal layers. Spend more time thinking! If you are working on Class I and other low end rapids, the thinking will be rapid and you will be able to jump into the actual development rhythm real fast. If I am doing another 4th generation language or meta language, I do not spend much time anymore designing. I just leap in and do it. Been there, done that. The automaticity kicks in. As noted in the international ratings for rapids, a Class VI rapids may be downgraded to a 5.x rapids after the Class VI has been run a few times. The analogy holds. The more you run them, the less scary and uncertain they become. Re-read the chapter on skill and expertise if you must.

A fourth concept is very important and is related to the above. As you proceed through the design exercise and build the layers and wrapping, you should be doing this from the holistic, organic view. This should build in a form of elasticity in the design. You do not build in a feature, but you do not design it out. This is not the same as putting in hooks and then stubbing the hooks. You cannot anticipate every place you will need a hook and a stub. You can however, design the system in a way that allows you to easily decouple, insert, merge, and create hooks when needed. Hooks and stubs take time to create and their existence must be tested to ensure that they do not interfere with the real code. If you do not add them, they do not exist. The capability to add a hook and extend the code is not something you can test. When I mention this do-not-create-it-but-do-not-design-it-out idea, many people think that I am talking about hooks and stubs; NOT! It is about the layering, abstraction and wrapping.

Part IV

Level VI Rapids & Mushing

Overview

Part IV is about software and system management, and the actual execution of creating a *mushed* system. This part could easily be another book, but it seems to make sense to combine the analysis and requirements thinking with the execution. In Part I, I brought up the analogy to white water rapids — rafting and kayaking — and noted that *ZenTai Mushing* was like a Class VI challenge:

- Class VI: **Extreme and Exploratory.** These runs have almost never been attempted and often exemplify the extremes of difficulty, unpredictability and danger. The consequences of errors are very severe and rescue may be impossible. For teams of experts only, at favorable water levels, after close personal inspection and taking all precautions. After a Class VI rapids has been run many times, its rating may be changed to an appropriate Class 5.x rating. (American Whitewater — www.awa.org)

Going into a new situation and doing the Part II analysis feels like this; for the low volume or unique systems I have done. The Part III ideas are like mapping out the river, designing a raft or kayak for the specific experience, picking the action team (if not a solo-run), and setting up a support group on shore. Not too exciting, but necessary if you want a decent run through the rapids with a chance of success.

Part IV is about how you can go about doing the actual *Mushing*, dealing with risks, team design, project management tricks, documentation needs, etc. When you do this stuff long enough, you basically become a card mechanic, flip a switch, and hit it. The cards appear behind ears; a royal flush appears in the winning hand, and magic happens.

Chapter 22
Management

What does management mean? Why do we have such things as team leaders, supervisors, managers, middle management, senior management, and boards of directors? Why teams? Why departments? Why organizations? Management exists because there is something to be managed. The something to manage exists because someone (for some reason) has decided that something should be done. Someone wants this; someone wants us to do something. Even if everything can be done by one person, there is some form of managing what is done, when, and why. The management problem just gets bigger when the tasks are larger and involve many people; there are things in life bigger than what one individual can do in its entirety. Whenever there are one or more people involved from the cradle to grave of anything, there is either implicit or explicit coordination and planning. Management, in one form or another, creates the environment in which the work gets done, creates the tactical policies that guide the operation of the endeavor, authorizes and acquires the human and non-human resources needed to get the work done, and performs various levels of planning of scheduling necessary to get everything mobilized and completed. Simple? No. It turns out that management of Class V and VI systems is not so simple.

22.1 The Management Challenge

One reason management is not so simple is the human factor, and it centers around how people learn and remember. For fun I do research on the history of management in the context of manufacturing, dating back to the early 1700s. When doing historical research on management, you get to do a lot of reading of musty, mildew ridden books. Not great when you are allergic to mildew. That aside, I also get to do a great deal of thinking, discussing, and debating about the cycles of management, what people did, why they did it, what were the results, why certain practices went out of favor or were forgotten, and then why were they re-discovered again a generation or two later. Consider things like drill presses or physical artifacts in which knowledge and skill are imbedded (e.g., once a twisted bit was invented, the idea of the twisted bit and what you can do with a twisted bit was not usually re-invented every generation). Unfortunately for civilization, people are not like twisted drill bits; twisted perhaps, but not drill bits. People move through management ranks, deal with the immediate situation, and newbies follow behind; not necessarily knowing how things were managed before, how the current management structure evolved, or what other practices could be used. There is no physical equivalent in management practice to drill bits. I remember reading one text from the late 1940s about the lack of smart practices and the author commented it was because the current managers had forgotten they could do something different.

Another thing that makes management difficult is that you are actually managing something. Sounds like a riddle? Well, it is — kinda. There are some general, transferable skills associated with management and these are the easy parts. This is the usual knowledge about what needs to go into a general budget, how to deal with a purchase order, government taxes, and that type of thing. These parts of the cognitive skill set we call management are transferable from one situation to another. Some people think that just because you can manage X, you can manage Y. If you are a good manager, you can be a good manager in every situation, right? What this line of reasoning is forgetting is the stuff that is actually being managed and the environment within which it is being managed. Just because I can manage a software project does not mean that I can manage a hardware project. Just because I can give advice and help manage an automotive supply factory dealing with metal does not mean that I can give advice or help manage an automotive plant dealing with plastic, let alone a food processing plant, or a phone company, or whatever. That is because these parts of the puzzle are context dependent, and skills that are context dependent are rarely, if ever, transferable. Occasionally, someone might be lucky (perhaps by having a bunch of people protecting them or picking up the pieces) and go from one domain to another, managing their brains out, but this is not going to happen too often. However, if you think management is only about numbers, reports, or making sure people are busy, then you might be right that anyone who can add a column of numbers in situation X can add a column of numbers in situation Y. If you think management should include thinking, participating in problem solving resolution (i.e., making reasonable and appropriate choices between possible options), and understanding what the people are actually busy at, then you have a different story.

So, I am amused when I see people going from one domain to another, sometimes in rapid succession, making decisions about things they really do not know. They just rely on the figures from a spreadsheet. In some cases, they do some quick actions, get some money moving, and get some quick praise for saving the day. But, was the coin moving a result of their actions, or something that the previous person(s) had set in motion? You hear about so-and-so taking over and within a fiscal quarter, the mighty ship has been turned and the company is now making serious inroads on market share and dramatically improving profit margins. Does anyone really believe that someone just arriving can affect contracts and supply chains and other factors so quick? I also get amused when I see companies rapidly rotating the up and comers through various departments; every 12–18 months. I talk to a lot of the lifers who babysit and do damage control. The new arrival thinks that anything and everything running smoothly and successfully is a result of their own esteemed skill. This month I am in shipping; last year I was in the fabrication shop; next year I will be in charge of information systems, and then I will be promoted to greater and more wonderful things. This is what I call hit-and-run management. The individual thinks that they are the cause of everything great, but in reality, there might be a group of people fixing things behind the scenes.

The problem with the hit-and-run type of management is that the managers do not really understand what they are managing. They do not need to do everything their staff does, but they need to understand what staff should be doing, what is reasonable, what is appropriate in terms of policies and practices, and so forth. Else, what are they actually managing? They should know what their direct charges do in a typical day, week, month, year, and life cycle. Since a manager is going to make decisions about how long something will be allowed to take (e.g., setting milestones and due dates), shouldn't they have a clue about how long things should take? How will they know if the person supplying the estimate is supplying a good estimate or just blowing smoke? This is one danger of managers managing things they have no clue about. But, there is a bigger problem. I have seen managers who were quite capable of managing an electrical engineering task, promoted to a role over hardware and software, and 100% unable to understand or judge anything about the software side. But, people being people, these ducks out of water decided that since they were promoted to being a director or senior manager that they were supposed to direct and manage the details. In the example I am thinking about, the electrical engineer who was promoted to a high level then proceeded to go forth and direct software activities, made decisions about the software development process, and the software technology used. In one place I worked, we as developers spent most of our time fixing problems created by these managers. They promised clients things that could not be built and even when they were lucky and proposed a feasible design, the product could not be built in the time promised or at the cost promised.

People get promoted and are made managers for various reasons. In my limited experience, I have seen some people promoted to management who were skilled in decision making, excellent at understanding the problem, able to coordinate others, good at working with and developing their personnel, knowing what their people were doing, and in general were good managers. They do exist. Unfortunately, I have seen others seemingly (according to gossip anyway) promoted to management because they were the last man or woman standing with the highest seniority, had the highest education, were the best-man or brides-maid at the higher boss' wedding, or were relations of the owner. I have seen people promoted because the company had a hiring freeze and were forced to promote from within from a weak bench. I have seen other people promoted because they did a super job on the technical stuff, so the bosses made them a manager; with no obvious management training or skill or aptitude. Read my lips: *not all good technical people can be good managers or capable of managing.*

Management appears to be a transient practice in which ideas arise, ideas are applied, ideas fade away leave, and then the same ideas arise again later. Over the last three hundred years, we have done a really poor job in understanding what management is needed when, how to introduce it, how to make it stick, and how to change it and evolve it based on the needs of the moment. This appears to be independent of the number of MBAs, business school graduates, industrial engineers, consultants, and super-stars who have got THE answer and THE method. Over the last 100 years, I personally think that things have got worse, not better. Back in the early 1900s, an unscientific bit of writing speculated that about 75% of the business bankruptcies were caused by incompetent managers. In the 1990s, I saw a business text book (on strategic management) that speculated that 90% of bankruptcies were caused by incompetent managers. I do not know the real number. Suffice to say that professional management has had a real problem in the public relations department at times.

22.2   Good Management

What you want to do is become a manager that is able to avoid bankruptcy. I have been watching and observing managers for a few decades now. Some very good, some average, some not so average, and some that were really terrible. My unscientific conclusion is that being an above average manager likely suggests the following:

1. You know something about what you are managing; the *what* and the *how*. Gantt pointed this out in the early 1900s and the idea still works for me. If you just know how to buy and sell, don't try to manage something that actually has to make something or do something.

2. You know that it is your job to make it possible for your staff to do what you ask them to do. This implies that if your staff cannot get the job done, it is your fault. Novel idea.

3. When the facts change, you are prepared to accept the fact that the facts have indeed changed. You might have to change some previously held assumptions and possibly change some former decisions.

4. You minimize or avoid problems in advance. Not all waste matter that will suddenly appear is random, and you are accountable and responsible for bad things that can be anticipated.

5. Your organization is adaptive and responsive to changes. Holding firm to the "this is the xxx way," or "this is the way we have always done it," is a way to become history. You might become an interesting extinct species to be studied in a business school case study some day.

6. You develop and work with the personnel, on their hard and soft skills. You help them develop skills that will be long lasting and not just be brains on a stick.

7. You know what you do not know and if appropriate, set up advisory teams of experts that can give you the pros and cons, the risks and the assumptions so that you can make a reasoned decision.

8. You realize that not everything worth doing is worth doing right; especially in an agile/extreme situation. No point wasting a great deal of time on something that is going to change in the next few days.

9. You know that just because you can do something, does not mean that you should. You can hit your head on the wall; but should you?

10. You treat people fairly, with dignity and respect. You are not Dilbert's pointed haired boss. You might be a hard boss, but you are fair.

11. You direct with purpose and while showing measured mercy and compassion, carry

through with a firm hand. The manager manages and the managed do not.

12. You realize that probably it is because of everyone but you that the project is successful and not because of you. You share the glory and the praise; it is not about you.

13. You know how to plan and orchestrate all of the necessary elements to keep your bosses happy, the groups you interface with happy, the downstream groups you are supplying products or services to happy, and your own team's activity. The who, what, when, how, and why.

14. You solve the problem once, and you do not solve the same problem multiple times.

15. You know what to worry about and when to worry about it. Everything has its time and place. There is no point worrying about things you cannot control or what is way out in the future.

16. You know how to focus on the critical path. You worry about things that are on the critical path and do not worry about other stuff.

17. You know how to read and write, and know how to communicate with people on your team and people outside of your team.

18. You know how to decompose, delegate, and then get the heck out of the way and let the people get the job done.

19. You know how to do all of this effectively and efficiently, making the fewest number of mistakes possible.

The above and more applies to almost all forms or situations in which management is required. This includes software and computer hardware. You will of course note that some of the points are context sensitive, and will remember that success in one domain does not imply success in another. I am not going to delve into all of the above topics in this book, just a few: those that relate to the specifics of *ZenTai Mushing*.

You can learn from good managers and you can also learn from bad managers. You might actually learn quite a bit from a bad manager. And you might find out sometimes that someone you think is bad, is not — and the reverse. Observe, reflect: what decisions are made, when, why, what are the outcomes. Was the manager close enough? If a decision was off, why was it off? How off was it? Is it your perception that it was off, or was it really off? It is probable that there is no *perfect* manager who would pass or demonstrate all of the above points. Very probable. However, you should be aware of what is considered better management techniques, assess yourself, identify strengths and weaknesses, and if you can deal with your weaknesses directly, manage them or hire people around you that will offset those weaknesses.

22.3   Strategic, Tactical, and Operational

Depending on the level of management — strategic, tactical, or operational — a manager makes decisions with a certain scope: *what is affected*, over a certain horizon, *how long will the decision take before it gets results and how long is the decision expected to have impact*, and involving certain resources, *how much and how many*. The big people typically play with the longer horizons involving more of the venture, deciding on more of the risk. This is the strategic stuff. Strategic does not just mean a plan. I can have a plan or strategy for tying my shoes, but that does not mean that tying shoes is a strategic decision unless you are an Olympic runner. Strategic decisions are those that will either make or break the project. If you make a mistake on a strategic decision, you will likely lose the war and possibly your head. The decisions in this category affect most of the project or task, and it almost impossible to say oopsy and recover. Recovery takes too long, costs too much; it is too late and the point of no return has been reached.

In software development, a strategic decision would be to enter into a certain market, to commit most of your funds to buy a company, to make instead of buying, to go single vendor, to scale up immediately, or to invest all of your coin and fortune on the assumption of immediate success. If you make a mistake with these types of decisions, you might be bankrupt. Tactical decisions are mid-level and involve hiring, policies, procedures, and medium-sized decisions. You might make a mistake at this level, but it will not kill the company. The tactical decisions depending on the situation might span several months or can be revised periodically. For example, a tactical decision might be the hiring policy about the junior and senior mix, training procedures, team design, policies for conference travel, when to order equipment, what documentation is needed, who signs what, … It might hurt a bit to fix a tactical mistake in terms of cost, time, effort, and pride, but it does not hurt THAT much. Operational decisions are the ones made daily. For example, you do not make decisions about a team design or documentation plan each and every day; what will it look like and what the requirements are. However, you might decide in a daily scrum what features are to be coded for a daily build. Operational decisions are the day to day things. If you muck up an operational decision, there might be some moaning, whining and some costs and extra effort, but it will not kill you. However, be careful. Continued and extensive operational problems can mess with your reputation, customer service levels, customer confidence, penalties for missing objectives, and such other nasty little things. When enough nasty little things accumulate, they can become tactical and even strategic challenges.

One observation I make when I visit or study situations is to see who is making what kind of decision. Are strategic folks making operational decisions? Are operational folks making tactical or even strategic decisions? If I see these types of mismatched decision making, it does not make me feel warm all over. Doesn't the big boss have something better to do with her or his time than decide the color of garbage containers on the factory floor? Why is a low level developer making decisions about end user functionality in terms of policies and procedures: "the user should not do this, we will not let the user do that"? This latter example is not to say that input cannot be sought, encouraged, or listened to from a lower level, but who makes the final decision? Directors direct. Managers manage. The big people get the big bucks and they should also have higher risks and expectations for performance and delivery. They should have a reasonable track record of directing and managing demonstrating that they know how to make decisions and having the decisions proven over time.

22.4 Management Skill & Training

Remember the writing about expertise back in Part I ? Managing can be considered a cognitive skill. You can hope or expect managers and decision makers to proceed from the level of a naïve novice to that of a journeyman with respect to decision making skill. Expertise in managing comes from repeated managing, with suitable feedback and reflection. Not all managers are created equal. You do not just learn good managing just by managing. Just like you do not learn teaching just by teaching. You need to study it, get mentored, be evaluated, get feedback, and learn it. Otherwise, you might just manage the way you were managed and you might teach the way you were taught. Neither might be acceptable! You might be amazed by the number of managers who do not get any real, serious training in management. It is assumed that if you can do it, you can manage it. This is not very enlightened thinking. Some people think that you can learn management by going to a session a few hours long, or going to a few weekend meetings. Wrong. It takes many hours of concentrated learning and thinking about the task in a conscious fashion. Not a group hug once a year. Too often management training sessions are presentations containing a bunch of pre-digested thoughts presented in nice checklists.

There is another problem with management training done in a vacuum. I have observed training sessions that did not address the inherent organizational design and how that may or may not match what was preached at the training session. The people nod and feel good during the training; they work together; they bond; they trust each other. But, then they go back to the old situation, the one that does not support the ideas being preached. Guess what happens. It isn't pretty.

Most managers do not purposely go out and try to mess things up. I would like to think that they are actually trying to be a manager. I hope that they do not try to put the firm into bankruptcy, create stress on workers, and create situations that need cleaning up. I believe that most are trying hard to manage in the way they have been taught, with the skills they have. If there is a less-than-desired soul in a management position, the fault can be placed at previous or higher management, or the HR department. In one example, one group had widely acknowledged poor management, and when I asked HR about their training policy, it was that they did not force any manager to get training. They had a few short courses (as in a half day or day), but they did not have any systematic way to develop and create a management structure. It was ad hoc. In these cases, you cannot really blame the manager for 100% of the problem. However, people should realize that they should get training, develop themselves, and if they can self-assess and realize that they are not suited for management, be adult enough not to be one. I also suspect that many managers are delusional and think that they are really good managers, doing what managers should be doing, and all is well. I am not sure what can be done with these managers.

Management should be considered a profession, which implies someone who is claiming to be a manager is in a profession and should act accordingly. Would you assume that you could be a professional actor or sports player without years of training, coaching, and development? Ongoing development and training? From my observations, many managers think that they can be a professional manager right out of the box, without any specific training and that they can immediately be good managers. They know how to manage. Everyone is a music or art critic. Everyone is a good manager. They know so. Neither is it a sign of weakness to seek advice and guidance from other people. There is also a cavalier attitude to planning and scheduling it seems. I think that most people spend more time planning groceries and managing what is in the fridge than when they play with large amounts of stockholders' cash, resources, and people's time. I also think that most people spend more time planning their vacations than they do their team's time each week/month/year.

So, imagine my feelings when I occasionally hear that one of my bright students has been immediately hired as a program or project manager by some firm. Wow, isn't that nice! The graduating student might be lucky and have a good mentor, but it is likely that the student has not really done much of substance. Unlikely that they have shepherded major pieces of software development through a full life cycle, dealing with user requirements through multiple releases, and having experiences with multiple technologies and processes. The graduating students are sharp as tacks, but that still does not prepare them for the subtleties of project management. This often suggests to me that the hiring organization is in reality poorly managed, or has lots of fat with inefficient processes, and has some management who love to gamble. Or, perhaps the project management they are doing is not really what I think project management is. Perhaps the definition of project management is different! Perhaps the new hires are strictly chasing paper or people, and are not really managing and making decisions. Hmmm...

Recently, I was talking with a large firm about the type of graduates they wanted to hire and where they thought the shortage was. They were doing lots of outsourcing of the *programming* and the *doing* to Asia. They were not hiring junior programmers any more. They did not want to hire people in North America to program, no no. They wanted to hire fresh graduates who could manage at the program and project level the people offshore. Interesting. I was curious to know what skills they thought project managers should have and how they thought these would be developed in the absence of jobs that would be needed to develop such skills. I suppose they thought management skills could be taught without knowing the *how*s and the *why*s. Very interesting. Guess they know something I don't. I hope that my pension plan does not invest funds in this company.

Part of management development and training is giving others the chance to make decisions, illustrate their powers of reasoning, supervision, and possibly develop into individuals you would consider making a manager or a team lead. The individual can bring fresh ideas and their experiences and this can strengthen and improve the project and effort. Long term, the individual may be a candidate for being promoted even higher and this is a good way to progress through the ranks. These are the positives. There are also potential negatives. This training can be tricky and you need to anticipate that all training will not be successful, and that you will need to have some slack in your timeline, and that you will need a small cushion in your budget. You have to take chances on people, assume that they can do things, have patience to see what they will do, and hope that things will work out in the end. If you tell people everything, and give the answers to every question, you do not know what the people think or can do. You are giving them the legendary fish and not teaching them to fish. It is also giving them enough rope to hang themselves. Once the people demonstrate their processes, methods, and current knowledge base, you can work with them on development and possible corrections, but this is not always possible and some people will not accept alternatives, and they will be blind to change. It is not a comfortable situation when you have to deal with a misplaced individual who thinks that he/she should be in a decision making role and who regularly demonstrates poor judgment. This has happened several times in my career. Depending on company policies and dynamics, there might be ways to find other tasks for the individual to do, or you might have to encourage him/her to find work elsewhere. Bottomline: if the individual does not work out, you might have to arrange for someone else to take over the tasks, or deal with it yourself. It is a difficult situation.

As a higher level manager, you are responsible for trying to develop others, but the developing does not always work out and this implies that schedules may slip, that some work is thrown out, and that some people will be unhappy. Developing someone is also a two-way process and the individual must want to be developed, and be open for development. You cannot develop someone who does not think that he/she needs development.

So, developing or training is not free or risk-free and your schedule or planning must include contingency planning for when an identified leader fails. Yes, you can try to test and interview to ensure that someone has the skills and ability. Unfortunately, there are some aspects that are hard to control. If this is a growth opportunity for the individual, you might not have solid evidence of management, team leading ability, or other decision making skill. You might have to rely on interviews, words, and talking. Talk is cheap and it is easy to use the right words without knowing what the words imply, or knowing if you know what the words really mean, or how to put the words into practice.

I believe that managing or leading in an Agile/Extreme project requires additional skills and personality traits. In the section on user engagement, I listed a number of user traits and the list is very similar for developers. First, a team leader, senior developer, or manager cannot be pedantic. Someone who is pedantic is excessively concerned and consumed by minor details and rules; to the point of getting potentially upset by the loose requirements, building, and pace implied by Agile/Extreme. Pedantic developers are busy looking at what others are doing, what is prescribed, and look for what others say that they should do. They will also argue and debate minor details about potential decisions before any of the major decisions are made. They will worry about what others think and whether others will approve of what they do. In the worst case they are likely to be paralyzed by their inability to see the bigger picture and ignore the details. They will also not like breaking rules or creating new ones. They might be academic in their arguments and cite texts, web blogs etc. as reasons or truths for doing something or for not doing something. They might also be dogmatic and cite certain principles or opinions as incontrovertible truths. They might be so concerned about what is prescribed that they might lack the imagination needed to create new solutions, new rules, or new best practices. I have encountered or observed such individuals during my career.

These characteristics are not handicaps for Class I or II situations. These traits might actually be assets in traditional or well-defined situations. But, these traits are not assets when dealing with Class V or VI. For the more difficult developments, I would also avoid people who are too literal and who are perfectionists. In a new and challenging situation, the verbiage and language will not be crisp and there are likely to be metaphors and allegories, and concepts struggling to find words. Someone who is literal-minded will not interpret and will not work with the meaning. Perfectionism is another problem with Agile/Extreme. As I was told once, excellence is a relative term and does not imply perfection. As you rapidly build up, refactor, and experiment, you will not have perfection on all fronts. You should do good enough code that bugs are not presented to the user, but the visualization, flow of functions, groupings of functions, names on buttons, etc. will not be final or perfect. These will evolve with feedback and experience.

Pedantic, dogmatic, literal-minded, close-minded, unimaginative, and perfectionist. These are six traits which are not well suited for Agile/Extreme development at the Class V and VI level. Everyone will have a little bit of each of these traits, but you want to avoid the excessive ones. There is no black and white and you should not be pedantic, dogmatic, literal-minded, or a perfectionist when thinking or looking at these traits either. People exhibiting these traits may be smart, be personable, fun to have lunch with, witty, may know technology secrets and magic like no one else. These traits are not all bad and they are very useful in small doses, or at specific times in a development life cycle. The problem occurs when they dominate someone's personality.

Chapter 23

Risk Management

A great deal of my academic research and applied consulting is about rapidly changing situations that require human judgment. These are repeated situations, not one-offs. It is about situations where people pick up clues, make predictions, anticipate risk, and do all kinds of innovative things to avoid or mitigate what might go wrong. This is what my PhD was about and subsequent efforts: looking at the people, trying to understand the problem they face, thinking about the system they exist in, building math models that capture some of the human's common sense, and creating software systems to help the humans in such situations. A common theme to all of this is risk management. If you look at what some managers do, or what planners do, they spend most of the time anticipating and controlling for potential risk. The idea is to be pro-active and create a situation for the workers or other value-adding resources that allows them to get the job done and to meet the objectives. Mind you, I have seen, worked with, and tried to teach managers and planners who do not think for an instant about risk and what might happen. They did not think that such thinking was possible. They seemed to enjoy the fire-fighting. They were in an organization that rewarded fire-fighting and saving the proverbial bacon from the fire. I also got the impression, right or wrong, that if people admit to being able to do pro-active thinking in a situation, they will then be held accountable to actually do it. Imagine, accountability and responsibility for the managers and planners. Needless to say, the ideas in this chapter have often met resistance.

23.1 Risk Analysis

I try to take things one thing at a time. When I am working on the *what*, I try to avoid the *how* for as long as possible. But, as soon as I start thinking about the *how*, risk rears its ugly head. I keep risk on a parallel track and multi-task between thinking, designing, and potential risks. When I do the big picture, I will try to summarize and itemize potential risks to the venture. I do this for several reasons. First, itemizing and thinking helps me think through the system on a realistic level and not a rose-colored glasses one. Second, it helps me remember and not forget. I do not rely on my memory. Third, it helps communicate with others and makes it transparent what is or is not being considered. I try to organize the risks and my first pass is not really detailed.

For each of the risks, the likelihood and impact should be briefly considered. Either the risk should be avoided (by engineering), or mitigated if it is not within the control scope of the client. The mitigation must deal with possible sensing of the risk before it is fully developed, timeliness of risk identification and isolation, and risk containment.

Risks will exist with personnel, hardware, and software, and will exist in the development, deployment, and subsequent use phases. In many developments and execution, there are many categories of software. For example, vendor software, software developed by you, and other possible software or components done by the user. It is not safe to assume that compilers will always compile, that relational databases are error-free, that file exchanges will be valid, and so forth. How much you worry and what you will do, will depend on how important the risk will be — e.g., for mission critical systems, most risks will be worthy of conscious analysis and forethought.

It is not necessary to have specific mechanisms in place for each and every possible risk, but there should be classes of strategies for how to deal with problems. It is also possible to minimize the impact associated with a risk event. For example, the Agile team approach of pair-programming minimizes the impact associated with single developer situations.

What kinds of things could go in a preliminary risk analysis list? In 23.2 are parts of a real list from a recent project; just to give you a taste. It was one of the first items written for the system's draft architecture document. Two major groupings were used, development and operations. Each entry usually has four points: what is the potential problem or risk; what is a way to deal with it; if the risk is considered high impact or not, and what the perceived probability is. For many items, nothing more than documenting it in the list is sufficient; basic recognition and the idea of a backup plan. For others, the risk has to be dealt with in the design, policies, training, group organizational design, etc.

The original list had approximately fifty or so risks documented. Many risks were of the usual sort. But, some unpleasant ideas were also listed. What about a malicious programmer? A rogue one? Perhaps not too politically correct. But what about a programmer who legitimately leaves or is sick for a long time? I have seen too many plans that are optimistic and candidates for the Sugar-Coated Thoughts Award of the Year. You can't usually discuss these points because it might make the programmers or workers upset; thus, best to ignore it. You can't raise the idea that not all workers are competent or want to do a good job, or even do enough work to warrant their wage. Everyone is A+, is competent at the task employed at, and puts in the appropriate effort for their compensation. No one ever tells another worker to slow down because they are making others look bad. Never. Everyone puts in a solid effort and cares about doing a good job. These are naïve statements. It is like being a manager and finding out how to hire someone without also knowing the process to fire someone. There is risk if there are not enough resources, but there is also risk if there are too many programmers and developers under foot. Close the loop, take your head out of the sand, and face up to reality. Look at the key factors from both ends — too much, too little, too early, too late, too much skill, not enough skill, etc. This does not mean that you have to be negative or politically incorrect, just realistic.

Some companies and institutions will regularly bring in outside consultants to perform risk audits and analyses on IT projects. This is a good thing. The outside view should not be the only thinking relied upon though, and the developing group should do regular self-assessments for risk and risk management. If this internal thinking is also done, the outside analysis can then be compared to the internal one and used for learning and improving the firm's own ability to identify and manage risk. For any complex system, the outside consultants will have limited domain and implementation knowledge. However, the outside firm can bring knowledge from other engagements and a different perspective, and this is very useful. Ideally, the developing group should be responsible for preparing their own risk assessment and having its quality assessed by the outside firm. It is easy in hindsight to say that you knew something and you had considered something, but unless you had it written down before you were told, you might have or you might not have. The developing group should be expected to know industry issues, trends, and practices with respect to risks and issues. I have seen cases where this responsibility had been abdicated and placed on the shoulders of outside consultants, allowing the developers to be lazy, myopic, and risk prone. The outside consultants cannot be around for all of the key decisions, and sufficient skill should exist within the group to do a decent risk analysis. You still need the outside look, but this is not an excuse for not doing it yourself in an ongoing fashion. If you can prepare an internal risk assessment and the external consultants cannot add anything of substance to it, then this is a good sign. Of course, the external consultants might not be that strong, but at least you will know that the internal team is as strong as the outside experts.

23.2   Development

There are a large number of development risks to take into account. Here is a partial list taken verbatim from a real project of what might be considered when first planning a project.

- Automated test s/w — Does it exist? Will it function?
  - o Automated systems will need discipline and effort to use.
  - o Automated systems might not be fully functional.
  - o Use trancode and virtual machine concepts, macros, playback.
  - o High impact. Medium probability.
- Version 1 Functionality — What if all of the functionality is not ready by the first release date?
  - o It is difficult, if not impossible, to guarantee that ALL functions requested for a release will be ready by the ship date.
  - o Ensure project plan focuses on main functions first, options second.
  - o Med-High impact. High probability.
- Unstable programmer workbench — What if the programming tools fail?
  - o Work might be lost; test and debug takes longer.
  - o Use mature workbench and tools.
  - o High impact. Low probability.

- Programmer leaves — What if one of the key programmer leaves?
  - Knowledge and experience is lost.
  - Use of paired-programming teams.
  - Regular part of project is show-and-tell — design concepts, subtleties.
  - Keep list of other local talent for short notice work.
  - Medium impact (controlled). High probability.
- Quality standards not adhered to — What if the problem is not discovered immediately?
  - Impacts code quality, user experience, credibility, etc.
  - Regression tests.
  - Regular reviews of code, tests being done.
  - Visible to whole team — some form of quality reporting.
  - High impact. Medium to High probability.

In a real project, the number of points to list and reflect upon can be in the many dozens; easily 50–60 points for a small to small-medium effort. Each project will be different in its entirety, but there are many common themes. When I do this and when I reflect back in hindsight, I will be right on some, wrong on others. I will try to analyze why I was off, why I did not identify other potential risks that did occur, and I will try to think about those risks that I was sure would occur but did not.

23.3   Operational Considerations

The risk list is not done yet. There are possible other considerations when it comes to the real system running in production. Remember the seven levels?

Ⅰ. hardware (e.g., physical device level),

Ⅱ. chip level s/w (e.g., imbedded s/w on chips),

Ⅲ. operating systems (e.g., base technology and drivers),

Ⅳ. systems level tools (e.g., servers for communications and database tools),

Ⅴ. supporting subsystems (e.g., email, search engines, reporting systems — using level Ⅳ tools, development environments and other toys),

Ⅵ. application (e.g., the main system — the cause for existence),

Ⅶ. user level logic (e.g., macros, work flow tables).

For each of these, you need to think about the risks and do a similar analysis to that in Section 23.2. This is more likely to be done as you get into the Part Ⅲ activities: the design and detailed architecture. My advice is to do the analysis in two passes. The first pass should be generic and work with the specie characteristics. E.g., for any operating system, these bad things could happen; for any communication server; for any db system; for any user interface driven by tables; ... You think through what can be a risk, what the risk implies, what would be a backup plan, and what is the probability of the risk occurring. You can then review this level of analysis and make sure that the biggies are addressed. The second pass takes it down a level and looks at the specifics of a chosen system: what is the risk with such and such operating system for hack attacks? What is known to go wrong or could go wrong with such and such database system?

You look for security, performance, sustainability, scalability, usability, supportability issues. And, are there risks for migrating from legacy, parallel execution, and verification? The more important the system is, the more time you spend on this stuff. For one project I was part of, the phase one risk document done by the team members was about 60 pages.

23.4   Risk Identification

There are probably three ways you can identify a risk. You can see a risk once it appears out of the fog and bites you. This is called reactive management. There are some things which are truly random and unpredictable, but there are other things that can be somewhat anticipated. This leads to the other two ways to identify a risk.

The second way is for someone else to identify it for you. You contract the consultant, bring in the risk audit team.

For the third way, we need to go back in time. Back in the good old days, in the early 1900s, the fathers and mothers of management advocated that one of the roles of management involved with planning and scheduling was to anticipate difficulties and discount them. This means, in plain English: avoiding or minimizing. This is a proactive view of the situation. And, it should remind you of various other things written in previous chapters. You start thinking about the temporal significance of events, looking for any change in the status quo, etc.

For software and hardware development, you have to think through the pieces that come together to deliver the goods. This means that you play through in your mind the human team, the development process, the technology used to do development, stuff to be used in production, how the operations are expected to execute, as well as the risks due to the changes from the user's side.

The lists you end up with come from a systematic review of the situation. You look for:

- The amount of experience you have, in the aggregated whole, with the tools and approaches.
- What might be first versions or releases versus more mature versions — identifying potential early life failures.
- The tools and methods which have been used for in the past. You look for similar functionality, similar scale, similar importance, and similar time line.
- The assumptions you are making about skill and competence. About your own skill, the team's, the supporting cast, the vendor, the users.
- What aspects appear to be rapidly changing with respect to releases, standards, and conventions.
- The aspects which are reasonably easy to fix or replace if the assumptions are wrong and what aspects are hard to fix or replace.
- What aspects are reasonably cheap to fix or replace if the assumptions are wrong and what aspects are expensive to fix or replace.
- Where in the process has the most value been added. If a problem occurs before this point, what is the impact, and what if the problem occurs after this point.

- The major interfaces and coordination points out of your control. With other groups, legacy components, major milestones, how many occur close together, and what else is happening at the same time.
- The major review or approval points. When do they occur; how many at the same time; what time is allowed for revisions; who has to sign, approve.
- The politics: who is likely to actively engage and help; who will be neutral; who will be passive aggressive; who will be actively aggressive.
- The stability of the vendors and supporting cast. How robust is their organization; will they be bought, sold, re-organized.
- The single points of failure. Not only at the technology level, but within the process and human part of the organization.
- What is likely to be on the critical path and who can work on the critical path.
- The legal and third party controls and interventions on the project. Who else has to approve, review.
- The legacy system issues for compatibility, migration, and integration. What are the number of systems, the quality of documentation, the type of wrapping and encapsulation on the interfaces (or lack thereof).
- The tasks and skills that require deep or broad or specialized skills.
- What knowledge exists in the organization about the functionality and implementation of any legacy system. Are the programmers still alive or on staff; does the source code exist.
- The techniques and concepts used to create systems being integrated with. Clean, modern style systems with nice flexibility and elasticity or 1960s–1970s application styles with rigid and unforgiving requirements.
- The levels of geekdom and cuteness in the systems being integrated with — how fragile and how understandable they might be.
- What aspects are expensive or time heavy for verification and regression testing.
- The aspects that are expensive or time heavy for users to learn, use.
- The quality of information coming into the system (human or system interface).
- The reliability of the interfaces coming into the system — with respect to up time.
- The typical life cycle and what can go wrong during usual developments (and why you do not think you will encounter them). The coding, the testing, system verification/sign off, roll-out, training, maintenance, legacy system integration, etc.

This is a partial list. You will have a different list for every project and as you progress through your career, you will add some and discover that some are lesser risks or non-existent depending on the situation. With progress in tools and technology, the risks change. Some risks from the 1970s and 1980s would not be considered a risk today. A risk today might not be risk next year. The risks are different if you are involved in a Class I system or a Class VI system. You need to have different lists and be able to figure out what should be on the lists. A master list might be a starting point, but you will need to edit it!

Chapter 24

Project Management

What exactly is project management? I am not sure if I know a precise definition. I am sure in my own mind that there is not currently a unified theory of project management that can be used predicatively or normatively. There are little capsules of theory that can help explain and guide elements of what project management might be, but there is not a grand schema that can be followed and applied. I have taught a graduate course on project management and I had a very difficult time finding enough good, solid theory and science to fill one-half of the course. There is lots of mechanical material on what terms mean, what forms should look like, that you should have something called a work breakdown structure, that you should know how to calculate the critical path, but these are mostly recipes that deal with the process, documentation, and recording and not with the real, intellectual challenges.

By theory and science, I mean ideas that have a theoretical base that can be tested and relied upon. I do not mean the thousands of articles with people's opinions, suggestions, or examples from one or two successful events in their careers. It is very hard to prove causal relationships, and to prove that an idea can be generalized into something that others can use. As a project manager, you need to know the terms and the tools of the trade, but that does not in itself make you a project manager. This is like Emerson's example used at the beginning of the book about buying a lawyer's library of books, sitting in the middle of the room, and thinking that you are now a lawyer. Just reading a medical book does not make you a doctor. Just reading a programming book does not make you a programmer.

I encourage everyone interested in project management to read and study at least one of the many books on project management methods, and to be aware of what is in the Project Management Institute's *A Guide to the Project Management Body of Knowledge* (PMBOK), and even get a certificate from a project management training program like PMI. Just remember though that this does not magically make you a project manager or mean that you are qualified to manage a project. I can read a book on white water kayaking and memorize all of the advice about what to do, but that does not mean that I can execute an Eskimo roll in a Class VI rapids.

You have to separate what people do versus what they are called. I have seen people called project managers that simply push paper, and document other people's decisions. They really do not make management decisions in terms of what the budget should be, who should be hired, or fired, or what the hiring level should be. They do not make decisions about what the risks are, or what should be done in what order. They have no authority or accountability except to make sure that people do the documents, that the right meetings are held, that dates are supplied, and that the chapter headings are filled in, and of course that the right signatures are on the document. They do a fine job maintaining the documents needed to satisfy the auditors, ISO standards, and making sure that the process is followed, but they are not really managing the process or making decisions about the process. To me, this is not project management. They should be called *Project Clerks*.

Project clerking is not to be dismissed in its entirety. The functions are important and someone must do them. There are people making lots of money doing these types of tasks. There are some people who do this really well. Based on my own experience, the majority of the people I have seen doing this type of project management did not have the skill or knowledge to know what was being done or what should be done. They did not know if it made sense, or whether it was appropriate or not, what an alternative might be, or what the long term risks might be. They do what they are told to do. They cannot problem solve and they cannot make suggestions about what to do next or what to do if something goes wrong. They make sure that there is a section called blah-blah, and there is a design document for blah-blah, and the design document has been signed off for blah-blah, and that the design document should be done before blah and after blah. This is like a postal clerk making sure that an envelope has a name and address. It does not matter if the name or address is correct; the postal clerk does not have a way to know; the postal clerk just cares that the fields are filled in.

This chapter is not about clerking.

24.1 Early Phases of Project Management

While there are parts of project management which are sequential and can be done via a nice checklist, there are many parts which are done in parallel — back and forth, back and forth, converging to a plan, decision, milestone, or other form of target. In the days of old, there used to be planning and project documents used in the early phases of project management called:

- Product opportunity (PO).
  - The glimmer in someone's eye about a market or need.
  - This would justify in someone's mind that the preliminary, high level documents should be prepared — Preliminary Product Requirements, Preliminary Project plan.
- Preliminary product requirements (PPR).
  - The initial wish list and set of concepts that would support the idea that someone might want it.
- Preliminary project plan (PPP).
  - An initial stab at how big the beast might be — basic scale based on specie and assumptions — 2 months or 2 years, about $1m, maybe $5m — the order of magnitude guess.

After these documents and review processes, approval would then be given for the next phase of development:

- Project plan (PP).
  - An overall plan that addresses the approved PPR, taking into account what is in the Functional Specification.
- Functional specification (FS).
  - A description of the *what* to be delivered — not the *how* unless it is a part of the PPR — considered a contract between the part of the organization or client wanting the *what* and the folks creating the *what*.

There would be lots of other documents as well depending on the type of organization, level of governance and adherence to standards (e.g., ISO), and there were and still are, many titles and different names for these types of documents. There is the intent and form for each though, regardless of name. These were the foundation of the Waterfall approach. The major key piece was the detailed *Functional Specification*. In the old days, you would try or would be expected to prepare these five documents and then the many other documents that follow. There would be detailed design specifications, test specifications, and so forth. If my memory is correct, you would do this if you were in a large corporation, or service company, or in the department commonly known as data processing. Not all development was done this way and if you were in a university group creating systems software or special infrastructure code, you probably never saw a formal document or a formal process. You went out, analyzed the requirements, and then built the software, possibly with user involvement as you went. You still had to pitch the idea, sketch out the requirements, get a rough design of the solution in your head, build it and get it working. If you were lucky enough in the 1970s to be working on interactive time sharing systems like I was, you were able to build things up quickly in small pieces and be intimate with your code. If you were doing the batch processing approach which was the dominant force of the day, it was larger modules and a different coding technique. Of course, you could use the interactive systems in a batch oriented, monolithic process, but it was possible to do things in an agile way and some of us did. We did not call it agile. We did not call it anything except programming. You might have also been lucky to have a boss who was user centered and believed in creating high quality software that would provide value to the user, who encouraged you to sit with the user and interact with the user to build what the user needed. This also did not happen to everyone, but it could and it did. We also did not have a name for this style or approach to creating software. It just made sense to a bunch of us. It was called programming. We knew that we were developing software in a different way and with different characteristics when compared to the batch oriented software and developers, and it was a little bit of them-versus-us for a while. We were programming. Everyone was programming.

This perspective may help with understanding the Agile/Extreme movement and what people are trying to do with respect to documentation, requirements, and specifications. As noted in the previous paragraph, you still need to consciously consider many of the infamous Waterfall components, but the difference is when, how, what sequence, and what form the component is documented in.

The *Product Opportunity* document is basically the "I think we should do this" discussion and could be verbal, in an email, on a napkin, or on a whiteboard. I do not care what you call it or how you paint it, but someone along the line came up with some idea that they wanted to pursue or have others pursue. This happens regardless of the Agile/Extreme versus Waterfall versus? debates. The more $ on the table or more risk or more complex, the more a formal process and document stream is warranted. If you are doing this in a garage or a two-bedroom apartment, no need to worry about ISO standards unless you are planning to supply to an ISO certified user, in which case, your project might also need to be ISO compliant.

The *Preliminary Product Requirements* is the part of the process that gives enough description about what you basically have to make; not the details though. You need to know it is a cat and not a dog. A guard dog and not a lap dog. You need to know the capabilities and expectations of what you are building. You need to know if the software will be used by one or two people versus the whole organization counting in the thousands. You need to know if it is a website for selling goods versus a contact management system for a volunteer organization versus an internal system for tracking job applicants. The more of a Class I –III system, the less documentation and preparation you will need. If it is a Class V or VI situation (and it is during this process that you get a feeling for what class it is), the more effort that you will need to scope and understand the basic capabilities you need to deliver.

The *Preliminary Project Plan* sets the expectations for, in a ballpark way, how long things will take and in general, what resources will be needed. No competent, credible manager will just give you permission without some planning and some scoping about costs and resource requirements. Again, might be verbal, might be napkin time, but the task must be done and should be done by someone with enough experience that you are indeed in the ballpark. If you do not have enough experience, you will be very lucky to be close to your estimate. It will be luck and not skill, and you might not be able to repeat the process a second time.

In a large sense, this early phase and three topics — the opportunity, the capability, and the basic plan — deal with determining the *what* and the *why*; the topics covered in the first two parts of the book. In light-weight Agile/Extreme projects of the Class I –III types, most of these topics do not need to be formally documented and analyzed to great lengths. You are building a blah and there is enough general knowledge available to you about blah to get you going and to avoid most of the pitfalls risks. There is enough knowledge about how to build a website for selling something that you should be able to quickly get in the zone and the ballpark. You are not on the edge and I would speculate that most small client software projects in Classes I –III do not approach the millions of dollars or consume many person years of effort. But, the same kind of thinking that occurred in the Waterfall approach is still done for these three topics.

The major, major difference between Waterfall and other looser approaches starts with the next set of tasks and related deliverables. The classic definitions of *Project Plan* and *Functional Specification* are not used in the Agile approaches. Instead of trying to get everything specified out in advance, the plan, the functions, the user interfaces the system interfaces, you iteratively and progressively plan, build functions, refine the user interfaces and get the systems hooked together. This approach was likely done in the 1960s; I know it was done in the 1970s and the decades following. The concurrent engineering or manufacturing approach that involves many parts of the system in parallel, including the user, was advocated in the early 1900s, and was a big thing in the late 1970s. This concurrent, user involved approach included the software side.

In the formal settings using Waterfall, this meant that you had the users and various parts of the development and production team working together from the beginning. It was not done by X and then passed to Y and then passed to Z with no involvement between the parties in advance. They worked on the detailed planning together and they worked on the functional specifications together and they worked on the user interface together, all be it sequential. Waterfall did not mean that you did not work with other groups; it was the sequencing and formality. And, if by chance you were in a different type of organization, the overall process was not sequential and did not look like the Waterfall model. Not all *old school* software was done in the Waterfall style!

24.2   Detailed Functionality and Planning

After the preliminary planning and concepts have been agreed to between the folks with the $ and the need, and the folks who will deliver, the fun starts. By the way, the former used to be called the **MAD** people or person — money, authority, desire. In Agile/Extreme, the functionality and detailed project planning is done informally and iteratively; in what next, how much will it cost cycles. You still need to give reasonable and feasible estimates for each cycle. If you blow it too often, the client will possibly become annoyed, frustrated, and start to think that you are milking the cow and are doing this on purpose. For bigger, more strategic systems, you might need more of a formal functional contract that can actually be measured and tested against between two parties and this implies a more formal project contract with clear deliverables and milestones.

If I am in a firm contractual situation with a fixed budget and specific goals, I try to delay a more formal or firm project plan until I have a good enough or rough set of functional specifications to work with. When the client says you have so much $ and so much time, you better have a reasonable picture of what you need to build. My argument has always been: I cannot give you the Project Plan until I know what I am delivering under contract: the Functional Specification. To me, this is an obvious fact. Unfortunately, this was not always so obvious to some of my management. Sometimes, they wanted a detailed plan and commitment before we knew what we were doing. In Agile/Extreme, your dates and budget need to be fluid.

In a high volume, low risk development where the effort is measured in people weeks or people months, and you have exemplars for comparison, you have the Class I or II rapids. This means that you have been on similar water, you know the tools, it is reasonably predictable, functions might be somewhat different, but not bizarrely so. You can do some additional functionality work, but if you have the right stuff, you can pull the trigger and start the

Agile/Extreme processes as documented in the manifest and elsewhere. Follow the bouncing ball and within an order of magnitude, you will probably be close to the original guesses for project costs and efforts, and the original requirements, assuming that you are skilled enough to do it. For the Class I and II rapids, you should be able to give a reasonable estimate to the client, time and cost, and be very close to it. The risks and unknowns are relatively low and you have done similar things in the past. There is no excuse for wildly misestimating, over or under.

For high volume development, there will be others who will know more than I do. They have written lots of good books for the bread and butter types of applications. In general, I do not keep repeating myself. I usually do not create the same type of application dozens of times. I might do something a few times (e.g., a parser for a language, or a symbol table implementation, scheduling systems), but I do not engage in repetitive developments year after year like a contract software house might. In some cases, I have made dozens of certain types of applications over the years, but that is not what I go looking for. So, I will leave specific guidance for Class I and II rapids to those who know better. The rest of the chapter will focus on the challenges with a greater degree of difficulty — say, Class IV+ rapids.

Doing Class V or VI white water kayaking all of the time is possibly a bit too stressful for anyone, but an occasional run is OK and regular Classes III and IV activities are good workouts. I think you have to think about software the same way. You need to understand that there are different classes of software development, some easier than others. You need to understand that a diet full of the extreme categories is probably not good for you, and that you need to slowly build up the skills and the knowledge to do extreme development. But, once you master it, what a rush. Starting up a Class V or VI project takes planning and preparation work. If you have not done something for a while at the extreme level, you need to get back into the groove and get your feet wet. Go slow and careful before letting rip. There might be new systems, new languages, new domains; I move like a snail for a while and do lots of little probing before wasting lots of people's time and resources.

24.3 Budgets and Plans

As the planning and preparation moves ahead, you have to think about a number of elements in addition to the basic functionality. You have to think about the:

- o Development phase.
- o Introduction or deployment.
- o Possible migration from an existing legacy system (manual or computerized).
- o Ongoing use.
- o Future evolutions and migrations.

You also have to consider future de-commissioning and how the system could be turned off and phased out in the future. In each of the phases, you have to do the drill: who, what, where, when, why, and how. For example, you have to consider the toolkits and platforms used for development and for deployment, and the costs of equipment and personnel. You have think about boring things like where people will sit, whether travel is required, how the client will be involved, what security is needed, what policies, standards and conventions might need to be followed, what policies might need to be crafted, and you have to do the infamous things most people I know hate: the planning and the budget.

In the initial plan, you have to identify the big tasks, major milestones, and the task sequence. This is the planning portion of project management. You need to know what tasks might be involved, what are reasonable and feasible durations for each, what milestones are possible, and in what order the tasks should be done. You might seek advice from others, but you should be able to doctor up the first draft yourself for others to beat upon. This is assuming that you are project managing something you are qualified to project manage. In the budget, you might be called upon to give a cost and personnel estimate for each of the major phases or activities. Even if you are not asked to prepare a budget in advance, have one in your pocket because someone is sure to ask you and it is always better to have something roughed out than to dream one up on the spot. Most companies and industries have rough numbers they use for personnel. They have guidelines for hiring costs, yearly increases, levels of skill, and ballpark average pay for each level. This you have to learn. There are also usual guesses used for how long it takes to hire and for learning curves. These are things you have to take into account when estimating how the first phases are going to ramp up.

You have to learn what to include in a plan in terms of the type of tasks and budget items. For example, what are the normal tasks at your firm, or at the client's when it comes to developing or acquiring a system? On the budget, do you have to account for purchasing licenses, yearly fees, phone, travel, hiring, office space, office supplies, training, vacation pay, salaries, overheads, home equipment, backups, security upgrades, consultants, external audit costs, etc.? If you are doing project management for a living, you should be able to size up a project and prepare a list of appropriate tasks at the ballpark level and the matching budget items. Items that make sense and budget items that are relevant. When you are a consultant or a senior analyst, I have observed that many debriefings usually end up in *what next, how do we proceed*, and *what will it cost* discussions. I make it a habit to anticipate this and have a one–two pager ready, not polished and very rough. You want to be able to pull it out of the briefcase and present it as some initial thoughts.

As the plans evolve and it gets more serious, it is typical to get others involved. You will need help to flesh out, review, and rationalize the planning and budgeting. This is where you get domain experts as part of a task force, coffee club, whatever. If you are relying on them to tell you everything and to give you guidance on everything, you are in over your head. You need to know enough to know if what they are advising is reasonable and feasible. They may know the details and more subtleties, but you need to know enough. If you do not, start reading; fast. A good project manager quickly assesses what they know and what they do not know. I do this first and put this on a project plan as my first task and the first thing to sort out. This is the first risk: the project manager is clueless. Hence, the project manager must figure out what they do not know, what the team might not know, and develop some kind of gut feel on how much effort it will take for the project manager and key personnel to have enough of a clue. This is not perfect knowledge, but this is the good enough level you see noted in many books — **you do enough requirements analysis to start**. Interesting that some authors who write about Agile/Extreme methods never define *enough*. For me, I need to get the Zen feeling in this first phase. Over time, you will learn to estimate how long it will take to get the Zen. You size up the current situation, note what is different or new to you, do some quick research, know your own ability to learn new material and technology, assess what you have researched, and voila you get an estimate for the first task on the project plan. Sorting out what you do not know and your confidence levels is very useful for risk management and helps prevent you from being caught with your pants down.

For large projects, there will be a head or chief Project Manager who decomposes the big puzzle into smaller parts and assigns the parts to other lower-level Project Managers. There should always be one final boss. There might be a boss for the functional and another boss for the technical. That works if the bosses respect and trust each other and can work together. If they do not, then you need the big boss over top. The aggregate or more senior Project Managers need more skill and knowledge as they have to see the big picture and do the decomposition. They need to know enough about the parts to know if what the other Project Managers say makes sense, judging the overall costs, durations, risks, effort, milestones, and task sequence.

24.4   Degrees of Certainty

In the spirit of *ZenTai Mushing*, you will never be a 100% perfect Project Manager. Shoot for 80% and cross your fingers. If you are caught in a situation where you do not have sufficient skill and knowledge to be the 80% Project Manager, fess up to this and strike an appropriate set of experts and helpers as a task force or advisory committee. Do not pretend that you can make a technical decision between A and B by yourself. Know your limits and ask others to prepare pro/con analyses with a risk analysis and keep asking questions till you are confident that a reasonable decision can be made.

In high velocity development, you will often find yourself holding your breath and jumping off the cliff. You will likely never be 100% confident that the design, approach, team design, etc. is right. When I talk to developers and they are claiming that something is 100% anything, I get nervous. This goes for users as well. When a user says that something will NEVER change, I assume that this might likely change in the very near future; and if the user says that something changes ALL of the time, I assume that more homework will be required to sort this out. It might never change, change often, change occasionally, or it might be wishful thinking that it might change in the future although it has never changed in the past. Jumping off the cliff into the unknown is not a cardinal sin and something never to do. If it is a calculated risk, done knowingly, with a backup plan, with early problem detection and exit or release clause, it is OK. If you just blindly toddle along in a daze without taking precautions, packing a parachute, and without looking out over the cliff first, then this is bad.

You do not have time to sort out everything in high velocity development and you will have to have patience. You will have to assume that you will do some things twice, or perhaps three times. But you should not have to do everything twice on every project. If you are doing this, you are in the wrong career. You have to assume that you will lose a week or two along the way, or more depending on the size of the project as the problems come out of the closet at night and mess with your project. If you cannot stand wasting ANY time, if you are often IMPATIENT, if you are usually INTOLERANT of yourself or others when mistakes are made, if you do not like doing ANYTHING over, and if you insist that everything is 100% right from the start, find another type of project to work on. Please. A high velocity, white water adventure will not be for you.

24.5   Slack and Project Elasticity

As project manager, you have to plan strategic slack at appropriate parts of the project, creating elasticity between key tasks and dependencies. This is an important part of project management. How much slack do you put in the schedule and where do you put it? There are four ways I have seen that describe how slack can be inserted in a plan to account for risk.

First, some people just take an estimate, and then do something like double and double again for each element. Or, they add a percentage like 50%. This is fast and easy to do, does not require much intelligence or thought. If you are in a sloppy business that makes money in spite of itself, this might work as no one will notice the gross inefficiencies and ineffectiveness introduced. Second, people seem to put random slack in, scattering it like pepper on a salad. This requires even less work and intellectual ability. This approach can turn a Class II rapids into a Class V or VI. Third, concepts like the ones proposed by Goldratt (1997) suggest placing the slack at the end as a big bucket. Compared to the random or fixed slack approaches, it looks like rocket science. You estimate accurately, or as accurately as you can, each element in the plan and plan with no imbedded slack just leaving the bucket to be consumed. This approach seems to make many assumptions (McKay and Morton 1998) and does not account for any domain knowledge the planner might have, nor does it take into account that not all perturbations are totally random.

A fourth approach takes certain knowledge into account and this can come from past tasks or experiences and guides where to put the slack. You plan most of the tasks tight as Goldratt suggests, but you take some of the bucket (not all) and distribute it strategically at key merge points or where high risks might exist. This creates some elasticity in the plan and probably makes the final execution of the plan match the plan. This allows more feasible and reasonable plans to be created and does not drive certain merge points or activities earlier than needed by pretending that everything will go right. Guess what this requires though. You need the experience and awareness to recognize what might be risky, what might go wrong, how long a recovery might take, and you need to be able to explain this to others. The plan and schedule will still chatter along as it keeps being repaired and adjusted, but there should be less chattering and less repairing when it is compared to Goldratt's method or a method that adds a little amount of slack everywhere. If you are doing the sloppy planning in a sloppy situation, then the chatter is likely to be low in any event. You would have to be really bad to constantly bounce into missed targets in those cases.

A similar situation about tight and loose planning existed in manufacturing in the 1950s and 60s. When business was tight and inventories had to be kept low, the planning was tight and the amount of effort invested in planning was high. In manufacturing, the inventory is often your major slack or safety mechanism. Tight planning is difficult to keep going and this involves managing people tightly: what is done when, how much is done when, how to get things moved quickly between tasks. If you have large inventory banks and big buckets of time, then you manage the inventory and you do not have to manage the headache bit, the people. For example, in the early 1960s many firms went to a material movement model that had each operation scheduled to be done on a week time estimate. The operation can start any time in the week as long as it completes by the end of the week. The inventory would be taken out of inventory at the beginning of the week, or thereabouts, and after the task was done, the inventory modified by the task would be placed back into inventory, ready for the next operation to be done the next week. This was the weekly bucket idea. Easy to plan and easy to manage.

One problem with this concept was that people had a tendency to leave the hard and possibly risky things to do for the end of the week. The work either did not get done on time as problems occurred, or did not even start. This bumped the work into the next week and spoiled this nice weekly bucket brigade. This process also broke down if someone wanted something mid-week, changed their mind, or competition appeared. It was expensive, required more space, and had the agility of a cement block. But, it was easy to plan and think about. It was also great for firms that rewarded fire-fighting and reacting, rather than pro-active management. At the tactical level of planning, thinking in ballparks — it is about a week, about a month — is necessary and OK. But, at the operational level, you should work on a finer level of resolution and flow work through the shop. Anyway, inventory was one way slack was accounted for in planning and scheduling. You can think of adding a set amount of time to every task, or using a set unit of measurement (e.g., it takes a week, everything takes a week) as the same idea. OK if you can afford to be sloppy. Not OK if you have to be efficient or effective.

The trick with smart planning and strategic slack placement is having the skill and knowledge to know where and how much to use. There is no magical schema that I have up my sleeve for low volume, non-repetitive situations. For repetitive situations, there are ideas (Aytug et al. 2005), but these do not help you much in software development.

To summarize the main points so far:

o To be considered a true project manager in my book, you have to be making decisions and doing planning and control, not just acting as the project's clerk.

o The first thing to consider is what you know and do not know, and think about what it will take for you to figure this out, or who you need as advisors; so that you can indeed manage the project.

o You will need to do some high level thinking about deliverables. What is feasible and reasonable? And of course, plans and budgets.

o You need to consider risks and put strategic slack into the schedule or planning.

These points are applicable for any software or system project I can think of. It does not matter if the traditional Waterfall or Agile method is being used. Another key concept to understand, again independent of the specific method, is that of critical paths.

24.6   Critical Paths

A critical path contains those tasks that if any one of these tasks is delayed by a day, the whole project is delayed by a day. Got that? You do not want to starve or block **ANYTHING** on the critical path. You do not put amateurs or novices on a critical path item unless you want a nasty adrenalin rush. You do not use the critical path tasks for training if they are to be largely done by the trainees. Luckily, there are software tools that at a press of a button can highlight the critical path. If you are a good project manager, you should have been able to get close to this realization without the project planning software, at least for small to medium sized activities. You should have known what really had to happen in what order and what were the key dependencies for the major or key features/functions/technology before pressing the button. The critical path highlighted by the software should not be a surprise! It should be a confirmation.

You use the critical path knowledge for a number of purposes. You use this to determine who should be doing what. You also use it to figure out where you spend your time and effort as a manager, and where to allocate and focus the resources of the team. Anything you can do to help the critical path, helps the project. There is always a critical path. And, there is always at least one task gating or timing the critical path. Why it cannot be shortened. As you address one task after another, another will rear its ugly head. This is another of the lovely water/rock analogies that is often used by consultants and book writers: as you lower the water, you see the rocks. As you continue to lower the water, there is another rock or set of rocks previously hidden. Same idea for dealing with the critical path.

You should not worry about things that are NOT on the critical path. That is money foolishly spent. The old industrialists knew this and Goldratt (1997) reminded people of this focus. Until something is on the critical path or about to become the critical path, it really does not matter. You can exploit this knowledge in various ways. You can focus the best and the mighty on the critical tasks. You can also use the non-critical path tasks for training. If a task does not become critical for say six weeks, I do not care if someone takes a few weeks when a hotshot could do it in a few days. I want the hotshot on the critical path or doing major tasks. This creates some problems though. I have had to deal with hotshots who could not stand to see others take longer to do something than they could do. Instead of keeping on their own task, they would constantly go interfere with someone else's task to show them the best and fastest way to do something. The individual was delaying the whole project by helping someone not on the critical path. Stupid. The individual was also not accepting of the idea that people learn from making mistakes. Could not stand it. I would like to say that I have had to only deal with one of these types of hotshots. Nope. I constantly bump into them. Sometimes they are a real negative factor and as a manager I have regretted hiring them, and have had to occasionally seek ways of doing some waste removal. The critical path also helps you as a project manager to know what to worry about and what not to worry about. You worry about starving, blocking, and tightening up the critical path and where strategic slack is needed on the critical path. You need to ignore the non-critical parts.

The people not on the critical path might also cause problems. They might not realize that their number one priority is not the team's priority. They might not realize that they can take a bit longer or stumble a bit compared to their ideals or own view of reality. It might make sense at times to optimize the critical path resulting in decisions that make life a little bit more difficult and awkward for those not on the critical path. This is hard to accept for people striving for textbook perfection in code and process. It is hard for some to accept the fact that not everyone's tasks are of the same importance. I have had staff upset by tradeoffs based on the critical path or major tasks. For example, I might tradeoff hours of someone's time off the path against an hour's effort associated with the person on the path. Not everyone's time is worth the same on the project, even if they are getting the same pay or think that they have the same skills. Some people understand this type of situation; others do not. People who believe that everyone's opinion has the same value in a meeting, will also dislike critical path tradeoffs if they think that they are not being used in the best way.

For the mechanics of documenting and tracking the critical path, please see the traditional texts on project management. Lots of good stuff there.

24.7   Resource Flexibility

Resource flexibility is yet another concept that is needed to be understood for project management. As briefly noted in the preceding section, not everyone is created equal in the area of skills and the use of the skills. In both Waterfall and Agile, I look for certain skill sets and use this knowledge when designing teams and assigning tasks. For example, there are problem solvers who are creative and there are people blinded by methods and process. I once had a discussion with someone associated with the space effort, and how they chose planners and schedulers and their experience echoed my own experiences in factories. You go to the higher manager or manager of planning and ask who to work with in creating a planning system. They inevitably point to so-and-so; he/she knows the process, knows all of the rules, is tidy and organized — go interview them and study them, the manager says. But, if you ask the manager who they usually end up asking for help when help is needed, they usually point to another individual and make a sarcastic comment about their poor methods or process, hard to manage, etc., but they know how to create magic out of thin air. This is the individual I need to know about when thinking about how a planning and scheduling system needs to function, what the real problems are, and what are the ways that the system can be bent, folded, and mutilated to get the job done.

Another type of person I look for are those who know more than one thing and can do more than one task. The more skills, and I mean near expert skills, the more valuable the person is to me. You can mathematically show that assigning tasks to the least flexible resources first and keeping any slack on the most flexible is the best way to go if you have perturbations and you need to react. The most flexible resource, if not loaded 100%, can be flown in, hit the problem, and run away again with the least amount of disturbance to the whole plan. They are also the person you would probably want to fix the problem anyway. If you have these people 100% loaded and tightly intertwined in the plan, any change to their tasks will ripple through many parts of the plan and require substantial juggling. Managing this way can be difficult if the team members are always looking at others and comparing their own status and activities. Petty jealousy and envy in those who are not as good as they think they are can create team problems.

So when I can, in both Waterfall and Agile, I target and use the space cadet type of hotshot and most flexible hotshot in certain ways. In both, I have a swat team capability (hidden if necessary) that can be easily mobilized to turn the sow's ear into a silk purse and to do this with the least amount of disturbance to the system. On one project I had about two dozen developers and kept four hotshots in reserve. Officially, they were loaded on paper 100% but in reality, only planned to be 20%–30% busy. They were always 100% busy, but the 70%–80% was spent stabilizing and doing lots of things that allowed the other twenty to just march along without headaches, interruptions. This was like having the four running ahead of the marching band filling in the potholes and crafting temporary bridges. The marching band just keeps on marching with a predictable rhythm and progress. Depending on higher management's enlightenment, I have made this capability visible and in some cases not so visible.

You get a lot of really good software created if you have twenty people just going along in a nice steady fashion! I kept the four dancing, but the others were just kept to an internal plan and agenda and lived happily under the umbrella created by the swat team. In the one case where this was not so visible was in a situation where management looked at the salary of the *top* developers and insisted that they had to be pre-loaded up and pre-planned for 100% of their working hours (if not 120%). Not so smart. I have also seen this in manufacturing where people look at the most flexible, and likely most expensive machine and want it to have planned utilization at a very high level. Management has to get their money out of it! Also not so smart, as the most flexible and most expensive machine is likely to be highly utilized at the end of any time period doing unplanned for things to stabilize the whole concern!

Key idea: schedule least flexible resources first, most flexible last and keep spare capacity on the ones that would do you the most good when something bad happens. And believe me, something bad will always happen.

24.8   Multiple Plans

As I was learning how to be a project manager I quickly became aware of the need of having and using multiple plans. As a researcher, I was able to probe this and discover that others also did the same thing. I have been lucky to study some really great schedulers. Once you have their trust and have a chance to work with them for a long time, you observe the following.

First, there is a political schedule or plan that is provided to management. This is what management wants to see and will be happy with. This will keep management out of the scheduler's way. In some firms, this political plan is close to the other plans I will mention, but in some firms, it is really an impressionist painting of the cubist form. The schedule has a three eyes, two noses, no mouth type of reality. The more the management level is not trusted or believed in, or the more games they play, the more bogus the political plan is. I am not recommending anyone create a bogus plan; I am merely reporting what I have seen.

Second, there is the plan or schedule that is verbal or internal to the department or team. This is what the project or area manager is really driving the people to. "Can you do this by this date — see what you can do, and I'll buy you a coffee." This may or may not reflect the higher plan. I once observed a scheduler really play the system, but it was shown that if he had not, the plant would have been in deep trouble. By his playing the system, he was within 1% of the monthly target instead of missing it by 30%.

Third, there is the plan that is in the planner's or scheduler's head that is the "this is what I expect to happen." They ask for 10, expect 9 or 8 and they are happy with the 9 or 8. That kind of plan. You ask for it to be done in five days, but you expect it will take six or seven and your master plan is really based on six or seven — not five. Yes, this is game playing, but it takes into account human nature and the organizational reality of setting goals and how people behave. If the people know that you are playing this game, it does not work!

Fourth, there is the plan that is in the project manager's head about an ideal situation. In this situation, the project manager can just get on with it. They will not sacrifice quality, functionality, or cost; but will just get it done independent of possibly arbitrary constraints or policies that on the surface, and possibly below the surface, do not make any sense and provide no value. This plan is often called upon as milestones get close or when problems occur. This is the "just get it done!" order from management. The better managers, schedulers, and planners I know have this plan in their toolkit and know what needs to be done and how to do it.

As a researcher or consultant, it is not easy to get below the political plan. If you visit the project office and ask for the schedule, they haul out the pretty picture and proudly display it and are quite happy to give you a copy. It takes a long time, perhaps months, to learn about the other plans and to know how these other plans work. This is why I do longitudinal studies with ethnographic methods. It takes me all of that effort to learn and understand the other three plans. There are probably other types of plans as well. I have just been lucky to observe and document the four. On the research side, it is always interesting to watch people obtain a political plan and then use it as if it demonstrates the scheduler's skill. They will critique the human's mathematical optimality. First, depending on the organization, the planner did not give a hoot about optimal; they were concerned with political. Second, the plan probably had many more realistic and feasible constraints in it than the researcher's limited mathematical model. That is, the planner's plan might actually be executable and not be pie in the sky. Third, the researchers did not get the planner's fourth plan which really shows the planner's knowledge and skill at crafting a good sequence of tasks and work. It is not a fair comparison to take a political plan and compare this to a mathematically crafted one.

The four plans above are in the first layer or ply of thinking. They are in parallel and in play. They depend on the current context or situation. A good planner will have at least three plies of planning. They will have the current reality, but they will also have some idea about the major risks and how to re-jig or regroup if one of the major risks, or types of risks, occurs. Call the current reality, Plan A. Plan B will be a plan or set of concepts for how to move forward if minor or medium problems occur. Plan C will be a scheme for how to proceed if a major problem occurs. A good planner should not be caught flatfooted or without a strategy. On some projects, I have been lucky to use Plan A all the way through. On other projects, I have had to resort to Plan B. On some projects I have had to bail out of A, try B, and accept the harsh reality of Plan C. Plans B and C might not be detailed till needed, but they have the big picture and scheme somewhat thought through. As you are trying B, you try to think of a Plan D and sometimes, there might not be a Plan D. That is life. A successful planner will have these contingencies in hand and will know when and how to move from one plan to another. A planner must anticipate and think ahead of reality.

All of the above applies to any form of software I can think of. What is needed for Agile/Extreme — above and beyond the above?

24.9   Dancing with the Devil

I am not sure what the Devil dances like, but there are days that managing a *Mushing* situation feels like something un-earthly. And, I do not mean something of the nice variety. Night of the living dead and zombies comes to mind.

Assuming that you have the right personality traits to even start, there are three things to keep together on a Class IV–VI development. First, you have to make extra efforts to manage the client and the client's expectations. You also have to keep tabs on any evolving functionality and avoid scope creep. This is different compared to Waterfall processes and different again from the Class I –III situations if we use the white water analogy. In Classes IV–VI, you rely more on the user and impose more because there are fewer reference points and existing systems to leverage. Each experience is very unique and you need the local guide to help navigate and understand the situation. You also have to be careful about what feature or function to reveal in what order so that the user does not get unrealistic expectations set in their mind. Nothing worse than a frustrated user. So, managing the user interaction is different. You also have to think about the state of what you show the user. I do not recommend that you show the user software with debugging aids or output visible as they create noise and confusion. You should also be careful how much is missing as some users have a hard time following an interface or flow that is ragged. Anything that makes the user excited in the wrong way will increase the tempo of the dance.

Second, you need the tools and the processes to carry out the *Mushing*. In some cases, your plan has to include the creation of the basic tools and features. You need a toolkit that you can use rapidly and with confidence. If you do not have the right tool for the situation, you have three smart choices. You can try to find one, you can build one, or you do not do the project. If you try to do the work without the right tools, do not collect $200; go directly to jail. You definitely do not get a get out of jail free card. If you have to build the quick prototyping toolkit, you might have some long, awkward moments with the client while things do not appear to be moving forward.

Third, you need the right people and be able to manage them. The right developers and the right client representatives. In the Agile/Extreme, the team is composed of both parties. For extreme development to work, you need at least some old folks. The ones who have been through the wars, seen the good, the bad, and the ugly. You need depth and breadth in the senior people and I cannot imagine doing a high quality *Mushing* project at Class IV+ without some old, grey haired people from the client and from the developing team. They provide the balance and maturity to avoid panic attacks as rocks get close and they can understand the subtleties of what a minor twitch on the paddle might do. You also need some young guns from both parties, the ones with the energy, curiosity, and desire to drive close to the edge. They bring new ideas, alternative thinking, and the ability to ask why and why not. They will not be blinded by history and will not be afraid to do something different. You need to balance the team. Some of the old and some of the young. And I have to repeat myself because I feel that it is very important: you have to avoid the pedantic, literal-minded, and perfectionist developers, and remember, age is not a determining factor in these characteristics. You should also try to avoid these traits in the client representatives. They will be disruptive and cause great pain. On some projects, you will have a choice, and on others, you will be told who you have to work with, both for the development and for client representation.

This team design of high powered thoroughbreds, young and old, can cause more grey hair for the manager. It is very possible that some of the people will have attitude; some good attitude, some marginal, and some bad. For example, many are probably capable of being a lone wolf and they like to be the lead dog in the pack and they want to know everything, be part of every decision, and provide advice on everything. They want to air their opinion and sometimes whine and moan when their opinion does not hold the day. When someone is using a foundation based on opinions and not evidence, they hate my favorite word — *Why?*. The more at stake, the more *whys* will be heard. If you cannot defend and explain a request, do not expect your opinion to be held very high. Just don't say "We need this" or "We need that." Be prepared to say why, what might happen if we get it, what might happen if you don't get it, what might happen if you get part of it, what might happen if it is delayed, and what might happen if it does not work as you think it should. This is true for both developer and client interactions. If someone has thought it through, they will be prepared to answer the questions and will not mind being asked why. If they have not thought it through, they might be speculating and they will not like being challenged. They will be defensive and have attitude. Unfortunately, it is a planner's duty to ask why and to challenge. Everyone. All of the time.

The reverse is true as well. The planner has to answer challenges as well. If I cannot answer a *"Why?"* challenge, then I have to question the decision or the task I have specified. If I cannot explain a user feature or function, then I have to question the need for it.

As lone wolves, the folks might not be very trusting in the other wolves' skills and advice, and will spend time checking, double checking, and looking into things others are doing. Again, true for client and developer team members. Most are likely to be wired and impatient with any and every delay. As part of the team, you need to find some steady people — young and old — that will help the team chill and avoid an implosion. You need at least one or two people who will be able to de-fuse the bombs and also try to herd the wolves in one general direction without too much wasted time and resources. Specifically at the Class V and VI levels, you will need the following on the team:

- Someone who enjoys the bowels of the system, is particular about system robustness and integrity.
- Someone who enjoys the inner workings of the business logic from the client side and will leap at the chance to probe, document, and explain.
- Someone who is good at detailed, process work.
- Someone who will force lateral and out of the box thinking.
- Someone who is cool-headed and capable of being the peace maker and glue that will hold the project together.
- Folks who will just put their heads down and get on with the job, ignoring all else.

All the while ensuring that the people do not have the traits I have warned about elsewhere in the book. The size of the team will dictate how many of the above you will need. If the project is calmer and has lower risk, you can dial back a bit. Each of the above will play a key role in a high velocity development. At various points in time, the lead dog of the mushing team will change depending on the current need. And, as with any pack of high spirited sled dogs, you will also likely have the occasional heel nipping, barking, and some of the folk getting on the others' nerves. It is not always pretty. You can try to mediate and keep people from each other's throats, but be prepared for an ongoing balancing act of personalities.

A high velocity project is an unholy trinity — the client, technology, and developers. You are pushing the limits, taking risks, assuming that some stuff will be changed (note: in some cases, most of the stuff will change), and relying on the overall skill of the team to create magic out of sawdust. If you create a system from scratch, built upon a bare stack, without lots of black boxes of functionality, you need people who can build and who have built boxes of functionality from scratch. If you have things that look like operating systems to build, better make sure that you have people who have actually built operating systems before. If you are doing a finite automata, have someone who has built some before. DO NOT do *Mushing* with a team of newbies, new clients, new developers regardless of chronological age. You need to be able to craft and make things from scratch or fake things out or do things with proxies if you are doing rapid development at the Class V – VI levels. You need to grab a branch and use it as a paddle to steer. You need to be able to arrive on site and quickly react and modify your craft to the conditions. You CANNOT be the type of team that just uses black boxes, fills in tables, or uses pre-existing patterns. At Class VI, you are inventing and creating solutions and this causes a very fluid and dynamic management style. The situation will be a combination of business logic, or requirements, and the technical elements that can be built or pulled together. The same software development dynamics will be echoed in the business logic and processes. It is easy to focus on the software development only, but many of the issues will also exist on the client side and since you are a team, you must help each other and work together. You will not find the best practices for everything you want to do on the web or in a textbook; remember, you are creating something that is likely to be non-traditional and not recommended. You will likely find warnings and checklists that suggest that you cannot and should not consider doing what you will have to do. You have to be prepared to take calculated risks and do things others will say are unwise and foolhardy.

To manage in risky situations, you need to constantly read the situation. You need to view the track or water ahead, feel the current vibes, and go with the flow. You have to have faith in the team and knowledge about how things will react given certain pushes and pulls, and you need to know the limits and boundaries of your skills and tools. You do not simply think about the immediate situation. You are simultaneously anticipating what might happen next so that your muscles are flexed and you are on the balls of your feet. You need faith, and copious quantities of it. While you are not looking for trouble, you are ready to handle it if it appears. You are not looking to write an operating system, but if you have to, you can and will. The best developers are lazy and do not program for the sake of programming. They will do everything possible to go with the flow and only if it is really necessary will craft something. They are not hung up on the not-invented-here (NIH) problem. Nor, are they hung up on needing to leave their mark on what they are doing. In *Mushing*, there are some things worth doing right and many things that just need to be done good enough, and that is the attitude you need to instill in the team and in yourself. If you cannot let go and go with the flow, you will be beat up against the rocks.

From a distance, it might not look like you know what you are doing. The arms look like they are flailing about and there is frantic activity at various points. Much of the activity does not look planned or has a premeditated path. It looks as if the drug dosage is off a bit, a big bit. People will try to predict and anticipate what you are doing and will do and usually not get it right. They are not sitting where you are, and they do not see what you see. They do not feel the same things. You are in the driver seat and they are in the cheap seats on the sidelines. It is not possible or reasonable to explain everything to everyone all of the time in a Class IV–VI situation and this will make the control freaks very upset. The scary thing is that they are partially right. There are some aspects not planned, nor predictable on these rides. This type of process has that chaotic rush of instantaneous response. Thrust and parry. You really do not have all of the answers or know what will happen and it is best to always remember this. However, it is not complete chaos and mayhem. It is planned chaos and controlled mayhem.

To manage a *Mushed* project, you have to do lots of planning and thinking. The difference is when you plan and think and what you plan and think about. You are planning your tools, your craft, and they are prepared well for the journey. Every pole, paddle, strap, rope, clasp, and bit of the toolkit is checked and made worthy. You have ensured that training has been done for all of the crew or helpers. You have run the water or course in your head a few times, thinking about what might or might not happen. You have done the risk scenarios and planning. You have to admit to yourself that there are some things you can control and others you cannot. Then plan accordingly. The war can appear chaotic, but the warrior is prepared, with reflexes on automatic and the mind focused on the task while sensing the future. This is not for the faint of heart, or people who like extreme clarity all of the time. You have to accept and embrace change without worrying about what was said a week ago or a month ago — *that was then*; *this is now*. You have to be ready to plunge into the fog, enter the snow storm, and like the legendary warriors, be able to fight in the dark or when blinded. Why am I going on like this? Best I can figure is that I think managing a *Mushed* project is 90% mind control and thinking like a warrior in a heated battle and 10% thinking like someone doing a Class I cruise.

If the *Mushed* project requires interaction and interfacing with other support groups, extra care must be taken to bring the other parties online. Bring them to the timeline and to the style. Every organization and situation is a bit different, so I cannot provide many useful insights on this. It depends on how old the culture is, how dynamic and open the culture is, types of systems and developments that you have to interface with, and so forth.

One of the things you have to do as a project leader or manager is to help identify what will be done when. This leads into the next section: the work breakdown structure. This is the bit of planning that identifies what will be done and in what order. If you are doing daily builds and hope to keep it daily without a meltdown, you should do some serious thinking about the build sequence, what a work unit is and the significance of these decisions. It should not be a random process!

But first, there is another dance that has to be done; again with high heels and dancing backwards. This is the initial dance or startup phase. You might be doing a project totally from scratch with new people, new hardware, new facilities, etc., or organizing a new project within an existing organization that already has a background in high velocity projects. The latter is not too bad as there will be organizational knowledge, personnel knowledge, and established track records. If you are doing a startup from scratch, the dance is more difficult. Loose guidelines and approaches should be used initially to see what each player brings to the table, how they do project management, how they approach mentoring, what their process is for software development, and so forth. This is another risk management issue. You have to have enough faith in your team members that you know that at the end of the day things will be OK as you weather through some stormy periods at the beginning. If you are the main manager, you should have a pretty good set of ideas of your own about how things should be done and what should be done. However, you can learn from your staff and it is quite possible and likely that your startup crew will bring new and improved ideas to the table. You should be open to these ideas if they are improvements. You need to give the staff sufficient time to demonstrate their skills and processes before introducing any control mechanisms to re-direct or alter their methods. Old geeks may or may not be open to suggestions and it will be a delicate balance between what each team member's strength is and what the manager might think. It is also possible that you have made a bad hire and then you have to figure out how to control the damage this can create.

At the end of the day, the manager will be responsible for the project and will be in the hot seat; the manager must have the confidence and feel comfortable about what the team is doing and how the team is doing it. In this case, the responsibility is somewhat shared: the team members must recognize this and take on the challenge to prove to the boss that the processes and methods will work and they must proactively work on creating the boss's confidence. If the team cannot persuade the boss that all is OK and will be successful, it is likely that the boss will insist on some other methods and processes that he or she is comfortable with. Everyone has a comfort zone and unfortunately, the boss's comfort zone is usually more important than the individual developers' zones.

Chapter 25

Planning vs. the Plan

In the general area of research on planning and scheduling, the majority of work has been on the physical output: a sequence of tasks, each with start and end dates, resources allocated, etc. I like to call it the *Theory of Sequences*, rather than a *Theory of Planning and Scheduling*. What people do is planning and scheduling and part of their job is to represent the final decisions as a sequence. Software systems typically support the sequence portion and fail miserably at providing the human with any help in actually planning and scheduling. I once talked to the chief architect of one of the leading planning and scheduling systems. He had been hired two years prior to the chat. He had yet to go visit a user or talk to a user. I also talked to the product manager for another leading scheduling tool. Off the record he admitted that they really did not know what users did with their tool. This lack of awareness did not prevent either company from making grandiose claims, nor making money for a while. They sold to the higher level management, who were desperate for any silver bullet, and who bought in on the dreams without knowing what was really needed, feasible, or reasonable. Initially, both companies made loads of money from their product; I do not know if that was the same for all of their clients.

One of the key tasks in planning is figuring out what to plan. This comes from forecasts in some industries and what you have to build or deliver in others. For software systems, this comes down to something called a work breakdown unit. A work item might be done by one or more people. There is no magic rule that sets the number of people. A unit might take hours, days, weeks, or months. No real magic rule for that either. The size of the unit must make sense and provide sufficient visibility for progress, issues, and feedback. In Agile/Extreme and *Mushing*, a detailed work unit better be close to a 1/2 day or a full day at a maximum. You have to be able to start and finish units if you want to deliver on the daily builds. Some units might take longer of course, but in general, smaller is better.

A plan is made up of these units, with interdependencies and constraints controlling the sequence. Work might be constrained by finite resources: how many different tasks can be done at once? Work might be constrained by requirements: what needs to be done before you can start? This thinking is after you figure out the actual work. I have seen project management books suggest that you (if you are the erstwhile project manager) ask the domain manager for the work breakdown units and sequence. Asking someone else for the work breakdown and blindly accepting it takes the fun out of managing, not to mention that you just turned into a clerk. You might need help with some of it, but you should be able to do the majority of this yourself at the level you are responsible for. In mathematical formulations, the so-called optimal plan is calculated by relatively sophisticated mathematics using a set of information about the task such as how long will the task take, who is needed, and information about other tasks considering what else requires the same resource, what needs to be done first, next, in parallel, etc. You might even include costs and penalties for fun. This information is processed by the math and out of the meat grinder pops a mathematically superior sequence of work based on some objective function: minimizing costs, delays, or whatever else you can quantify. The main inputs are the unit and the relationship between units. The act of taking this minimalist set of input and creating a sequence of various granularity is called planning or scheduling by the majority of literature. The creators of such tools love to call the sequences *feasible*. That they are. They are mathematically feasible using a limited set of inputs. They are rarely feasible in the operational sense. There have been very few studies where mathematically generated sequences have been followed in a longitudinal study to see if they could indeed be executed and followed as the project or process evolves.

To a planner or scheduler, if the plan cannot actually be executed, it is not feasible. As far as I know, science is not at the stage of creating operationally feasible schedules yet except in very few situations which are either exceptionally trivial or almost 100% automated in some way. The situations I work in cannot use math. Math has a hard time creating feasible schedules in the real-world sense, let alone optimal ones. What do I mean by operationally feasible? I might have four junior programmers and on paper, a junior programmer is a junior programmer and is plug compatible with other junior programmers. They might have had the same courses and the same training. In the database, they look the same. But, I know that one is a party animal and Mondays are not a good day to ask this particular programmer to do anything critical. I would have the other programmers do critical work on Mondays. The one programmer is OK the other days except Friday afternoons when he is thinking about the parties on the weekend. If an automated planning system assigns the risky item to the party-oriented programmer on Monday, I would not consider it feasible. Another example? I know that the compiler will be updated next week and there is no option about the timing as the upgrade fixes a number of critical flaws. The compiler is not really a task on the plan and the planning software does not know if this is V1 of the compiler or V2. The planning software does not even know that there is a compiler in most cases. I would not schedule any demos or important tests for the immediate few days after the compiler upgrade.

Thus, the plan is made up of units representing the tasks or operations sequenced by the act of planning. The key are the units and their characteristics. What goes into a unit? One aspect is size and effort. For this, I refer you back a chapter or two. When I get down to the close horizon, I like to start thinking in the smaller chunks of time, especially while *Mushing*. Remember: design, code, debug, test, integrate, and cap. This controls the size. Another aspect is the *what*. This is more tricky.

Try to think it through like building a shed or a house. It is easier to build the foundation first than to build the top floor, raise it up, and then build the foundation. You think about the major structures — beams, supporting walls, and such. You should not be thinking about wallpaper or rug color while you are still determining the snow load. Try to sketch out the system like an architect would, with different layers and overlays. One for plumbing, another for wiring, one showing doors and windows, etc. This will help you figure out the *whats* you have to build, and the major sequence or phase in which the work should be considered.

There might be some work that must be done that you have no choice over. You need to schedule this in as appropriate. The work might be the key elements used by many other bits and you have to do it first. No way around it unless you want to stub out everything. You should still think through risk and we will deal with that in the next paragraph. If the software component is not a key feature, there might be discretion. For these, you could do it on Tuesday or Wednesday; it does not matter. When you have discretion, how do you figure out the trade-offs? I use risk to sort this out.

It is a casual observation, with absolutely no scientific basis to support it, but when I look at some plans and work breakdown structures, they do not appear to be created with any rhyme or reason. They look somewhat ad hoc with work bundled together like a dog's breakfast and units that did not make sense. I do not think that it need be so random. I worked with a Master student, Jennifer Jewer (2003), on some ideas for guiding what goes into a build and the sequence of things when you have discretion. Discretion means that you can choose the timing and sequence. Her research used a risk oriented taxonomy and then she did an experiment to see how easy it was to teach the ideas and be understood by the typical Computer Science senior undergraduate. Turns out that figuring out how to decompose a problem into a series of daily builds was not so easy, even after being exposed and taught about how to think it through based on the taxonomy. The major difficulty is the old experience one. Not only does the planner need to know what is flexible with respect to sequencing constraints, to craft a high velocity build sequence requires the builder to appreciate, recognize, and then respond to risk factors. And, these risk factors are hard to explain and hard to understand. It is like telling someone how to ride a bike. Or, how to kick a carefully arching soccer ball over the head of the defender and just catch the upper left hand corner of the goal. Please provide a recipe for this, including the situations in which it is reasonable to attempt it, and please explain how the kick actually feels! It seems that it is hard to see a risk or appreciate a risk until you have encountered it once. Or, you have had a similar base case that you can extrapolate from. Why shouldn't you touch the hot burner? Why should you wear protective gear when playing hockey or American style football?

Jennifer worked with me to develop an approach that should be somewhat obvious if you have understood all of the stuff so far. We tried to quantify it, but the ideas can still be used informally. For Class I or other simple systems, you can do most of this in your head without much effort.

There are three big criteria in deciding what to do when.

o The first is essence — what the function or feature means to the user. This provides a relative benefit measurement. From page 44 in her thesis: "The Essence criterion is used to determine the relative benefit that each requirement would potentially offer to the end product. It indicates the degree to which the success of the project depends on each requirement. The *essence* of the software represents the minimum requirements necessary to satisfy the basic needs of the users and clients. An important aspect of the Essence criterion is that the essence of each requirement is determined independently of the technological constraints of that requirement."

o The second is a risk assessment — the relative risk associated with a feature based on the technical and project risks. <u>Project Risk:</u> refers to the risks related to the management process or business risks of implementing the requirement. <u>Technical Process Risk:</u> refers to the risks inherent in the development methodology or tools used to construct the software. <u>Technical Product Risk:</u> refers to the risks inherent in the functioning of the software for that requirement.

The figure below illustrates this:

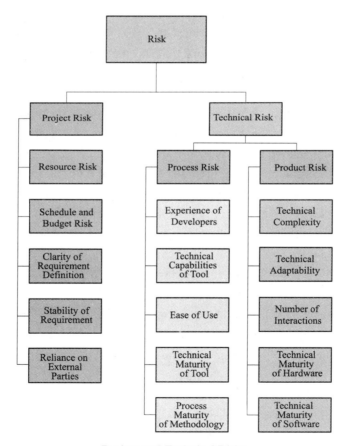

Project and Technical Risks

o The third is an independence criterion — how the pieces are interrelated based on benefits and technical interplay. From page 60 in the thesis: "A requirement that should be implemented *before* another either because of technical or benefit interdependencies would be assigned a higher priority than an 'independent' requirement having no benefit or technical considerations constraining or promoting its implementation. For example, a requirement with the most dependants, and thus having the biggest effect on the rest of the system if it changes (i.e. infrastructure) should be implemented before another requirement that does not affect the system functionality or require other aspects of the system in order to function."

Again borrowing from Jennifer's thesis, the prioritization approach is summarized as:

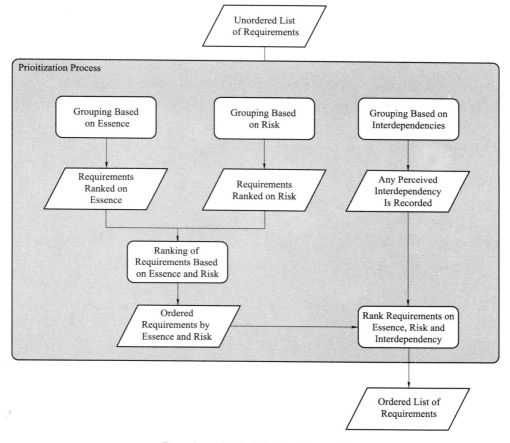

Overview of Prioritization Approach

The steps are:

1. Determine the requirements to be prioritized.
2. Select the developers to participate in the prioritization.
3. Individual scoring of requirements against the *Essence* criterion.

4. Individual scoring of requirements against the *Risk* criterion.
5. Ranking of each requirement based upon *Essence* and *Risk* scores.
6. Estimate the *Interdependencies* between requirements.
7. Assign each requirement to a development build.

I will not be repeating all of Jennifer's work on how to quantify and assign priorities. Suffice to say that to get to the end of step 5, you can assign simple rules like low–medium–high for tri-state levels or go for five or seven levels. Try three first. In a spreadsheet, itemize the points, assign the weights, calculate the priorities, do some aggregation if you want, and then sort by the three criteria — essence, risk, and interdependence. Step 6 can be done using paired comparisons of the features and for a large project, you can do some math and logic to determine the path and what depends on what. Step 7 is the hammer — where skill and expertise really comes to play. "The output of this process is a development schedule. The schedule outlines the order in which the requirements should be developed, so that the requirements with the highest priorities are developed first."

This process will help you figure out the key pieces of software needed to deliver the product, and the risky elements. You want to deal with the risk ASAP to allow you to sleep at night and to give some warm and fuzzies to the client. You use this information to inform your decision making. There is no single way to look at all projects. In some projects, the interdependencies will override other considerations, or the functional risks are higher than the technical risks, or the technical risks will be higher than the other risks. The point is to have a systematic way to look at the work to do and to create work units that can be defended and explained.

It is not a magical, brain-dead approach. You need to think through if a technical risk must be resolved before you can deal with a functional risk. You need to think about the number and types of risks you will try to address in any work unit or build cycle. I recommend one major risk at a time, thank you. You do not want double jeopardy. Double jeopardy is perhaps OK on a Class I or II run, but NOT on a Class VI. If nothing else, it helps you understand and think through the development and minimizes the risk that you have forgotten something or that you have delayed something that should have been done sooner than later.

In a Class V or VI type project, you might have all kinds of risks. You might have risk on everything dimension and you have to consciously assess the risk and manage it. You should not blindly ignore it or pretend that you will escape unscathed.

In general, you need to be open-minded while planning and executing, to the point of using evidence based reasoning or decisions that can be justified. As a high velocity project proceeds, you will find initial assumptions changing and you need to adapt and change — *when the facts change,*

change your decisions and opinions! This goes back to what true experts do. They do not hold onto something past a best-before date and they do not become emotionally attached to anything in the project. Beware of opinions though. As a team lead or manager, I will rarely change something based on someone else's opinion unless they are the client and even then I have a strategy for how to deal with that. I have a deaf ear to "I think" or "I know" or "I like" statements which are not supported by some form of logical, supported reasoning. Everyone has an opinion about everything it seems in software development. Why is one person's opinion, based perhaps on a singular experience, better than another person's opinion? If I am accountable, I know whose opinion I will listen to if I have to listen to any opinion at all in the absence of evidence and reasoning — mine. I can probably appear to be a dictator or a wet blanket to people when they come up to me, give me some advice based solely on opinion or thirty seconds of thought, expect me to follow that advice just because of who they are, and I do not. But, give me data, give me examples, and give me facts and I will change decisions and plans.

I think that planning and designing go hand in hand when doing Class V or VI developments. You have to plan the journey and know enough about the tools you will take with you to design your *package*. One of the topics that will rise up is best practices and common usage. This will refer to the tools you choose and how you use tools. Most of the literature will be talking about Class I or similar situations and that is fine. They are common and people need guidance for them. They are easy to describe and since one Class I looks a lot like another Class I, comparisons and benchmarks are possible. However, in Class V or VI, there are few benchmarks and you are likely to do things people consider un-orthodox, not acceptable, weird, stupid, and crazy. Because: there are few Class V and VI situations, relatively speaking and there are no recipes to follow!!!! This should be obvious to all involved. You will have to bend, fold, mutilate, staple, glue, and do things others have not done just to survive. You will use tools in ways that the creator of the tools has not imagined. You will not be on the travelled path. This is the nature of the beast. Everyone does this. Everyone says this. The gurus say this and that.

You need to assess all of these things in the context of your Class V and VI situation and do what needs to be done and what makes sense. You need sanctioning of course, and you should not be doing different things just to be different. People paying the bills and the people who will have to support the system need to be part of the process; we are not talking stupidity. But, you need to have a thick skin and sufficient skills and knowledge to make these types of decisions. You need to know when to get off the traditional path and you need to know that you have the skills and knowledge to create a new path without killing yourself or the project. People will make predictions, take cheap shots, and possibly say nasty things about you and the team. I have had other senior developers go complaining to the senior engineering management about me and my team of idiots and how our ideas would not work and that we were doomed and wasting the company's money. In one case, we had working examples in the back room of everything the other senior architect said was not possible to do. My common response to most criticism is that I am too stupid to know what I cannot do.

Back to the client who has an opinion that you do not think is quite right. In this case, I rely on my toolkit and rapid prototyping processes and in most cases I can build A and B and let Darwinism take its course. I try to not get into arguments or debates with the client. This is not wise and not good for long term relationships. It is better to quickly build two versions and let the client play with both. Let the client decide. Often you might find that the final idea has a little of A and a little of B. It is hard to get to this point debating around a conference room table. If you have the right tools, you can actually demonstrate the choices in less time!

Remember, you need to be able to take a plan and re-jig it while it is executing with a minimum of disruption. This is the planning part. You take a task and make it into several; you re-assign people; you delay tasks; you bring tasks forward in time; you cancel some tasks, take several and make one, extend an estimate, shorten an estimate. You constantly review the horizon.

Here is another suggestion. I often recommend a 2-4-6-8 policy for project managers or any kind of manager. The precise measure will depend on the project and what is being managed, but the idea is the same. Each week, you have a specific sit down with your war council and think about the future. What is happening in the next two weeks, four weeks, six weeks, and eight weeks? Any special events? Milestones? Any changes in the status quo? If you identify a possible situation, react to it in advance, not after the fact. This is *planning*. You anticipate the future, identify difficulties and discount them. You take responsibility for the future and do not play victim.

Part of planning is knowing when to stop doing X, when you have learned enough from X, and knowing that you are climbing a tree trying to reach the moon; you are making some progress and showing value but it is not going to work out in the end. This is what I referred to as the *stopping criteria* when discussing expertise. When building a Class I–III system, you should not be throwing out much code. In these well understood situations, doing major refactoring should be seen as possible incompetence. In the Class IV–VI situations, you need to be prepared to do lots of prototyping and building, and plan to throw out various pieces of code. In moderately risky areas, you might only have to do something twice. In very turbulent areas, you might have to take several runs at the problem. You gain value from the attempts and the value should be realized in the next refactor, but you should not assume that you will make the right choices the first time and that the code will survive. For this style of exploratory software development, you cannot have an overbearing ego. You have to look at the situation philosophically and think about value gained, and not think about the code that is on the floor. As a planner and manager, you have to assess the code and code development; you need to know when enough is enough. A programmer might not like his or her code not being used and consider it a sign of disrespect. A programmer might not like all of their coding and their effort not being used. They have to remember that the project is not about them, it is about creating a system for the client, and that they should feel good as long as their effort had value along the way. This is hard for some egos. For the high end projects, your planning needs to take into account more refactoring.

Chapter 26
Aversion Dynamics

Aversion Dynamics? This is a term I coined in the early 1990s with a colleague — Thomas Morton. It refers to a category of heuristics and ideas based on a simple observation that most skilled planners and schedulers spend a great deal of their time managing risk in their assigned activities (McKay and Black 2006). They have an *aversion* to risk! They do not like risk. Most do not do things risky for the fun of it. They do not like the extra work and blame that comes their way when something risky blows up. The best, the absolutely best, planners and schedulers are lazy and they want to create plans and schemes that will just work without any sweat and anxiety on their part. It turns out that they actually do work hard and do lots of work, but it is an advance type of work, to avoid the not-so-nice work that might occur when everything goes wrong. They do not like being contacted in the early morning hours to sort something out; they like their sleep; they like to eat their lunches un-interrupted; they like their social times, etc. It also turns out that smoothly flowing enterprises are often very effective and efficient. You can show mathematically that this idea of being averse to risk and doing things about risk is a good thing.

The *dynamics* implies that the topic is not static. It is more like a dual stage control mechanism. Each stage in the control mechanism is actively sensing, using control rules and ideas to alter the system to control the output. The control logic of the lower level is the output of the higher level controller. Risk is not static and the control mechanics are also not static.

I did not invent the concept of being aware of risk and being smart about it. People have been doing this for a long time. I did observe the practice, did it myself, and did some *science* on it. I think I was the first to really study these precise ideas, but that is by luck and not by design. I created some math with Professor Thomas Morton (Carnegie Mellon University) in the early 1990s and have continued with this stream of research ever since with some other colleagues. The basic idea is to take a conservative posture around risk through sub-optimizing, and then return to an optimistic outlook using traditional optimization ideas once the risk has passed. The current *Aversion Dynamics* set of heuristics look at scheduling, dispatching, and inventory control, to illustrate the concept and issues. The planning math is a double mathematical heuristic. One level determines the base priority for a task using Bottleneck Dynamics (Morton and Pentico 1993) and then an *Aversion Dynamics* function is on top of that base function, modifying the priority. The second order math uses ideas from heat loss theory to decay the modifiers. Lots of exponential decays. The math is designed to function in cases such as: a machine was just repaired, but there is an identified risk that the technician did not fix the machine correctly, so anything that could be messed up and cost a fortune should be delayed a bit and other work performed for a shift or two. Or, if you have not made something for awhile or have done something, you realize that things might have changed such as patches, upgrades, different people, or documentation has been lost. It makes sense to make a little of the product before you have to satisfy the whole order. The more risk, the earlier you start and the more tests you do before you start the main part of the task.

If you sub-optimize and are careful for a bit, and the problem does not occur, the cost is not too big. If you sub-optimize and the problem occurs, you get a big gain. If you do not sub-optimize and instead decide to lead with your glass chin, you will suffer. In a sense, this is similar to creating toothpick code or creating key chunks of code to test an idea before investing more effort.

Without going into the math models, some of which mathematically model the bunny hop diagram introduced earlier, there are some basic observations about the ideas behind *Aversion Dynamics*.

- o There is risk. Surprise!
- o Even if you do not like risk or want risk, there is risk. Even if you mandate that risk will magically be absent.
- o There is potential risk with any change in the status quo. It does not help to pretend that there is not.
- o Some risks and issues can be predicted reasonably accurately, some not. Not all perturbations and risk are 100% random.
- o You might be able to predict an event, but not the impact in advance.
- o You might not be able to predict something accurately to the day, but you might know it will happen sometime in a certain week or a certain month.
- o There are some things more sensitive to risk than others. A rubber ball can take a hit from a baseball bat; a pane of glass might not survive.
- o Depending on the situation, you can try to avoid, minimize, or otherwise mitigate risks. You can do things ahead of an anticipated risk, and you can react to risk that you are experiencing.
- o You can take exploratory excursions to test the waters and see if the risk will bite you. If you have not done something for a long time, do not assume that the world is as you knew it and that things are the same.
- o A great deal of risk and damage to one thing is a lot better than the same amount of damage done to many things.
- o Risk and damage occurring during peak loads or critical time seems to multiply like rabbits as people become frazzled, start making quick decisions without thinking, and run around like chickens with their heads cut off.
- o You can usually find things to *absorb* risk. These are things that you can do or work on when you suspect that there might be risk. These might be items which are cheap and

bountiful, or do not damage accumulated value-added, or items that will not take a lot of time and effort to fix if a problem does indeed occur.

o You can consciously establish policies for avoiding risk. You do not introduce changes on Monday, Friday or evenings, or afternoons. Introduce changes on Tuesday, Wednesday, Thursday during the day shift! And, remember to do it when the key personnel are around, not on training, vacation, or sick.

In a nutshell, it is commonsense. It is not rocket science. But remember that there are two types of commonsense. There is sense that is commonly held about a situation — the world was indeed flat at one point. The other kind of commonsense is good judgment in common situations. I vote for the latter. The best quote about commonsense I have seen, but cannot remember where I saw it while doing my PhD was: *if commonsense is so common, why is the lack of it so often lamented?* This highlights the difference in meaning.

Why so much personal interest in risk and how people deal with risk? My Master's research looked at the information flows across schedulers' desks and this set up my long term research agenda. The planning and scheduling looked like a cognitive skill, and the majority of the time spent by the schedulers was focused on risk management, not making sequences. Hence, the twenty-five years of thinking about risk and risk management.

Chapter 27

Reliance on Technology

This is another brief chapter. I have observed project managers and developers rely too much on technology and methods for planning and for doing. They rely on it to do their thinking for them.

There are some tools and processes you do need. Do not get me wrong. But... for Class IV–VI situations you should be able to plan, think, and design without fancy gizmos. For example, in a high velocity development without formal requirements, hand drawn diagrams will suffice and fancy computerized artwork provides limited added value. Or, when you are debugging and you hit a bug, what do you do? Do you sit back for a bit and think about the situation and think it through or do you immediately start banging away with the fancy debugging tools. Personally, I prefer the former and I like to isolate a problem based on the behavior and situation. I try to find the problem with a few strategically placed probes. Sometimes I have to use the debugging tools, but I do not rely on them to do my thinking.

Debates about the software development process have always been about deciding what to do when in the process, and the tools to be used. There have also been lots of fancy methods and tools developed to help the process. You have to be able to assess a situation and figure out what a tool or method will actually do. Will it save 30 minutes a week? Two hours a day? An hour a month? What is the cost and time effort required to create, acquire, deploy, and train on this tool? Is it worth the required effort to save the 30 minutes you might have to expend once a year? So, I usually challenge people. Is there a manual way? Are there other things to do while we get the tools in place? For example, in small teams of three to four people, you can manually coordinate version control without much effort. You can even put the key information up on the wall. Who is working on what, who does it go to next when it is done, etc. to avoid one person clobbering another's effort. When you introduce a tool or new technology, it will take time to learn and this time can take away from key activities on the critical path. You need to be smart about when you introduce new platforms, new environments, and other tools. There is a cost!

A junior developer might have to rely more on technology and methods compared to an older co-worker. This is to be expected. I have more patience with the young. I have little patience for the supposed experts who say that they cannot develop, or do something **WITHOUT** their favorite programming language, their favorite operating system, their favorite email package, and so forth. Zero patience for people who refuse to use a common tool or team convention just because "My friends will make fun of me" or "I don't like it" or "It is too tedious." There are some good points about being the prima donna when it comes to basic skill, but you will not likely find your favorite sparkling water and flowers in your room every morning when working with me. You have to be prepared and able to work out of a knapsack or briefcase occasionally and use whatever tools are available; just get on with it. There is no time for whining and moaning about some piece of technology that might not exist or is not up to the latest best-practice edicts from a specific web-site. I have observed people whine and complain longer than it would have taken to do the thing that they were whining about. Some prima donnas are like babies and I do not change diapers. No one ever spoon fed me or babied me on a job either. I doubt that I would have got done in my career what I have got done if I required tender loving care and regular diaper changing. My mentors, teachers, and bosses did not have time or the temperament for it, and neither do I.

For *Mushing*, you have to be agnostic and be willing to rough it. You need to know how to handle the basic tools, the hand tools versus the power ones, and if you know how to use your head and fundamental concepts, most of the basic tools will get you very close to the fancy ones. Possibly even faster and with less effort. For long durations, you do need some creature comforts. In a start up mode, you might have to rough it till you have the collaboration suites, automated testing tools, nice libraries, and all of the gear.

You definitely need technology and other aids to help prevent eye strain, strain on the joints, bad backs, and other considerations. This is smart. However, there are times that you will just have to do without some luxuries for a while and deal with the bare necessities knowing that a few weeks without an extra monitor, or special chair, or special keyboard, or special software will not kill you. It is nice to use those fancy Japanese toilets with heated seats and artificial intelligence, but when you are in the woods, you will just have to do without. However, there are some programmers who think that their brain is attached to a piece of furniture, or a special light, or a special piece of software that they just cannot do without. They just can't get into the programming or the task unless the situation is perfect. They cannot do good code unless they are in a nice warm cocoon. They are not code warriors — sorry. In the early 1990s, I remember seeing grad students at MIT using labs and office furniture from the 1930s. At least it looked that old; there were new things, but in some of the older buildings, it was anything but modern in terms of nice office space. They were able to do great science and work independent of the type of chair or desk they were using.

If and when I have to deal with people who require *things* before they can deliver the goods, and have lots of excuses and obstacles why the work is not proceeding, I will work through the excuses one at a time. Oh, you need that chair? Here is the chair. Now what. Oh, you need that software before you can code? Well, here is that software. Now what. What is the next excuse? I will use their excuses to paint them into a corner. I have talked with many other managers who do the same thing. Luckily, there are few developers of this ilk, but they do exist and love any excuse not to do the work expected of them. They talk the talk, but can't walk the walk. They ALWAYS have a reason why they are late; the job is not 100% done yet, whatever. They might have the concepts and be able to debate and criticize, but they cannot deliver.

You should use technology when it really does something that your brain and body cannot. For example, I do not rely on technology 100% to help me with the rapid coding for *Mushing*. I also rely on my ability to touch-type to quickly produce code, and I rely on my memory for chunking and regurgitating in real-time: algorithms, patterns, and code sequences. Remember that I have been doing this rapid coding approach for many decades, way before fancy development environments. I rely on tools while doing rapid coding that help me see and think about code behavior and I diagnose problems from a high level.

There are other things that I will rely on technology for. I will rely on technology for version control on larger projects. I will rely on technology for things like code-coverage statistics on testing. I will rely on technology for team collaboration. But I have been doing that since 1978, so that is old hat. But I do not rely on technology as much for small teams or informal developments. As noted above, I would prefer to rely on technology that helps my thinking and that helps reveal software behavior and issues. I do not rely on software that merely generates piles and piles of useless data, containing little or no information content. I do not rely on technology that tells me the obvious. I am not a Luddite, but I will not use technology for the sake of using technology or just because everyone else is. It has to provide value. I do not let other people do my thinking for me. I do not get excited by technology that helps me save a few minutes a day. I do not upgrade my systems to just get 10% improvement. I am interested in the tools and technology that will save me hours a day or many hours a week. That gets me excited. Double the speed of the computer's CPU and double the memory? You have my interest. But not when the applications I use seem to run at the same speed. If the software is double the size and twice as slow, where is the gain?

When assessing tools, you have to think about how often they are used and what they are used for. If something is used only once a week for a half hour or so, a 50% improvement only means 15 minutes! If you use the tool many times during the week and each time you save 15 minutes, now we are talking. And, don't forget: it does not make sense to have a bunch of highly paid developers debate something trivial for hours that will take minutes to do, or that will have low impact. The next time you are in a meeting, count the people and do a rough estimate of their salary; how much is the meeting costing per hour? Is the meeting worth it?

Using libraries of standard methods and routines is also a form of technology usage. I will use ready-to-use libraries to bootstrap a development, and then thin-wrap them. I will also use libraries without any alterations. It all depends if I can get what I want done. If I have to, I will create my own libraries. How much I create and modify depends on the mandate I have and the risks that will be implied. Sometimes you have to use the plain version without altering it. Sometimes you can take more liberties and roll your own. This is part of the project management strategy. I try to avoid libraries that are changing rapidly and seem to be fads. I also avoid libraries that I cannot understand or that hide things; I like transparency.

Chapter 28

User Interface Principles

I can read your mind. What is a chapter on *User Interface Principles* doing in this part of the book. Should not this have been back in the design part? In a more traditional situation, I would agree. However, this book is not for traditional situations.

In a *Mushing* project, you are doing the UI design during the actual coding phase and you will have to be thinking about good UI practices in the heat of moment. The first part of this book should give you the big principles and concepts from which to design. This section is a little bit more detailed. And, I am not going to repeat all of the UI principles that have been written about since the early 1970s. I will give you my general checklist for the detailed level. This is above the value, comfort, experience, and evolution notes from earlier chapters. You need to think about the UI design each and every day you do a daily build that affects the UI!

In addition to the obvious suggestions of minimizing clicks, not cluttering, and not making the screen look like a Las Vegas casino, my top dozen or so list is:

1. Be consistent. The system should look like one person designed it, not a horde. Consistency on user interfaces was noted a long time ago. One of the first (if not the first) interactive systems in the early 1960s (I think it was the Atlas system out of the UK, but I could be wrong) discovered that users got frustrated by inconsistent times. So, they built in a delay to slow down the fast to match the slow. The user frustration dropped.

2. Organize functions by major function and then minor function within major function. This is simple grouping, putting things together that work together!

3. Place the stuff on the screen that matches the basic cognitive processes of the culture you are creating for — e.g., flow from upper left, down to lower right. And, have the same types of buttons in the same place on different screens.

4. Think about where you are coming from and going to. Where was the user looking, where were the fingers, where was the mouse; on the previous screen, on the next screen. Screens do not exist in isolation.

5. Specifically, considering the above points: design the mouse movement and options that reflect common patterns and repetitive sequences: minimizing extreme left/right, or top/bottom movements. Try to avoid repetitive stress situations, not induce them.

6. Design the screen flow based on a user scenario and common functions so that the flow makes sense. Set up for the most common options and settings. If the vocal and influential few want something odd or exotic that will be used very infrequently by the average user, make the few pay the price of extra clicks and menus and visual clutter, and not the masses.

7. Ensure that the user knows where they are, where they have been, and where they can go. You should never be lost when looking at a screen.

8. Be transparent and check your assumptions. If I go randomly to a page, can I tell what I am to do and how to do it? What the user can do should be obvious. If you have neat features that are not commonly found, better make sure that the user knows what they can do.

9. Remember that a significant number of users are color blind. Avoid instructions like "Press the red button." Do not rely on color codes to help guide users.

10. Not everyone will know your cute short forms and acronyms. Not everyone sits with you and shares a coffee chat.

11. Use the real estate wisely. Allow things to shrink or collapse if they are not used. Do not put something somewhere just because there is some free space. Think it through.

12. Do not keep sizing and moving things dynamically, making the screen bounce about. Hold the visuals firm when possible. Do not keep forcing the user to hunt for the buttons and components as they float around the screen. Screen stability is important, especially if the use implies that the user might park the mouse over the button and use it repeatedly while looking elsewhere on the screen.

13. Remember that people can become blind to things on the screen and will ignore items. Do you really see banner ads, or changes in a message when there is a message on the page, always in the same place, in the same font and color? Personally, I will pay attention to things when they appear where nothing is expected, and in a color or font I am not expecting. I will not see messages and ads that are always there. I can block those interface components. Put normal information in one place, medium stuff in another, and big, important stuff where you cannot miss it!

The style of user interface and usability should depend on who is using the system. If the user is periodic and for short periods of time, this can imply one type of interface. If the user is periodic but uses the system intensely for a period of time, that is a different style. If the user uses it daily for many hours, yet another style. If the user wants to use the system and comes to it without mandate or force, that can imply another style. For example, if you are building a system that no one really wants to use, you need to design the system to allow the person to get in, do their job, and get out as quickly and as painlessly as possible. You can have general concepts and principles, but you need to interpret them in context. Just because a certain style is used by a social networking site does not mean that it will work or is appropriate for another purpose.

You also need to be careful about assuming what other people did before creating or arriving at what you see. You might assume that a vendor or company did due diligence, or analyzed the situation, or thought about the user, or did user testing, or that the interface was compared against others and considered the best. It has been my experience that many software interfaces and other products are designed with little thought and evidence; they are usually designed based on personal opinion and sometimes with a wrong set of assumptions.

Chapter 29

The Toothpick

At the beginning of the book, I introduced the concept of the toothpick and iterative construction. The design and use of the toothpick idea will vary with the type of project you are doing. For example, you might do an initial toothpick of some very basic things. Then you might do a number of other toothpicks (some based on the original, some not), and then merge them back together again in a super toothpick.

What should go in a toothpick? Hard to say. It needs to support the Zen of the system and create the strength needed to counter the initial risks that might arise. You need the basics for the foundation, the main supporting beams, and the interfaces that will allow the components to work together but in a decoupled way. The toothpick is not just about functions and techniques; it is also about how the code is structured.

The goals for each version of a toothpick should not be too large or ambitious. You should always have a toothpick you can use as a demo and the goal should be a working toothpick each day! In terms of code coverage for testing, my target for a toothpick demo is about 60%. The code does not have to be perfect and not all functions operational, but it should be good enough that you can follow the major path and give a demo without crashing. You should design the user interface with a minimum of back and forth to background files, or manual typing to set different options. If you want to show A and B in the toothpick, make the switch part of the UI. It is important to remember that during a demo, any extra geek work will just confuse and possibly worry the client. They will not be able to interpret or understand the impact or meaning of your actions; what to worry about and what not to worry about. In the worst case, they will doubt and worry about everything they see that they do not understand.

In a Class I–III situation, you should be able to knock up a pretty good toothpick in 1–2 days. If it takes longer than that, you are not in the right situation to do Agile/Extreme, or you do not have the right tools, or you think you are better than you really are and you should never have accepted the job in the first place. In a Class IV–VI, a toothpick might take four to six weeks, or even longer. It will depend on how large the system is and how dangerous the journey is. Branches of a high-end toothpick may take 2–3 days to create, but there should be a daily target for the main base and the core of the toothpick.

If you are using the toothpick idea, I would suggest that you have a larger diagram of the whole system and as you do the toothpick, you illustrate to the people doing the work and seeing the demo, what parts of the system are being crafted and tested. This includes both technical and functional components. Personally, I like the first toothpick to take a vertical slice through the system, top to bottom. Then, additional toothpicks address the major limbs of the Xmas tree.

I have found that most users can appreciate and understand the toothpick idea. It helps them think that the task is indeed possible. It also helps the user loosen up and not be afraid of asking for things that they might have thought (or were told) were impossible. Once the user can start to feel the juice, they will start giving new requirements and providing additional insights. This is why I like to quickly create a toothpick, sometimes within a day or two of first meeting the user. The toothpick will have the basic Zen elements and demo the building blocks at a macro level. Then I can ask the user, what they think they could do with such things. They might not have been able to dream up the blocks or craft them, but I have rarely met a user who did not quickly think of ways to exploit them! They just need a little help to start the process.

Not every client or user will be able to understand, appreciate, or work with a toothpick approach. This will make life exciting and interesting if you have chosen this path. It comes back to picking suitable users. If you have quite a number of users that you have to engage, you might have some who are OK with this approach and others who are not. Given a mixed group, you might have to put your head down and just proceed, knowing that some users will not be very happy with you or the project. The same may be true for the development team.

The development team must understand the philosophy of toothpicking and the need for delivering to the client certain types of functionality in a certain order, to a certain level of completeness, and to a certain level of quality. When you are showing the client the work, this is not the time for debugging, hand waving, and testing. You may have the wrong functions, or the wrong functions on the wrong page, or the function needs tweaking, but you should be able to show the client the function without bugs, crashes, and excuses. The development team must have a single definition of what is good enough and what quality means. Yes, you have not done 100% testing, and there is likely to be a bug, but you should be able to do most of the client interaction without bugs appearing rapidly and repeatedly. That is sloppy development and demonstrates to the client that you, the developer has a cavalier attitude towards their time, effort, and concerns. This is not good and can be interpreted as you being ignorant and arrogant. Before you show the client the function, you have to do enough unit testing and functional testing so that you do not waste the client's time. Their time is ALWAYS more important than your time. This reinforces the need to incrementally build up a solid base that you can little functions to quickly, accurately, and reliably. As you design and reflect upon interfaces and methods, you need to think about how the methods and interfaces will be used in a toothpick way.

If you promise a toothpick in a day, you better deliver. If you say two days, deliver in one or two days, not three. Even if the user is not able to see the demo immediately, let them know that you have it ready for them and have-laptop-will-travel is ready and willing to show them some nice functionality. You cannot blow the user's confidence during the early stages. You need to do everything possible to meet your objectives and your promises. This implies that you have reasonable skills at quickly understanding the Zen, knowing the tools you use, and being able to estimate how long it will take to create the toothpick. I am very serious about this. If you fail to deliver what you promise, the user's confidence and trust is lost and is hard, if not impossible, to regain. Sometimes you will lose the client's trust and involvement through no fault of your own, or because of a weak team member, or because of other processes or politics; regardless of cause, you should always plan or be ready for the situation of client engagement or client disengagement.

When you deliver the software when expected and as expected, the upside is that you have a trusting client who probably has just been given a big emotional boost. Routine delivery will also help you to develop a reputation for pulling the rabbit out of the hat, consistently and reliably. I am very careful about what I promise and this is a key to being able to keep promises. I do not widely inflate estimates either. It is hard to inflate a one day toothpick promise! However, if I think it will take a week or two to deliver the toothpick, I tell the user two weeks. The one week part is what I think it will take and I have a few risk factors to deal with; I itemize them in my head, think about how long it has taken in the past to sort out similar risks and add that to the estimate, giving two weeks. For my own personal deliverables, I have been reasonably successful delivering toothpicks, iterations of toothpicks, and having good client involvement. When I have been working in a team or managing, I have had mostly successes, but some failures too. It is hard to predict if the Agile/Extreme and toothpicking is going to work. You need the right client, the right development team, the right processes and tools, and the problem to work on.

You need to schedule or deliver a new toothpick at regular intervals. The frequency will depend on the size and type of project. It might be every two days, leading into a daily rhythm of refinement. Or, the toothpicks might appear monthly until enough functionality and toolkit exists for more frequent iterations. The toothpicks typically are more complex than what you do in daily iterations, but they quickly become something that you can focus the daily builds on. If you have to build your own libraries or key technology, as you might in a Class VI case, the toothpicks are more about the libraries and tools than they are about the nice user interface. Once the tools are crafted, the attention can turn towards the user's value chain and interface.

At some point the toothpick disappears either into the forest of code or is just thrown away. This brings us back to the programmer's ego and attitude towards their work. It bears repeating: do not become emotionally attached to any toothpick code or design. That can be fatal. You need to be able to sit back and critically examine the good and bad aspects and deal with them. That is part of the process. You will do things twice and sometimes three times or more. If you are hung up on trying to do something once and doing it *right*, stay with Class I –III development. Always be prepared to light a match to the thing you call a toothpick.

Chapter 30

Factoring

I have mentioned refactoring and pre-factoring at different points in the book and this chapter brings the various ideas together. This is an interesting topic and I suspect that many pages could be dedicated to the topic, but I will keep this relatively short. The term factoring with refactoring and what I call pre-factoring is all about what goes where, what the interfaces are between the software components, and what the code looks like within a module. Refactoring is going back and changing something that exists to be smaller, quicker, cleaner, or functionally re-grouped. Pre-factoring is doing extra factoring work in advance, minimizing later refactoring. At the end of the day it is about system and code design. How I factor, refactor, and pre-factor depends on a number of project characteristics.

The closer the project is to a Class I–III activity, the less refactoring and more pre-factoring is done. This makes sense. The project is reasonably well defined and the patterns and major concepts known with considerable confidence. You can anticipate future functions and interfaces, and when you are in a specific module or zone of functionality, it makes sense to code up the extra functions. There are obvious benefits to doing a number of similar functions in a concentrated way, and not going back to the module later. The story line or immediate use is not there, but you know that certain functions are very likely to be needed. This is what experience brings to the game. A junior developer does not have the knowledge base to know what functions might or might not be associated with the project and this can lead to later refactoring as the functions expose themselves. If a seasoned professional is doing a great deal of refactoring at the Class I–III level, this suggests to me that insufficient analysis was done in the early stages, including perhaps some homework on the part of the analyst.

For the Class IV–VI projects, there are some things you can do to reduce the refactoring. For the Zen elements, extra analysis and thinking will lead perhaps to a better understanding about future functional requirements and interface aspects. The extra analysis, if done the *ZenTai* way, will help you understand the life cycle functions, as well as what might be expected of the system objects. Another technique is to identify the Zen subsystems and decouple them via a pre-factored *soft* interface. Decoupling has also been mentioned frequently in this book; it is an important technique to master. The Zen to Zen decoupling I am talking about here is the type that provides earthquake proofing. You take a little bit longer and do a little bit of extra work on the interfaces between major subsystems and try to protect one system from the other. You want to be able to easily and quickly change the characteristics of one system without equally or substantially impacting the other. The need for a soft decoupling might not be apparent in the beginning but I have yet to be involved in a system where large components do not eventually take on a life of their own. As one major component changes, you do not want grinding and rupturing on the interface.

The harder elements to deal with are the Class IV–VI features that can easily change in the user's domain. Equally hard are complex elements which are algorithmic or bleeding edge in using breakthrough technology. I expect that most of this code will have to be done twice. Or three times for the real tricky elements. If you program expecting this, then you are not as frustrated or disturbed when you decide that effort is needed to refactor and that code needs to be thrown out. You might be lucky though and not all of the code needs to be redone. This creates a gambling situation. When you are doing the initial code, how much effort do you want to expend on complying with standards, making the code clean, documenting the code, and possibly pre-factoring?

With the high risk code, I will try to do several layers of abstraction, one working directly with individual objects, a second layer that is more of an object utility library that performs meta functions between objects or with groups of objects, and a third layer that introduces application dependencies (e.g., business logic). I find that three to four layers sufficient in most cases and there is a danger of confusion if you introduce too many layers.

I will also try to do some pre-factoring and take a chance that some of the code will remain for the long term, balancing out the chance of refactoring this pre-factored code. And yes, quite often I find myself refactoring pre-factored code. However, I also find that lots of pre-factored code sneaks through in its major form and retains value for a long time. I will pre-factor based on my gut feeling or on the system analysis thinking about object life cycles and task flows. Part of the pre-factoring might be to create clear prologs and epilog sequences that allow normalization of inputs and flows, and then a corresponding fan-out on the return. This takes a small investment when designing and coding, but the rewards are usually worth the effort.

At the detailed level, my code typically goes through three stages, especially for the code I suspect will change rapidly and frequently. The first stage is a wild west approach where I do not really care if I bloat the code, do things inefficiently, or do things which others would not consider totally clean and compliant. I will try to be consistent, but I will be focused on code creation so that I can get the function to market so to speak and get the user's response. That is my goal — get the function done as fast as possible with a little bit of concern about the future at the interface and structural level, but not too much attention to the best coding or language practices. I code for quality but it might not be pretty. It will not look like textbook code and this is OK. It is not intended to last and is intended to be re-factored later. This quick and possibly bloated code is more efficient and effective to use in the early phases when you might not understand what elements are common, what can be generic, or what variants might exist. This is similar to the early days of a manufacturing industry where waste is efficient and the winners do not worry too much about systemization and process optimization. I need to get the functions to the user quickly in the Class IV–VI situations as they are high risk by definition. We do not know 100% of their usage. We do not know 100% how the user will respond. We do not know 100% how they should be grouped or what the work flow is. There are no exemplars and we are in the barrens and wilderness mushing through a serious winter blizzard. In these extreme cases we do not know a great deal and it does not make sense to me to spend too much time on the pretty aspects when I am hitting rocks and guessing at what might be around the bend. I still need good layers and a good design that allows me to respond, but code elitists will hold their noses when they look at my first versions of functions. I do not care what they think about my code or coding ability at this stage. I care about the functions and what the user needs. Pretty will come later. If I worry about pretty on everything from the beginning, then I will likely run out of time, user patience, budget and not deliver the product. It would be pretty code and make a purist smile, but the project would be dead. In the early stage of a Class V or VI, it is about compromise and knowing when and where to make the deals with the devil.

The second stage corresponds to the functional layout phase in manufacturing evolution where you group things together and gain expertise with specific tooling and features. You gather up a bunch of functions and perhaps drive the logic with a 'for each' type of structure going through a list or class. The code does not do many types of objects in a general style, but the code is not bloated and there is better grouping in the code. I have observed my code going through the same evolution from the sloppy wild west phase to a systemized refactored version. The third phase resembles the process flow aspects of industry when you have gained sufficient expertise and you have enough usage to justify refactoring the code again to create meta functions, clean logic, generic objects, pretty code, and better, softer interfaces. The refactored code that reduced the bloating at the functional level is refactored itself to group objects sharing certain characteristics with a corresponding reduction in the code base. If the code is very complex, there might be several passes at the functional stage, and several passes at the process level.

Just like in manufacturing if you are driving for speed and functional delivery you have to be careful to pick the right style of code design. If you try to systemize the code too early, you are wasting effort and possibly painting yourself into a corner. If you try to create an optimized flow process through the code before you understand the tools and requirements, you are also possibly in trouble.

There are developers who insist that everything should be done right every time. I disagree. History is full of failures where inappropriate processes were imposed too early and in the wrong context. Part of the experience equation is knowing when to be sloppy, when to refactor to the functional version and then again to the process model. In all cases, you need the time and discipline to go back and clean up sloppy initial efforts, but that is part of the process. Sloppy might mean using some hard-coded values initially on an interface until the time is taken to create a library for the literals. Sloppy might mean copying code and bloating instead of creating something that might do multiple objects or functions. Sloppy probably means that you will not get certification for this code effort. Sloppy does not mean buggy. Sloppy internal code does not mean a sloppy external interface to the user. Sloppy in an early phase does not mean sloppy forever. However, for software in the Class I –III categories, you have no real excuse for being sloppy or for doing the functional approach. Class I –III situations are repetitive and should have almost no sloppy or compromised code; there is no excuse for it.

In the simpler systems, it is likely easier to freeze, or at least consider freezing, the function sets and what will be delivered. This is one reason that there might be less refactoring in the Class I –III situations. If one does not understand the inherent difference between the Class I –III and IV–VI cases, one might expect that the IV–VI cases can be treated the same way, with the same expectations for freezing interfaces, functions, and requirements. In the Class IV–VI, it will be hard to freeze anything and you might be lucky to have a week or two of stability. The Zen should be relatively firm, but the user side will be like an amoeba. Rigid, rule driven developers will not enjoy the fluid nature of the requirements and how the functions will come together. You can design software to be fluid and have flexibility; this is what is needed in the Class IV–VI cases, but this will not help the rigid programmer's need for knowing with certainty what is to be done for the next week, month, or half a year.

Chapter 31

Coding

What are my coding conventions and style? This will depend on what I am coding and who will maintain the code. For example, if I am coding for myself I might use exotic libraries, explore recursive algorithms, and not worry about the skill of the person who might have to maintain the code. However, if I am writing code that people with less training and skill than I have will end up maintaining, then I will code in a simpler, less sophisticated fashion. Other choices depend on the use of the software and life cycle expectations. If I expect the code to last hours or days then I might allow myself to use the latest libraries and functions that push the software engineering envelope. I might also do this if the system is not considered mission critical. I will avoid certain types of libraries and code if the system is required to be functional for many years, perhaps decades, or if the system is mission critical.

In general, here are my practices:

- I will almost always use long variable names with verbs and nouns. And, I do mean long. I will use underscore characters to create white space and assist with readability. I use this naming convention because it seems to make the code clear and readable. It also helps to find code during searches. It reduces the amount of inline documentation that is needed.

- I will try to avoid having a line of code do two things at once. I will try to avoid having nested or imbedded methods. I will try to avoid having a method on a return statement. The method uses a variable; the variable is returned.

- I will try to have consistent naming conventions for local variables, methods, and classes. The naming might be grouped or hierarchical so that I can immediately tell the scope and meaning of a variable or method when it is seen in the code. I will try to avoid using namespaces or techniques that hide where something is coming from (yes, you can use the development environments and move your mouse over something to learn something about it, but that potentially requires me to do something while I am reading and this is distracting).

- I will try to use Michael Jackson's principles for hypothesis and posit setting and testing.

- I will try to avoid any latest, neatest features or functions that the super geeks suggest to use. I might try them and experiment with them, but I will avoid them for any long running code or mission critical aspects. I will wait for them to stabilize and be proven.

- I will try to avoid any libraries or function sets which are rapidly changing and which may break my code on a regular basis. If I need the functionality, I will create an equivalent (if feasible and reasonable).
- I will try to code assuming that the module will have to be debugged and will have to be refactored. This possibly implies imbedded and specialized debugging aids built from the beginning and used throughout. I try to think about what makes code easy to debug and fix, versus what makes code hard to debug and fix, trying to do more of the one and less of the other.
- I will create special debugging tools for any non-trivial recursive routine. These will be imbedded but controlled externally so that triggers and logic can be activated in a quick and easy way.
- I will try to code that allows the invisible to become visible if needed. This might imply thin wrapping of things such as an empty string constant.
- I will try to avoid using punctuation or other short forms that increase my cognitive load when reading and interpreting code. I want to read the code quickly and parse it without effort; I let the compilers do that. I consciously avoid syntax like ++= or =++. I can touch type and I like to see things spelled out, not using cute, insider short forms or short cuts. Some are valuable, but short forms or short cuts should not be used solely to show off and demonstrate how much of a geek you are. And, if the statement has to change later, I would prefer to change the text on the right and leave the operator alone.
- I will use object models and use polymorphisms to extend base classes. However, there might be times that it is better to create a parallel class and not overload an existing one. I might also create a parallel class or create a mini virtual machine to establish control over certain functionality if the component is mission critical.
- I will thinly wrap and build up libraries for functions that might require memory or more cognitive load. For example, for date functions I will create libraries that provide methods to compare dates in different ways that allow me to forget what a -, + or = means (e.g., Is_Date_Between(d1, d2)).

The above list focuses on what the code looks like and not so much on what the code does or how I group the code. As I build up the code base, I will try to create reasonably functional libraries and subsystems in 2–3 days. Some of the code will be sloppy, but the basic points above will have been followed. I might spend months building up lower subsystems to the same granularity of functionality and quality before being able to plug them all together. There are some systems which are like an automobile engine. You cannot run the engine without pistons or cylinders. You cannot run the engine without a sustained fuel source. You cannot run without some form of ignition. You can bench test and exam each piece separately, but it is not until you have all of the pieces together and you turn the key that you discover if the engine will turn over, fire, and run. For Class I –III projects you can buy the engine, buy the transmission, and so forth. For Class IV–VI, you might be designing the engine from scratch with each of the major subsystems being new, without an existing, running, version to look at. If you have enough time and resources, you can do very sophisticated and realistic simulations, but most software projects do not have this budget and the specialized software simulation tools do not exist. In one system I had to build a dozen or so subsystems independently and then wire them together. This was about 80–90,000 lines of code coming together. Each subsystem was unit tested like a bench test, but the systems had not been connected together in any fashion. They had to be connected in an eighteen step logic flow to match the business logic. It took two days to wire the subsystems together, make the final glue, and debug the interfaces. The engine turned over, fired up, and executed at the end of two days. It was not rocket science. It was a case of certain coding and interface conventions combined with behavioral code testing.

For complicated systems I will use some additional techniques when I code. I will code so that my risk exposure between subsystems is very limited and flexible. This might mean using parameter lists and loose coupling between services instead of exposing contracts and doing tight binding. A second technique I use is that I will make the external functions at the service level available to the inner logic so that I can reuse the capability and features. For example, when creating a work flow language or capability, I ensure that the work flow engine with its verbs and nouns can access edge functions and trigger meta level user or service functions. A third technique I also try to remember is to code these types of engines so that they can be used to create dynamic logic and code — verbs and nouns that allow a work flow program to create another work flow program, or that allow a series of web interface calls to create a work flow program. A fourth technique somewhat related is to keep spare values and types in the data structures and to make them visible to the internal logic (e.g., work flow engine). A fifth technique imbeds meta data into the data objects for versioning control and self-identification (e.g., type, semantic purpose, version, variant). Sixth, I will expose my internal tracing, debugging, and logging functions so that they can be called via remote access, and make them dynamic based on triggers if they are associated with recursive logic. All of these coding and detailed design techniques assist with stability, robustness, and long running. They are not always obvious or considered valuable when you start out on a project. During one design review, a number of these techniques were being explained and as one reviewer commented, they would not have been thought of if this had been my first time doing the type of project being reviewed. This is what experience can provide. I had indeed not thought of them on my first project like the one being reviewed. I had learned through the school of hard knocks that the ideas had value and should be included from the first day.

When I am working on a subsystem I will also try to contain or control all inputs and outputs, allowing the software to turn on or off any or all of the interfaces as they become available. This requires the isolation and tight control of any functions that escape your subsystem, or that call your subsystem. You need to say to each one, quickly and cleanly, fake or not to fake. I also put logging and tracking on each input and output interface point.

If I am coding something like a web service or an imbedded subsystem, I will create a simple form, full of buttons hooked to the interface. The button/form interface will access the system I am coding just below the communication point and mimic calls to the service or subsystem. This allows local testing to be done and test cases to be created before introducing more unknowns. If someone reports a potential bug in the subsystem, an existing test can be used to see if the problem is internal or external, or to examine if assumptions are different between the external and internal call. If a new test is needed, it can be quickly added. This is the first level of interface debugging I do and it quickly helps to isolate the problem between caller and callee. The testing at this level also shows potential callers a working example to copy. The next level of debugging is at the edge on the caller, before the caller's logic kicks in. You want to make sure that the caller gets what the caller thinks they are getting before wasting any time or effort chasing dysfunctional elements. Part of the coding and debugging design has to allow edge checking! You code for it and build it in. I learned these techniques in the late 1970s and have never forgotten them.

There has been quite a bit of work recently on patterns and leveraging existing knowledge, libraries, and sharing of code. There is quite a long history of code sharing and the code sharing contributing to the base systems. Companies such as IBM encouraged their user communities to explore their code, modify, and suggest improvements. This dates back to the 1960s and was an early form of what is now called open-source. The user communities would gather, share, and discuss the extensions, modifications, and enhancements. One IBM group was actually called Share, and it is still functioning. IBM developers would be active participants and many good ideas that originated with the user community would find their way into the formal releases. Then as now, some companies did this. Others did not. The open source type of activity is old school, not new school.

One way to generate code is to pick up open source code, or code samples from the web. You can pick up whole applications this way and it can be a good way to create a base foundation. There are pros and cons with picking up code. Not all shared code is good code or quality code. I was once helping to move an open source application from one operating system to another. We actually had two versions of the open source application, one from location A, one from location B, both designed to do the same end user functionality and to interact with the same external interfaces. It was interesting to see the internal design and coding quality from the two locations. One was nice to work with; the other was not. The applications were very sophisticated and were actually applications designed to run on Arpanet. It was a great training opportunity as it is rare to see two complete, different solutions to a non-trivial problem. You get to see the results of various design decisions and coding practices. One was a good open source code base, and the other was not.

You need to really understand and analyze any open source code you are using if it is to be used in a Class IV–VI project. While there are still risks at the Class I–III levels, the risks and damages can be better controlled. Part of the analysis has to consider the sustained use. That is, if you adopt the code and you have to support it, what are the skills you need, what development environment is needed, and how do you get the software to your desired level of quality and keep it there? What is the cost? Some companies have been successful picking up an open source base and building upon it; some have failed. The companies I have personally talked to who relied on an open source base for strategic and mission critical purposes have had to make substantial investments and have a large development staff of above-average developers. I will use open source code for some elements and vendor supplied proprietary for others. When I do use open source code, I am careful.

The web has become a great place to find code chunks and solutions. It seems odd but I have encountered developers who rely almost 100% on finding solutions on the web and seem unable to create their own solutions. If it is not on the web, it should not be done and cannot be done! They are assemblers. They assemble code. They do not create. They will cut and paste portions from X, some from Y, and some more from Z without refactoring, understanding, or analyzing what they are getting. I do not think that this is good nor wise in all cases. I do not think that they have what it takes to be a high end developer on a Class V or VI project. They may know the language in and out, be certified, or have great skill with the magical libraries and short cuts. They might be great technicians. They may be excellent developers for a Class I –IV activity. Developers on a Class V or VI need to know how to create tools and software, not just use them or glue pieces together.

There are times that I will surf and find techniques and starter code. I will then use the ideas or refactor the code to fit my situation, not change my situation to fit the copied code. I do not like to copy or use code that I do not understand or trust when working at the high end. If I am prototyping or exploring a new language or system, then I will copy and assemble with the best of them and get something up and working. I then analyze it. I do not take this code and blindly use it in production for a mission critical system.

Chapter 32

Testing

I prefer to test and develop in a certain way. I think that a good testing strategy is important for delivering high quality code. The early toothpick and core code approach means that certain parts of the system will be exercised many times as you move forward. This is a key element. Another key element is getting high coverage on key technology or supporting features. Thin wrapping will force a concentration of execution flows and you will have stronger blocks of code. This helps you figure out how to use the method or interface in one place, get it right, and then you do not have to worry about getting it right in multiple places in the code. I create the virtual machine layer, get that debugged, and then put it to bed.

By forcing myself into the design, code, debug, test, integrate and cap rhythm, I am focused on the code I am doing and that helps me set up test flows and concentrate on one bit of code after another bit of code. I do not have to think about code I wrote days or weeks ago to set up tests. I do not have to think about large hunks of code either. If you create the basic tests and test as you go, it is more efficient and effective. You should also try to crack your own code and confuse it. In one test stream of about 5,000 cases I helped create for a relational database system; over half of the test cases were errors to be detected or odd situations designed to confuse the parser and language. These 5,000 cases were not created all at once. The list kept growing as we *Mushed*. This test stream was used at every build (i.e., every day) to ensure that we did not break something that was working. This project was in the late 1970s.

Tests should be deterministic and it should be possible for you to replicate and analyze a test. Random ad hoc tests are not usually a good thing. However, if you can log and track user interactions, it is possible to recreate the ad hoc sequence.

At the code level, here is an idealistic, short checklist of what to look for and test for:

☐ All methods or subsets should have unit testing.

☐ All major modules should have a performance analysis at the transaction level to know how many transactions can be performed per time unit.

☐ Coverage via unit testing should exceed 95% and near 100% at time of final system testing.

☐ Test for linear or non-linear degradation if queues or functions can pile up — test at orders of magnitude.

☐ For data driven areas, have test with seeded '1's — easy to check for counts.

☐ For data driven areas, have test with unique numbers — ensure that boundaries are checked.

☐ For any value or field that has a range allowed (a <= ? <= b), tests should include a-1, a, a+1, ?, b-1, b, b+1.

☐ For any Boolean test, both true and false paths should be tested.

☐ For lists, ensure that the first and last nodes can be reached.

☐ For sorted or manipulated lists, ensure that something can be added/deleted at the front or end of the list.

☐ For sorted or manipulated lists, ensure that something can be added/deleted in the middle of the list.

☐ For sorted or manipulated lists, ensure that the complete list can be deleted and something else added to it.

☐ For trees, ensure that every leaf can be reached by the processing logic.

☐ Every error code should be reached and tested.

☐ All code that calls a function that returns a status code should be checked to ensure that the code actually checks the status code and behaves accordingly instead of assuming that the function works or that a status code can be ignored.

☐ For loops, ensure that the loop can be bypassed — the empty, zero, or null case.

☐ For loops, ensure that "1" pass is possible, without iteration.

☐ For loops, ensure that "2" is possible — the first iteration.

☐ For loops, ensure that "3" or "n" is possible — this is the first general case.

☐ For any "if"/"then"/"else" — make sure that both paths are set.

☐ For every function that assumes something exists (at least one of something), test to ensure that the zero case is tested for.

☐ For every array or vector, test for a buffer overflow.

☐ For any recursive call, ensure the recursion stops.

☐ For every function that assumes something does not initially exist or is empty, test to ensure that the populated case is tested for and caught.

☐ For every throw-catch-finalize situation, ensure that the throws are done and indeed caught and processed.

☐ Create one test that takes the module or class for a life cycle spin — creation, functional life, various functions, termination. Make this test exposed and capable of being accessed remotely as a health check. Ensure that a predictable and deterministic result can be obtained.

☐ If the result of a method is composed of processing multiple items that might or might not be visible (e.g., returning an empty string), there should be a way to ensure that you have processed all of the strings — a way of seeing what was not included or a way of seeing where the empty values are.

A similar approach can be used for testing work flows, table driven systems, and so forth. Think the system through and make sure that you consider the special cases, and the general case.

For a larger system, you can implement self-tests and build in functions like health checks. The tests should be done automatically, periodically, and at the request of other software. This latter is live testing during execution. You test during development. You test with the user as you move forward in daily builds. You test. You test. You test. You never finish or stop testing. Things will change in the system and for critical systems, you need to have systems that are self-monitoring and self-testing.

In a formal situation, you will need to write a test plan, have a suite of tests you need to run and sign-off on before going live, and a way of knowing later if something was tested or not. If tested: when, by who and if possible be able to replicate and re-test.

A regression test capability is important for a number of reasons. There are too many things that can be accidentally altered as you make changes and there could be data dependencies, flow dependencies, or technology dependencies. If the system has a life cycle, the regression testing should cover the cycle as well. It is possible that certain combinations will only occur during a certain time of the year or month, or a certain code flow, and the system must have the ability to artificially run the clock and pretend it is running anytime you want it to.

Stress testing is also important. This is the worst case scenario in terms of the most complicated process or task flow. This is the maximum number of blah-blah that could happen in a time window or in an aggregated way. These are the classic load tests. I think you get the idea. You need to plan and have periodic stress tests as part of your normal process.

I also have a theory about who should lead the testing and if you have a formal Quality Assurance process, who should be responsible for quality. When I do consulting or give advice on development processes, I like to ask who is in charge of testing and who does the testing. There is of course the testing that the developer does or that is buried in the system, but who does the high level testing? OK, there is a user, but who else? I usually recommend that they put their best analyst in QA and make this hotshot in charge of quality. Yes, I think that this should be your best developer and designer. This does not mean that the best developer is doing both product and quality work. The best developer is dedicated to quality. You will get enough out of the best developer on product design through the review processes in any event. Do not worry that you are not using your best resource wisely. In a larger group, the QA folk are looking at multiple projects and are multiplexing. They have to be pretty good and probably better than the development group to keep it all together. They might also have to look at something in days or weeks that took the developer much longer to craft.

If QA is pro-active, the people in QA have to know how to design the system, have to know alternatives, how not to design the system, where typical problems might be found, and other subtle points. Hence the need to have top quality developers in QA. This does happen in many firms based on my own experience. Usually, I deal with organizations where they put fresh graduates or juniors into test mode and rely on random or superficial testing to find problems. An experienced geek will second guess and think about what the programmer might have forgotten about and then drive the software to that edge point to see if a failure will result. Hmmm, what might the internal code be like; what might be smart and not so smart ways of delivering the exposed functionality; what kinds of things might the programmer have overlooked or made assumptions about, hmmm.

Chapter 33
Tool Smithing

This chapter is about engineering the solution. If you are doing a small system of the Class I – III types, there are likely to be many tools and building blocks for you to leverage and it will be rare for you to create a tool yourself; you will be using tools others built for similar applications. This chapter is about the types of tools you might need in a Class IV–VI development.

If you remember, there was some minor discussion in the chapter on reliance on technology about development tools and environments, but I really did not deal with creating your own tools in that chapter. While you need to be careful about using technology blindly, there are many things you can use or do during the actual development phase to make life easier, for you and the user.

There are two kinds of tools for the IT developer. First, there are what I would consider to be passive tools. These are elements that you build in or utilize as part of the product. For example, there are functional libraries that supply you ready-to-use code, or design patterns that help you with techniques and algorithms. Second, there are active tools you use as you craft your solution or you use while your solution runs. These will be discussed in the following sub-sections.

The thinking about imbedded tools for development and deployment in the Class IV–VI systems is iterative. You need to know what tools are available from vendors and public sources, and how to use the tools. You also have to be prepared to make some of the tools yourself. For example, after having a rough journey through very rough water you might have to build a new boat and build the tools to build the boat with. Or, you might want to build some special tools in advance to help you through the rough water. In the ideal team design for the extreme Class VI type of project, I like to have both a dedicated blacksmith for the infrastructure and one or two tool smiths. The tool smith must consider languages, development technologies and topics such as:

- Coding and project/solution standards.
- How to incorporate licensed and open-source libraries.
- How to decompose the actual code into useful and meaningful groupings for libraries.
- How to decouple and protect one layer of code from another.
- How to create the code base so that it is maintainable, sustainable, and good investment.
- How to imbed testing and self-protection into the system itself in an easy and painless fashion.
- What aspects of the technology are rock-solid and which parts are like cooked pasta — you

want to exploit the one and avoid the other.

This list is not exhaustive, but illustrates what must be thought about. In general, projects will be different, but the general topics will be the same. At the start of the project you might not know what tools you will need or what tools will be built, but you know that you will likely need to create some tools along the way. Very true in the Class V and VI cases where you are bending, folding, stapling, and otherwise mutilating convention or tradition, going where few others have gone before. Just because a tool does not exist on the shelf does not mean that it cannot be built. Just because other people do not use a tool or structure their similar applications does not mean that you should not use a tool or engineer a better process. And, remember that all of this 'Best Practice' stuff is very subjective, perhaps even self-praised and self-promoted. What tool is best for someone does not mean that it is best for you or your situation.

One of the things you learn over time is when to start considering a tool because you do not want to over systemize or exceed the optimal mix of third party tools, own tools, and final code. History is littered with firms who have misjudged how much to invest in the system and infrastructure versus the parts that actually make the money. In many cases you start off without the tools and reach a logical point that says that not only do you do a major refactoring, but you restructure the system and create some additional tools to make the remaining part of the journey more efficient and effective. This implies that you need to know when to stop, know what does not make sense to continue with, that you recognize what tools can be created, and you know how to create tools.

However, there are other cases that, as the initial design and thinking is being done, it becomes obvious that tools are needed immediately. In several of the high-velocity projects I have been involved in, the tools are the first bits to be implemented. Since you also often work in a team setting, it is also possible that various team members will not value a tool or a tool concept until the project is over. There is no point pushing a tool or concept before people understand the value of the tool and what benefits arise from using the tool. This is not ideal, but seems a harsh reality. You can raise the idea early on, but if there are no takers, you might as well sit back and enjoy the show. Let nature take its course.

One way to reach the tool point is through the technology toothpicks. In a new situation, you just do not know at the beginning what should be done. You proceed, craft some code, perhaps good code, perhaps poor quality code. You learn the language or platform and through this, you discover how fragile solutions are created and where the weaknesses are. You can then think about tools and create them. I suggest that you do small test examples to tune the ideas in the tool before going full out on it. This might be a few dozen lines of code or hundreds or more. It will depend on what the tool has to deal with and what you expect out of the tool.

33.1 Passive Tools

Passive tools are usually easy to make. They include things like compile time options, macros, includes, prolog/epilog logic, and non-code concepts such as standards and conventions. These are all tools. A tool does not need to be a piece of software per se. A tool is a mechanism that helps get the job done. Standards help get the job done, but a standard is not itself a piece of code.

The passive tools and conventions will help you keep values out of the actual code, the visual manifestation separate from the semantic function, data from the logic, the highly volatile data and logic away from the stable components. This sounds like obvious statements again. Anyone who has studied software at a college or university knows that you should wrap, isolate, decouple, encapsulate, and layer code. They know that you should not hard-code things, expose inner workings, imbed and mix semantic entities together, and other such stupid things. Of course not. There might be times in rapid development that you cut corners on interface code, but the main code base should adhere to good coding principles once you are out of the wild west phase. Many of the inherent benefits of object-oriented programming are built upon such principles.

However, there are always segments of the coding public that think that they are special and that solid practices need not and should not be followed. The assembler programmers of old thought their low level service libraries and operating system modules were special. The Cobol programmers thought applications were special. Simulation programmers (e.g., factory simulations) thought simulation programming was special in the 1980s. Webpage programming and scripting is special. Everything is special. They are all so special that they are very much alike in many ways. They are not so special!

Unfortunately, programming is still considered by many developers to be an art and not subject to systemization and engineering. In my view, design is the art and coding is a trade. There is a danger of too much systemization or systemization at the wrong time. For example, if you are expecting to refactor some volatile code several times in rapid succession, does it make much sense to exhaustively unit test the code, document it, etc.? Does it not make sense to wait until the code stabilizes before investing heavily in the module? If you keep doing the same secondary tasks over again, this can amount to a great deal of wasted effort. When you leave the code unclean for a period of time, you must have the discipline to go back and finish it off. Some organizations have problems with people returning to modules to clean them up and they make the decision to have clean software at every iteration. You can do this at the Class IV and possible V levels. I doubt that this is very economical in a Class IV case where you will have lots and lots of major refactoring.

To build the passive tools that help decompose and layer the code, you have to recognize what should be where, and how to get it from point A to point B; data and logic. For example, I do not think that you should have hard-coded numbers in the body part of a webpage. They should be variables and substituted. This goes for text as well. You create a library of text strings and access them dynamically into the code part. This avoids changing the body code (and the associated risk) when you want to make simple word smithing changes. You try to separate the logic elements. We separate the GUI widget needed to do blah, from where the widget is on the page, from what the widget looks like in terms of form and structure, from what goes in it, and from how it interacts with other things on the page, with the server, and with other pages. This allows you to change various attributes of the problem globally and without having to hunt down each and every instance when you change something. Even if a programming language or tool does not immediately give you these capabilities, can you fake it out, and fake it out in a robust way? You do not want to create tools more fragile than the original problem; do you?

Part of this thinking helps you identify groupings of functions, functions for the major objects in your system, and guides you to the modules and library files. For example, if you have a task oriented design you can group functions by task and have task specific variables and values in one place accessed by all pages associated with the task. Often this type of thinking is done on the server side, but it can also be done on the client. Why not? This thinking also helps you identify layers for thin-wrappers. For example, you might be using an open-source library from blah-blah. You are smart enough to keep blah-blah in its own script file. That is the first step. However, I would also create another file between the library and my main production code. I would not reference the third party functions directly because I might want to intercept, replace, and do other things that happen during the course of long-living software. Yes, it is a bit of a pain to create these thin-wrappers, but I do it consistently in many kinds of situations because I have learned the hard way that it makes sense. Again, I do not always do it, but there are many times that I have to and would do again.

Another category in the passive tool area is the prolog/epilog concept. You might or might not know about these ideas. They have been used in languages and low level systems in the past. I use them at all sorts of levels, including work flow. You might also recognize these as constructors and deconstructors. Name is different, but the game is the same. You isolate the setup and normalization logic in the prolog and if necessary have the prolog customize the body of the logic. You have the epilog deal with any cleanup and possibly any last minute dynamic logic to alter the flow. The prolog and epilog is also a good place to put tracing, logging, and other goodies. Even if the language does not support the concept directly, you can program it and use other features of the language to mimic the desired functionality. This is where you turn the pig's ear into the silk purse. This is what a tool smith has to be able to do.

33.2 Active Tools

Active tools are usually harder to make. These are the monitors, analyzers, loggers, tracers, interceptors, test harnesses, health checks, and other logic elements that execute while your production code is running, either in development or live modes.

I cannot think of a case where I have not built or crafted passive tools in a project. Once I find I do things repeatedly or packaging is necessary, I start the library and interface routine. It kicks in automatically and the tools get built. They save a lot of time and effort in the long run, and are vital for quality and stability. I do not always create active tools. This depends on the problem and the cost/benefit trade-off. Not only will you have to support the main product, but you will have to support and maintain the tools. This has to be a conscious decision made by the organization. The passive tools imbedded in the product are easy to justify. The added tools on the periphery are not.

The active tools are quite likely to be unique to the specific system being built and while there might be some generic tools to help with the development, the generic information is not sufficient. The tendency might be to spend a great deal of time and effort on these secondary tools, and go beyond the practical return on investment. The tools are fun to create. Possibly more fun than the final product. It is tempting to spend too much time and effort on the active tools. All you need is the 20% that does the 80%. Do not over develop!

I am impatient about some things. I like passive and active tools that help me find bugs fast; real fast. If it takes more than five to ten minutes on a regular basis to isolate and identify the cause of a problem, you should consider some tools. Or, possibly consider a new career. The monitors and active tools help you understand the code's behavior and if appropriate tracing is active, you can think through the system in an organic fashion and analyze by thinking and by behavioral analysis. You should not need brute force code level tracing. The passive tools in the previous section leverage different aspects of the development cycle, such as the effort associated with code re-use. They are enablers. This is also true for the active tools if you think about them as exposing clues and hints about what is going on. It is likely that you will use a combination of the special active tools to identify a certain part of the code and then invoke the development environment's code level tracing.

I cannot tell you what tools you will need or that you should write. If a system is heavily recursive, you might need special tools to trigger at recursion level 245 and start exposing behavior — XYZ at recursion level 124, is interacting with ABC at recursion level 245 and this is what the scope of ABC–245 looks like before XYZ–124 called it and afterwards. In another situation, you might want to red tag transactions at the client side and trace all major conditions through the complete stack that relate to the transaction — creating human readable tracing information. All I can suggest is that you budget for a little tool time on Class IV–VI projects and be prepared to write the tools. You need consider the tools early in the process. If you wait too long, you might not need the tool at all, as the project you are working on is no longer exhibiting any life signs.

Chapter 34
Documentation

Agile, Agile/Extreme, and *Mushing* are not excuses for avoiding documentation. You still have to do key, important documentation. Assuming that someone is paying you to develop the code, you need to practice appropriate due diligence and protect your client and their investment. I have taught many students who hate to write and I have been told on occasion that if they were good at it and enjoyed it, they would have gotten an English degree instead of doing Engineering or Computer Science. Well, guess what. You will write and you will write lots if you are going to be successful in this craft. You do not need to be a Shakespeare and wax glowingly over a sonnet, but you better be able to communicate. Communicate clearly and concisely; and reasonably quickly.

What kinds of documents you will need to prepare will depend on many factors. You should be knowledgeable enough and skilled enough to do the full set; matching ISO standards for example. You could be doing high velocity development in a very structured environment with auditors and institutional policies that must be adhered to, or you could be doing this on a couch watching television, crafting a small system in a very informal fashion for one or two people. The two extremes are reality. There are some common concerns that must be addressed. The form and structure might be different, but the intent is the same:

- o There should be something in writing; it could be a napkin, email, or form document, that sets out the scope and general nature of the engagement. Setting the ballparks and understanding. This should address guidelines for cost, time duration, and assumptions about user engagement and dedicated time. If you do not do this, your best friend better be a lawyer.
- o In storyboard format, scenarios, or lists of capabilities, there should be something that describes what the software is expected to do. Enough so that you and the client know that you are both talking about a dog, not a whale; a cuddly lap dog that would fit in a knapsack and not a ferocious guard dog that does nasty things to anything that come close to it.

These address the messy issue of managing expectations. If you do not do this, you will have some interesting meetings as the project proceeds. These should be done for every project. A big challenge with Agile/Extreme is expectation management.

One other document that should also always be done if you are a professional developer and not just a hack is:

o Something on how to maintain and upgrade the system. You cannot assume that you will be the one fixing it or upgrading it in the future. You need to leave enough guidance behind so that someone can spend a reasonable amount of time learning, and then do something for the client. This should address any key, critical rocket science code, and general steps for the person to follow if they want to enhance the UI, the database, etc.

If you are creating a 100% self-standing system, you do not need to worry about interface documentation. And, if it is a 100% new application, you do not need to worry about legacy or migration documentation. If either of these 100% assumptions fails, then you will have to write some more documents:

o An interface document should clearly describe what is on the channel or communication path facilitating the interface, what the data is used for in your application, when it is communicated, what form it is communicated in, and if there are any assumptions being made. If you pass information back, this should also be documented thoroughly and carefully. Do not just say a flat file is passed. Address when it is passed, what are the rules generating its creation (is it a subset, is it a complete list of all members, when is the data considered accurate and complete, etc.). Do not just say that the field is eight bytes. Say what is the format and edit rules for the eight bytes. Are there rules to know if the file has been updated or not? Are there checks that can be made on the interface file to ensure that it is valid? Do not just repeat the internal variables names EX_@JH$1 and other short forms. Use semantic and meaningful descriptions for the interface document, in addition to the internal variable names!

o The migration document is important. It will map what is being moved and what is not. It will document how and when the migration will be tested and then rolled out live. It might also document how the old, legacy system data is kept up to date by the new system in case the old system has to be re-started on the new database. The migration document will also highlight any problems with data integrity; completeness, ambiguity, missing. Any rules for the translation or extension of data will also be documented.

So, we have covered two forms of documentation needed to manage user expectations, one for future maintenance and support, and two possible forms of documentation for interfacing and migration. The form of the documentation will depend on the legal, contractual, and organizational requirements! You might need to do these documents using standard formats and get specific signatures. Or, you might be able to survive with handwritten notes or emails. In either case, the key data needs to be recorded.

The bigger the project, the more critical the project, the more documents you will need to help capture and communicate with. There will be governance issues, levels of approval, auditors, and many people on the team that will rely on the documents for guidance and awareness. For larger, say 10–20 people or more, developments that are substantial investments, I would recommend the following before you line up everyone and pull the trigger:

o Some type of initial needs or opportunity analysis and statement. Is there a problem, what is the problem?

- o An initial set of requirements or capabilities that are needed to address the problem. Not a detailed functional specification, but something that says what the user or system must be able to do. While you think that this might not be risky, what happens with requirements understanding creates a risk. Your analysis may be data and evidence driven but people may not believe it or agree with it. In software systems with heavy user interaction everyone will have their opinions and desires about how they think they should use the system and how others should. If the client group and/or developers do not agree with the requirements analysis and the implied *general* solution, there will be problems with the development. There could be credibility issues, or just the reality that some people want to do everything themselves and will not accept another analysis.
- o You will need a rough organizational view, a rough budget, and a general set of milestones. If you are asking for $, you need to support your request with some type of information.
- o If the system is large or critical, create an agnostic, abstract architecture document that does not dwell on the minutia; but the big picture. Document the major systems or subsystems, general schema for what and how they have to function to satisfy the key abilities identified in the rough abilities document. But, this can be dangerous if your developers are used to doing architecture themselves and will not willingly accept another's analysis or design.
- o You will need coding and testing conventions established. Either using an exemplar to point to, or a manual.
- o As you get set up, you will need a more complete project plan, analysis of resources (equipment, space, people) and a general agreement with the client that the preliminary ideas captured in the above list of bullets make sense. Get or use official signatures if necessary.
- o If the system is mission critical, you might need a robustness analysis document, or an internal risk audit document.
- o If the UI is critical, you will need some initial ideas sketched out on the UI approach and concepts. What the themes or main considerations will be, and what assumptions are being made about the system usage.
- o Along the way, you might have made a toothpick to test out the basic assumptions and this is also part of the documentation package.

I think it is best to have the above documents prepared before you have a horde rushing about. In a *Mushed* project, you need a single, strong vision of what is to be built. It does not have to be detailed, and it can be in a very rough way; like if you are myopic and you take your glasses off — everything is blurry, and you cannot tell eye color, etc. but you might be able to tell a person from a dog. As much as you need the client to agree with the general requirements and concepts, you also need the development team to accept and work with this vision, set of requirements, and general solution strategy. This might sound obvious, but not all teams and team members will accept or adopt pre-determined designs and solutions. It seems that some will take the ideas under advisement, but they do not feel obligated to use them. In some cases, this lack of compliance is a company or institutional culture, or in other cases it is a clear case of insubordination.

Once the team starts, there are other documents that need to be done. And, again this will depend on the organization and formal requirements. You might need a formal project plan, documentation plan, test plans, design documents for each service or major component, security analysis documents, etc. If there is sensitive data, you will also need access and authentication policies and documentation to track who has what password, what privileges, etc. Many of these documents are not directly related to the activity of daily builds and high velocity iterations with the user! You need them for other reasons.

At the detailed level, documentation is needed and the degree or effort is sometimes an issue for debate. How much code level documentation is needed? This is another case of *it depends*. If the code is going to be a long term system and people will be expected to be supporting it full time, then I might expect the developers to invest in a learning curve; digging into the code and understanding it. Complex systems might take weeks or longer to figure out and I would not expect them to have first contact with code and have it figured out in minutes or hours. Even if there was lots of documentation, it would be rash and unwise in a complex system to assume that you have understood the system well enough in minutes or hours to effect a change. You have to document algorithms, pre-conditions, and you should provide recipes for routine tasks like "if you want to do this, here are the steps." Part of the documentation trade off is who is doing the writing and who is doing the reading. Remember, a senior developer's time should not be valued the same as a junior developer's. I value documentation that gives information about assumptions, subtle issues, and explains the non-obvious. I do not value redundant documentation that simply repeats what the code says.

I try to prepare documentation like a series of maps. There is a continental map that shows the major geo-political subsystems similar to countries in their power and scope. Once you pick a country, you should be able to see the major states or provinces, major cities and major connections. Then you can drill down to the province or county or city levels in terms of interfaces and functionality. There may or may not be pseudo-code or specific code level documentation. The truth is the code and you do not want to repeat the same description twice. The documentation should be guiding and help you understand the behavior and design rationalization.

Standards and conventions are important topics to talk about and to establish within a project as they relate to documentation and code. The standards and styles will depend on the type of software being written and one can expect standards at the browser level being different than in a web service, or database. You should expect that any significant subsystem is consistent and does not look like it was coded by a number of individuals. It should look and feel like it was coded by a single individual. This will require members of the team to agree and to perform as one. This is similar to paddlers in a boat; you need everyone in synch else the paddles will collide and chaos will result. If you are updating someone else's code base, you should make it clear that you were modifying, clearly state what you modified, and explain why you modified the code. Someone coming along later will need to know this information. Some people will hate standards and conformity. Some will not accept it. I have been successful and I have failed on this topic. The company culture will matter, as will the individual's attitude, as well as the penalty for non-adherence.

As noted earlier in the book, standards can be tricky if they are created prematurely, are used inappropriately, or are faulty by design. Think carefully about your standards.

Chapter 35

Client and Developer Build Cycles

High velocity development should mean daily, or more frequent builds. However, in the beginning of a project, the major cycles might be weekly or bi-weekly until processes and methods are ironed out. This should be expected. I addressed the question of how you can think about what new features and functions should be in each build in terms of risk management. I also mentioned earlier in the book about the bite-sized bits that should be considered. It is time to remind you about the bite-sized approach to programming that I advocate. The bite-sized concept directly relates to project control and project quality.

Consider a morning or afternoon. Either has about three hours of useful time in it. I like it when programmers think of two–three hours as a work unit. You do enough design work to break down your problem into relatively small chunks of code. You basically design and code for about 1 and 1/2 hours, then unit test for about forty-five minutes, integrate and test the code with other parts of the system, and possibly do some documentation. This is not the way most people do software development — that I admit. And, I have to admit that I do not do this in all developments either. I do a very tight control process for mission critical systems, or when customer perception and user engagement is sensitive. For these types of systems I believe that the extra care is warranted and that the programmers should slow down and do a more holistic approach before releasing the software to the client. If you do the three hour drill and take it seriously, then you will have fewer life threatening experiences and you will probably deliver something that is very close to bug free for each build.

If the system is easier to craft (e.g., Class I–III), is not mission critical, perhaps has a compressed time line or budget, and the user is very understanding and accepting of daily bugs, then you can reduce the amount of testing pre-release and extend the coding. This should be discussed with the client before you start as they need to know what to expect. If you detect that the client is uncomfortable with strange things on the user interface related to debugging or partial functionality, or with functions that fail, then you need to do much more testing before releasing the code. You should not dismiss the client's concerns and impose your own definitions of quality and good enough.

As part of the scrum tasks, you should have specific stories or assignments that take the code for a life-cycle test spin before the user sees it. In many cases, you should explicitly test any new code in a user environment, not your computer or your development world, and you should try all of the key tasks with suitable 'live' data. Then, you can give the code to the user and do the demo. This is the way I try to work. I try to break things down to two–three hour bits and I do the design, code, debug, light test and integration testing before considering it done. In some cases you have to code ahead or pre-factor the code and you might not be able to do the integration and test phase. When you cannot integrate and test as you go, you need to create the scrum stories to ensure it happens before the user gets exposed.

For a relatively easy project in Class I –III, the tools and basic infrastructure should be solid and the libraries used by the developer should not have a high failure rate. This implies that less integration testing and care needs to be taken before involving the user. When working with the user in rapid builds, you will have to be able to isolate and fix any bugs encountered quickly, so that momentum is not lost. We are talking minutes or hours to isolate, fix, and return to the user. If you do not do sufficient or adequate testing before engaging the user you might have the delusion of making progress and getting things done quickly, but the overall process is going to be longer than expected. If you take more time and make the most out of the user engagement, it will be like the rabbit and the tortoise — in the classic story, the tortoise wins. I think that this is also true in software development.

Even when I am just coding for myself without a client, I will do the three hour drill. I might reduce the testing down to an ultra-lite version, but I will still do the life-cycle spin and test the extreme conditions; the most complicated example within the three hours. My goal is to cap it and not to return to the code again unless it is for a design refactor. When I finish the two–three hour cycle, I want no known bugs and I want the regression test executed, and I want a working demo. I do not get this 100% of the time, but I do get there close to 100% of the time. It is a rare event if I do not do this. I will be very careful about the base code that delivers the function and I usually hope to be bug free for those modules when I finish my cycle. I consider it a personal failure if a bug is found in what I consider my solid code base. If the code I am writing is the test or simulation or scenario code used in the demo, I might have a bug occasionally and while I am slightly embarrassed, it is not a blow to my ego. If I am just coding with absolutely no debugging in advance of other functions being created, then you will find a healthy number of bugs; I expect them, I anticipate them, and if you find them, it does not bother me. In almost all cases, if there are bugs, they are usually isolated because of the short coding cycle and are quick to find and fix. The short cycle constrains the overall size of any encapsulated function and it is usually easy to analyze in such cases. My own goal is that most of my bugs should be found in five – ten minutes and it should be rare for a bug to linger an hour or more. If the logic is very algorithmic with heavy recursion, then I will test more and extend the integration test with special edge cases, perhaps spending a whole two–three hour period on this alone. I will also create special debugging tools and imbed them in algorithmic code to assist when testing and debugging. If everyone and all parts of the system are working on this cycle, you can reduce the full build cycles down to days and deliver to the client a nice system to interact with every day or two.

Unfortunately it seems that most developers are more tolerant of bugs and schedule delays than I am. If you believe that you cannot build almost bug-free code quickly, then why even try? This seems to be the assumption of some developers I know. They assume it cannot be done, so they do not even attempt it. Another group of developers seem to actually enjoy the thrill of fixing bugs in a fire-fighting sort of way. I have been often told by programmers that you have to expect bugs on every cycle, and that you have to accept the fact that there will be lots of debugging during the cycles, as well as functional evolution. These points about bugs and code quality are true about any size of coding effort, but quality is not the only issue that will impact the build cycle and client happiness. The size of the work unit is also important. I have been often told that breaking down the work into the small units I suggest is unnecessary, stupid, or cannot be done. I suppose that if you believe that you cannot do something, then you will probably be reasonably successful with that prediction. It seems that many people will make up their minds before giving such a concept a try, and possibly not even give it a try.

Rarely in the last three decades have I found it necessary to code longer than two to three hours to create something that I cannot wrap up, test, and integrate in some fashion. I need to finish things when I start them. I need to leave them in a good state when I leave them. These are two reasons for the smaller code units. Nothing to do with software engineering. They are personal. I have trouble focusing on a single thing longer than two–three hours. That is a third personal reason for breaking up the work. There have been many times where I worked on a number of similar functions or libraries that have taken days to complete. In those cases I still try to do the coding in a way that the library is built incrementally and where I can start, stop and finish self-contained chunks in two–three hours. I can remember a few times, and they are few, where I have had the code open and in a complete broken, torn apart state for a day or two. I have anxiety during that time and I do not feel good about having that much code floating around. I cannot sleep well if I suspect that there are any bugs in the code I have written. I have never liked those software products that supply a list of known bugs or problems with every version or point release. I am not saying that there are not any bugs in my code, just that there are not any known bugs. You are probably thinking that I might not do too much testing and that users encounter the problems later. I am also very embarrassed when a user discovers a bug after the product is provided and after I have said that the software is ready to use. They occasionally find a problem, but I then do everything I can to isolate and correct the problem. In a sense, my coding and development style are in synch with my own personal problems. I have done this drill quite often on dozens of software projects. It works for me. Small block by small block. High and focused attention on each block, slowly creating a larger unit like you would with a child's set of building blocks that snap together to create something of meaning. It is hard for me to understand why someone wants to program or deal with large pieces of open code at once.

There is another major aspect you need to consider: the feedback you will get through the build cycle. For some features and issues, you might get feedback overnight or in the morning and be able to reflect a rapid change by noon or in the afternoon for the users to experience and re-evaluate. In other cases, some of the changes might be in a short pipeline. The change must be thought about, some users consulted, others engaged, before a large number of users are exposed to the daily version. Some changes will require days or weeks to think through and implement and not all feedback can be dealt with quickly. This will frustrate some users on a large project. The feedback might be given in one week or one month and not be acted upon till many months later. This will not be the case for the smaller and simpler Class I–III projects, but may well be the case for the larger and trickier Class IV–VI types.

In the very extreme case you are doing daily or hourly builds with the client. In other cases you might do weekly or bi-weekly build cycles with the users. What you will do in terms of a daily process will depend on when you get feedback, the number of users giving feedback on the versions, and how you want to manage expectations and the feedback process. If you are dealing with one or two final users and no one else is involved, the process can be rapid, somewhat informal, and very quick. Given the right tools, the change can be made in front of their eyes, or while they are going for a coffee. Being able to make a cosmetic or minor functional change in real-time can really get the user excited.

But, there are times when there are many users, many types of users — perhaps hundreds and thousands. There might be dozens of major stakeholder roles and functionality groups. What do you do then? You need to manage expectations and you need to manage the communication flow. You try to make each build useful to multiple stakeholders, with a minimum of noise. It is doubtful in these cases that you can run a daily cycle and the best you can do is bi-weekly or even monthly.

If possible, some key, insider users might see changes within days and provide some initial feedback. Then others might be asked their opinion on a more complete iteration before allowing even more users access. This pipeline implies that people will still continue to provide feedback on the old while the new is being crafted. Features or functions that cannot be changed in a day will also continue to get feedback from irate, confused, or concerned users. This is hard to control and I have no magic solutions or ideas. The better and more understanding your users are, and the better your own client contact personnel are, the easier this can be. However, it will never be perfect or easy.

Your versions may appear spastic and confusing unless you think them through first. You need to design the process and understand who will provide the feedback and how you will disseminate the versions to more of the stakeholders. You might need several dedicated, or partially dedicated, members of the community to help vet feedback and make the initial sanity decisions. You will not be responding to every suggestion, and in addition to suggestions that should be noted, people will be beating on symptoms and not the problem, their most recent issue, and their own personal likes and dislikes.

During the development cycle, the different versions might not reflect or expose new functionality to the user. If a major system is being crafted, some of the versions might be more internal than external and the user may become impatient. You can hit points in time where more work is needed on the existing infrastructure or on new foundational elements. However, users will have become accustomed to regular versions and will not be amused if there is not something to look at with the same frequency. They will want to see new and wonderful things. They will want to see their suggestions and complaints addressed. To help control this, each version should note what is new, what has been fixed, and if possible what has caused a change (e.g., so and so suggested this). This user cycle of expectation further illustrates how important it is to create enough of the underlying toolkits and functionality before going live to the users!

If the system is big enough, a number of independent toothpicks can be crafted in rapid fashion and then integrated into a common development environment. Once you start integrating and joining developers together, you will need a collaboration environment, versioning control, and what not. You will need the processes and methods to:

- Check-out and check-in modules and subsystems.
- Build — compile and package the new version.
- Automatically run a number of unit tests and health checks.
- Automatically run a regression or a life cycle test.
- Publish or release the version.

For a small team, you can do most of this manually. When you have a larger group, you need to make sure that you have the environment set up before you start. In theory, you should have taken a toothpick through the normal process and a sample cycle, environment before getting a bunch of people involved. You do the same thing for the process as you do for the product. You develop a toothpick process to illustrate and to train people. This constrains the initial "what is this," "where is this," and "how do I" to one person while others are doing more useful work. Having multiple people doing the same probing is not a good use of people. Have one person run ahead and sort out the build process and then use this as the process for all others to follow.

A few thoughts on the first version...

The first delivery to the client — where the user is going to start interfacing with the partial system — is a very important point in time. While every release is important, the first exposure will haunt you for life. It is perhaps the most important point in terms of stakeholder involvement and engagement. The first deliverable should be very robust and survive the random testing it will receive.

If you are not sure of a function, do not expose it. Often, it is better to deliver one function in a better fashion than two, each in a half-finished fashion.

If you are careful and think about the exposed functions, you can present to the user software that has a nice fit and finish to it. Remember, you are not delivering a full, final product in this approach. There is nothing forcing you to expose or deliver to the user partially working or buggy features. Do not do it. Wait till the software looks and feels a bit polished. Wait until you do not have to warn users or tell users that "this does not work." The counter-argument is that the user will be tolerant, will accept bumps, etc. This is true in very few cases. You better know your clients really well before you take this approach. Some users will stumble over the nuances and get caught up in spelling, word smithing, etc. and that is their prerogative and nothing you say or do will change that. For critical users, never deliver shoddy merchandize.

If the software is critical, freeze the software a day or two in advance and dedicate the team to testing and bug fixing. No new functions should be added, and the software should not be tweaked at the last minute. If in doubt, have a clean twenty-four hour period where there are no bugs found and no code modified (and I mean no bugs and no modifications). You can bounce and bump along later, but the first access by the adopter is a very serious thing. A bad experience will take a long time, if ever, to fix! What might appear minor, irrelevant, or a fly speck to the geek might be felt to be very serious by the client and their feelings are to be respected and indulged.

Chapter 36
At the Helm

If you are the one in charge, you need to create and maintain a tricky balance of system process and avoid the rocks in the rapids. For example, you need to have clean authority and management tasks for your senior team members; what they are supposed to do, what decisions they are to make, etc. You need protocols and you need to respect the proper chain of commands. You cannot usurp their authority or interfere with their delegated tasks in a disrespectful way.

In calm water such as the Class I –III you have some clear visibility for the journey; you can probably let things go along smoothly and not inject yourself. There is time for people to discuss, debate, plan and react. Everything can be orderly because there are no sudden risks or issues. At various parts of a high-velocity process, this is not true and there are frequent encounters with risk. When you, as a master helmsman, sense an immediate danger you have to focus your resources to avoid or minimize the risk, hoping to get something done very fast. This is where egoless programming and an egoless team really helps. You do not have the time to send the message, person-by-person, down the length of the boat to the one person you need to do something different. You do not have the luxury of arranging a meeting in a day or two, nor can you take the chance of having port confused with starboard. It comes down to a quick "Hey you, yes — YOU — do this — **NOW**!" It is not meant to be rude, ignorant, or disrespectful of the individual, or others in the chain of command. It is a harsh fact of the reality facing you. If you do not do something immediately and directly, you have a major problem. You do not do this for minor issues. But, if you perceive it as critical or threatening, you have to have a team that understands this and does not get upset by the sudden change of direction and the sudden injection of orders top to bottom without warning.

I have seen whole corporations who seem to have this as a culture and it is really impressive. For example, in one case the CEO and others will obey certain rules and protocols that follow chain of command — to the letter. However, if the product is perceived to be threatened by a credible issue, kaboom — the CEO sends the message directly to the designer cutting through the organization, or gives a specific instruction based on his or her expertise. The organization and culture is focused on constant improvement and high quality. There is no wallowing about with things when it comes down to why the company exists. From what I have seen, no one gets upset when top-down hits. It is not finding fault with someone or firing someone; it is about the product.

This particular company works as a team and knows that such a top-down intervention is not to be taken personally or critically. Things must be addressed, addressed quickly, and it is OK. It is not the norm and the CEO does not do this daily. However, when it happens, it happens and people get on with it.

Doing a high-velocity software project is like this too. At the helm, you need to be able to operate in a normal fashion without much direct contact, but when necessary you need to be able to get into the middle of it and take charge — in person. This is another reason why you best be able to actually do things and be capable of hands-on duty. You need to be able to do the work if you have to. Roll up the sleeves and hit the code, design, development tools, etc. Not the norm, remember that. But, you need to be able to actually develop, design, and code if you are heading a *Mushing* project.

Chapter 37

Operational Control & Tracking

In an intense project, tracking and planning are very important. There are four levels of planning that might be needed in a substantial effort:

- Systems level — as captured in a high level project plan.
- Sub-systems level — as documented in the plan for each major sub-system and web-service.
- Monthly — team level — prepared by each full-time team member, incorporating their small team planning objectives and plans for the next four to five weeks — updated weekly, week level granularity of effort and tasks.
- Weekly — individual level — 1/4 to 1/2 day granularity for all team members.

The planning can be done via text and spreadsheet documents, or other fancier project planning and tracking tools. Getting people to do effective operational control, tracking, and planning is not an easy task. Some people will do it, understand the benefit, and learn from it. Others will not see the value in it and fight the exercise. They might only see the superficial aspects and not understand the bigger picture. They might not understand that the information being provided helps the next level up manage them and support them. They might not understand that the information helps the next level up balance resources, pick up clues, and anticipate problems. They might just think it is a list of tasks and dates for recording purposes only. Or, the person might not have the skill and experience to think through tasks, decompose them, and develop an orderly plan. Not everyone can do this.

As with many topics in this book I have had successes and failures when it comes to having people understand project management and engage. Possibly more failures than successes. I always try though. From a manager's perspective, there are additional benefits as it becomes clear through the exercise who might or might not have any project management skills and at what level that might be. Some people might be fine with day to day operational management duties, but not have the ability to think at the tactical or strategic level. Attempting to do the activities in this chapter help identify those people who can from those who say that they can.

The following subsections provide general guidelines for the types of project management reporting I prefer.

37.1 Tasks — Who Does What

The following is an example set of tasks that could be expected in a larger project:

- Systems level.
  - o Prepared by Director/Senior Designer/Chief Architect.
  - o Updated once or twice a quarter.
  - o Tracked monthly to senior management team via monthly status reports, altered accordingly.
  - o Reviewed and distributed to senior management.
  - o Subsequently distributed to the team via a collaboration site.
  - o Should require 1/2 day to prepare.
- Sub-systems level.
  - o Prepared by team leads.
  - o Updated once or twice a quarter.
  - o Reviewed and distributed to team leads, full-time permanent staff.
  - o Should require 1/2 day.
- Monthly — team level.
  - o Prepared by team leads.
  - o Updated Mondays — rolling plan format.
  - o Reviewed and distributed to team leads, full-time permanent staff.
  - o Should require 1/2 hour to 1 hour.
- Weekly — individual level.
  - o For current week — by 11am on Mondays — submitted to team lead, Director.
  - o For past week — update/reflection — by the end of Friday — submitted to team lead, Director.
  - o Should require 1/2 hour.

37.2 Detailed Plan Contents — Start of Week

The following are the types of things that I expect to see (or not to see) in detailed weekly planning by team members:

- Scope — does not deal with the past, strictly the current week.
- Sequence is not detailed unless it is important and is a cause for tracking or monitoring or relates to someone else's activities — e.g., where your task feeds someone else and they are expecting it on Tuesday afternoon.
- Any activities with interdependencies are also documented with best-before dates and expected completion dates.

- Usually, activities are identified in 1/4 or 1/2 day efforts/durations.
- Activities are tied back to a larger monthly goal/objective/task — via key name or code.
- For each task:
  - Task grouping.
  - Name or one–two word description.
  - Priority.
  - Short description of work to be done — the *what*.
  - Short description of the goal or intent — the *why*.
  - Estimated time start and duration.
- For the week at an aggregate level:
  - Note how any deviations from the previous week will be addressed and recovered.

37.3   Weekly Updates and Reflection

I like to have team members reflect on the progress each week. This helps people learn and it helps to put things in perspective. At the end of the week, I like to receive the following from team members:

- For each task:
  - Update the actual start and duration.
  - Note the status — if complete, postponed, in progress, cancelled, etc.
  - Note any deviation and reason for the deviation.
  - Note the results of the activities.
- For the week at an aggregate level:
  - Note any issues or questions that the team lead or supervisor should be aware of.
  - Note and reflect upon the week relative to your learning objectives and learning process.
- For any missed tasks, identify recovery plan or what is planned to address the challenge.
- At the end, identify any risks or issues to the monthly plan.
- For each task listed in the weekly plan, indicate any +/– variance in effort or duration and identify reason.
- For any new task not in the weekly plan, indicate why it came up.
- For any task dropped in the weekly plan, indicate why it dropped out.
- Provide any appropriate suggestion for what management can do to make your job more effective or efficient.

37.4   Monthly Level Details

Team leads and supervisors are expected to summarize and tie the weekly planning to the bigger picture:

- Monthly plan has reasonable detail for the next four to five weeks.

- A short exception note is included addressing the past week and how any deviations impact the plan and will be dealt with.
- Sequence is not detailed unless it is important and is a cause for tracking or monitoring or relates to someone else's activities; e.g., where a task feeds someone else and they are expecting it on second Tuesday of the month.
- Activities are given a key phrase or code for tracking.
- Activities are tracked at the week effort level, or 1/2 week — not at the hour or day level.
- Key milestone reviews or dates are tracked.
- Activities are summarized in two or three sentences. What is to be done, why is it being done, expected effort, expected elapsed duration, and any key dependencies (from others within the team or external). Activities are also rated in terms of priority and confidence of successful completion.
- Any activities with interdependencies are also documented with best-before dates and expected completion dates.

Chapter 38
Team Design

To use the analogy of a dog sledding experience, there is a driver and a lead dog — everyone else is in between. You might also have the equivalent of the pit crew providing supporting infrastructure during a automobile race. The analogy also works for a team effort in rafting or boating where someone steers and someone else extols the team and focuses the team's rhythm. I have briefly discussed the types of people you want to have or not have on the development team at various points in the book. This brief chapter looks more at the team as a whole and the team organization.

You need someone in charge who has a holistic view of the problem and an organic understanding of the user model(s). This person sets the general feel and philosophy for the finished product. This is the functional architect. You might also have a technical architect. I have worked on teams with the architecture role split, and on teams where one person was able to do both. It is likely that the functional architect's dream and promises sold the product or investment, so at the end of the day their word has more weight to it. Their job is also likely to be more at risk. The technical architect's role and duty is to make the functionality happen, not question it per se. The technical architect can point out inconsistencies and risks, but they should not waste time with discussions about what the user needs or does not need, or what the user will like or not like. Not their job. Similarly, the functional architect needs to focus on the end value and unless the technical decisions impact functionality, should not waste time debating the subtleties of one server disk array configuration over another.

Both architects, technical and functional, should be open to challenge, debate, and rational decision making based on evidence based reasoning. Do not expect to sway them with opinions or a group petition based on likes and dislikes. As a team member, your job is to convince and persuade a decision maker that something different is something better. Not worse and not just the same. The decision maker needs to be confident that the product will deliver on the promises made; promises that they might know and you do not. There can only be one boss and while most decisions and ideas might be arrived at through a consensus process (note: consensus does not mean unanimous), it is sometimes necessary for the boss to make decisions that only the boss can and should make. They may take a calculated risk that others do not agree to or support. That is the prerogative of the boss. However, this should not be done lightly and it should be clear that the boss is taking the risk and is not holding the others responsible for the decision.

It is common for the architects to present their solutions and have to defend their choices. One way to do this is for the architect to start off with the requirements and solicit ideas "This is my problem, what would you do?" This is different than presenting a solution and possibly getting defensive. This also allows the architect an opportunity to see what the critics know or do not know. After getting several ideas on the table, the serious discussions can take place about suitable solutions, if the proposed solution is suitable, if there are ways to make the solution simpler or easier, what are potential risks, problems, and so forth.

That covers the driver of the sled or the person steering the ship. You need others in the team. If you can, you have a separate project manager, and someone else to head up quality assurance activities. And of course, you need developers. As noted earlier, the overall mix should be of the old and the young for anything substantial at the Class V or VI levels. At Class I, you can get by with many young developers guided by a seasoned professional. For the more difficult projects, you probably want a tier of expertise and by the definition of expertise and the time required to gain expertise; this implies that you will have an inter-generational team. For example, in one Class VI venture, I wanted a seasoned mechanic or blacksmith to deal with the infrastructure and the various tools that will be needed. This individual needed to know or learn many tools and be able to configure, debug, and wire up a system with their eyes closed. They also needed to know how to deal with vendors, purchasing, and the developers who will want everything yesterday. The blacksmith had to be able to create a steady development environment that will allow others to work. It is the management's job to provide the blacksmith with sufficient resources to do this with. The blacksmith needs to be a real expert and have breadth and depth, not just depth on a single idea or system. It is hard to find someone with all of these traits, but a close enough approximation will do.

In a smaller team, the head designer might be the only designer and the roles are clear. However, in a larger team, there might be several designers and they might be the team leads. The role of the chief designer is a bit different in this case. The chief designer is responsible for the basic building structure and gross capability. It will be a factory building versus a house type of decision. The chief is responsible for setting out the guidelines and general ideas that the software has. This might include technology selection at a macro level, the identification of major subsystems, interface requirements between components, and goals for each subsystem. These requirements which might include concepts for flexibility, migration, evolution, encapsulation, robustness and testing are then interpreted by the next level of designer. It is not feasible or reasonable to assume that every designer is cognizant of all requirements and all assumptions. The higher level design sets the borders and boundaries and the rules for drawing within the box. The lower levels then have reasonable autonomy in design and implementation. If there are multiple teams, there might be requirements that all teams follow the same coding, interface, testing, and documentation conventions. This will require coordination between the teams and the team leads. In most cases, the chief designer should not and will not worry too much about the specifics of such conventions unless they will interfere with goals and objectives of the team's product.

The team leads must provide sufficient visibility via project management tools and other techniques so that the driver understands. The driver needs to see and comprehend enough to know that the right direction is being followed, that the assumptions are understood, that the end result will be the factory building desired by the client and not a townhouse development. The head must also provide sufficient information to the lower designers to basically get close enough, or closer, to the target.

In a Class I —III situation, this interplay between the head and secondary designers can be relatively clear cut and smooth. In a Class VI situation, especially one where the team is built specifically for the project from scratch, the journey can be expected to be a bit rough. In a rapid development, it is not possible to convey all assumptions and have a long learning curve before work has to begin. You have to work with patient team leads and know that certain work will be launched before all is ready, before conventions have been set, and so forth. You can control this by doing independent toothpicks and slowly putting together an integrated framework. Remember, it is possible and very likely that portions of the system will have to be redone in a Class V or VI situation. I have mentioned it before: plan for it, tell people it will happen, and do not be surprised when it happens. This can be a challenge with highly motivated individuals with high standards, possibly with limited patience for re-doing things or for doing things that they suspect will have to be re-done. There can be many lessons learned by early launching and taking the risks though and as long as it is two steps forward and only one back, it can be justified. If the team leads have been in this type of situation before, it is a good thing.

I like to have a multi-skilled team. As the team fills in, I usually cannot afford two people with exactly the same skill sets at the expert level. I like to overlap them. I might have a super super person on a skill and another just at the super level. This prevents the team from exploding if someone is sick or departs for places unknown. I also like it when the hands-on folks can do a little bit of everything. This helps with critical path activities when everyone might be needed on deck, and when design reviews are held. The team members must understand that on a high velocity project, decisions will be made if they are there or not, and they will have to live with the consequences of someone else making a decision or doing a task. Hopefully, everyone will respect each other's ability, and will respect the result. I also like paired activities. I prefer having a senior developer work with one or two juniors. You do not want too many juniors working together, or have one senior developer responsible for too many juniors. For mentoring and for tight project control, a lower number is preferred.

As part of the team and execution model, you also have to consider team dynamics and training. If the team is intense and working above the call of duty, I will sometimes specifically schedule non-project time; things like no-work-Friday-afternoons. I will have the team do right brain things such as drawing lessons, photography, watching movies, going for a hike, and other such things. If the team has lots of junior people, I will schedule software engineering classes for the team. And, if you add up some other things, I will try to have about 20% of my budgeted time dedicated to developing and training my staff. This is not always possible, but it is my general goal.

Even if the staff are transient, as in students, the investment is worth it. There will be other training for supervisors and others, but 20% of the whole project may have to be in weekly training and learning. For students, I might have little competitions to see if they have been listening and are able to apply the lessons being taught and this is worth the effort as you deal with what they learn and how they go about applying the learning. However, I will not babysit on the learning. I expect people to keep up to date and do professional development. I believe in lifelong learning and when I interview, I try to understand if the candidate knows how to learn to learn and is a self-learner. I will not teach lessons in topics that I think people are quite capable of learning by themselves, or that they should be getting via a certification process. This strategy of training and learning is part of what I usually call Plan A, when everything is going well and assumptions are valid. As a project evolves, it might be necessary to change this strategy if things get late and other mistakes have been made.

Egos. There are some good things about egos and lots of bad things about egos. It is impossible to have a 100% egoless team and 100% egoless developers who are qualified to be on a Class VI journey. The personnel need to be confident but not arrogant. They need to know what they do not know and know what they do know, and how well they know it. They need to take pride in what they do, but it is a different type of pride. Not everything will be perfect; there will be mistakes, and there will be bumps. The pride comes from the end result of having users getting value from the system. This you can be proud of and do a little bit of chest thumping about. And, the chest thumping should be delayed for a few years, to make sure that the system is really giving value. You should not chest thump during the development or while the system is on a honeymoon status.

On an egoless team, it should not matter if I design something, another person codes it, and a third does the final testing and integration. It should not be my personal property and I should not refer to it as MINE. Using the 'I,' 'me,' 'mine' words should be avoided. Ideas are put on the table and the team owns the ideas. If the coding and development is done to the good enough level as defined by the plan (where some stuff needs to be great and other bits not so great), it should not matter who does it as long as the final code meets its goals in terms of functionality, quality, and time. You should not care if someone takes your design and re-designs it when it is justified and for clear benefit. It should not be just-as-good; it needs to be better. It should not be redesigned just because it is not liked for personal reasons either. If egos are permitted to flourish, and people change existing designs and code for no good reason there will be many problems along the journey.

The team should be encouraged to use the royal 'we' and 'us' and 'our' words when presenting work associated with the team. In an egoless team, it should not be apparent to outsiders who did what and who contributed what. They may know who was involved and what people were generally responsible for, but they do not need to know that Joe designed and wrote blah. There are other ways to receive credit and get the praise you deserve. You do not need to crow yourself or fluff yourself looking for pats on the back. It is best to present a unified front. Of course, there are times that some identification is needed if credibility is questioned or needs to be established. So and so is an expert in blah-blah and he/she did the design for blah-blah. But, unless it is that type of meeting, the team needs to play it cool and unassuming. There are also times to highlight exceptional effort or contributions, but it is never your own; it is of others.

The ability of the team to bury its collective ego is important. By design, a high powered *Mushing* team is capable of living off the land without any help from other supporting groups. In a pinch, it can do all of the installing, configuring, wiring, operating, acquisition, and trouble-shooting. It is like a focused factory situation that has its own electricians, own receiving, own shipping, own planners, and does not have to wait for shared resources or for a place in the queue. This might cause other groups concern when and if the team has to do things that might normally be done by others. If others can do it within the necessary time period, the others should do it. You should also try to set things up in advance so that the others can help you. You should not do something that another part of the organization can. But, if the team has a mandate to deliver ASAP on a strategic system, it might be necessary for the team to occasionally do things in advance of others. For example, depending on the project, the *Mushing* team might be investigating new technology or using technology in a different way and it is not possible to wait for the supporting teams. It is the duty of the *Mushing* team to provide help and training for the infrastructure groups so that the project can be sustained and that other similar projects can be supported. It is important to remember that the team is not *better*, it just has different skill sets and is able to do different tasks if called upon. The *Mushing* team will not replace other teams or threaten their jobs, but the visuals may be hard to control. A controlled ego will help with the interfacing and inter-team dynamics.

It is best to have the lead dogs and pit crew on board reasonably early. It is too chaotic if you have everyone start at the same time. You need the leads to understand the product concept and the team philosophy before the junior staff arrive. On a Class VI project with new technology, it might take several months to set up the infrastructure and then another couple of months to bring the team on board and ramp up. It is easy for Class I —III situations. You get a call, look at what is needed, grab things off the shelf and get the job done. You do not need months to set up. You do not need months for a learning curve. Unfortunately, this is not the way Class VI projects proceed.

Hiring for a Class VI job is difficult if you are starting from scratch. You might have a whole team of people who have not worked with each other before. You might have people all freshly hired and all you know is based on the resume, interview, and reference checks. In an ideal situation, everyone is great and can deliver on their promises. This is not always possible, but it is the goal. This is Plan A. Plan B depends on your own skills and other people you can call in to help when hiring does not turn out right.

Who do I like to hire or work with for a Class V or VI job? I have a set of criteria that, when possible, I like to run with:

- o I do not like to hire problem children (see Sutton's book on the *No Asshole Rule!*). I like to hire people who are fun to work with and are nice to work with. I do not like people who are disruptive by design or by accident. Technical brilliance is only about 20%–25% of my hiring criteria. If you are a problem, forget working for me. If one is accidentally hired, you have to deal with the situation. In a new organization and asked to staff up a team, I ask two questions — how do I hire and how do I fire — close the loop.

o I like senior developers with demonstrated results and junior developers with demonstrated potential.

o I prefer hiring developers who have been with projects long enough to see if their judgments and decisions were reasonable and good enough. If someone moves jobs and projects every eighteen to twenty-four months, their resume ends up in the round basket at the end of the desk. If you do not get feedback on your mistakes and decisions, you will keep repeating them. There is no proof that your idea was the greatest thing unless someone uses it. There is no proof that your design can be sustained, can evolve, can be maintained, is flexible, or of high quality until the test of time has been applied.

o I like hard working, but lazy developers. The ones who will use existing code and structures if they are good enough.

o I want to hire people who get their hands dirty and are willing to do anything on the project — "I don't do that. It is beneath me." Does not cut it. I do not want managers and team leads that just point and direct. They need to write code, get into systems, and be capable and willing to do actual development. A *Mushing* team does not have space for people who just push paper and manage.

o I like hiring people who code for a purpose of providing someone value. I do not like to hire developers who code for the sake of coding or use technology just for the sake of using it. I will try to find out their very first programming experience and why they did it.

o Senior developers must be able to communicate and work with users. In written form (emails and documents), in meetings, and be able to lead a meeting and give presentations.

o If the project is ramping up with high expectations for speed and quality, I want developers who already have a clue, not developers who have to be taught and trained for many months.

o I want developers who are confident and work with evidence based reasoning. I do not want someone who lives by macho statements, rhetoric, and bravado.

o I like developers who know when to say yes, no, and "I don't have an answer — let me look into it and get back to you." I do not like developers who are quick to say something is difficult, is not possible, is possible, is easy, is trivial without some basis for the statement.

o I do not want all of the team members to be perfectionists who are not happy unless everything is 100%. They need to know when enough is enough and not be hung up on idealistic icons or concepts.

o I want developers who do not gossip, whine, or constantly complain. I want ones who don't just do what is asked of them, but do what needs to be done, without all of the complaining.

I might have to compromise on some of this list, but when I am given choices and options, I will go with my preferred profile. In various projects over the years I have been lucky with some teams and not so lucky with others. I have hired great developers and I have hired people who smoked me during the interview. Hiring is not an easy process. In some cases you can try to set the tone or team expectations. For example, I have zero tolerance for some things and it is best if team hires understand my view of life. Below is a code of conduct inspired by reading Sutton's book. I am not sure how practical it is, but it does set the tone:

The team executes under tight timelines directly with clients and relies upon a strong working relationship with other support groups. In most cases, an Agile/Extreme methodology is expected and the work will be done in a very visible and transparent fashion. This approach places high expectations on all team members in terms of professionalism and productivity. It is not only doing the right thing, it is being seen to be doing the right thing, and in fact exceeding expectations of the clients and interacting groups. All members of the team are expected to behave and conduct their activities consistently and in line with standards set for the team.

The following guidelines form the basis of the Team's Code of Conduct:

1. *A strong client focus —respect, support, and listen to the stakeholders. The special projects team exists to create value for the stakeholders.*

2. *A strong team focus —respect, support, and listening to all team members and supporting groups. Understanding scope, authority, and appropriateness with respect to policies, procedures, and personal interactions.*

3. *A high level of professional communication (verbal and written) —fluency, grammar, content, tone, and presentation.*

4. *Members of the team will help each other in accordance to the team's standards for learning and training as established by the team leaders.*

5. *Perceived barriers will be dealt with in a positive, professional fashion, involving team leaders when and where scope and boundary issues dictate.*

6. *The project will be ego-less and results will be left to speak for themselves.*

7. *All team members will adhere to team standards for project management, documentation, code, and testing.*

8. *All team areas and members will be professional in appearance and conduct at all times.*

Examples of behavior deemed unacceptable (from a recent book by R. Sutton on workplace behavior):

Personal insults; invading one's personal territory; uninvited physical contact; threats and intimidation, both verbal and nonverbal; sarcastic jokes and teasing used as insult delivery systems; withering e-mail flames; status slaps intended to humiliate their victims; public shaming or status degradation rituals; rude interruptions; two-faced attacks; dirty looks; and treating people as if they are invisible.

Chapter 39

Mission Critical Systems

This is almost the last chapter. At various points in the text I have mentioned something called mission critical systems. What is a mission critical system?

I have been dealing with mission critical systems since the mid 1970s and I have learned that there are different types of mission critical systems, and that mission critical systems in general have to be handled differently than normal IT tools and systems. If a mission critical system fails to deliver its value, there are potentially long term or costly repercussions, or both. The impact of a mission critical system failing can result in the death of individuals at the extreme, death of companies, projects, or products, or the crippling of organizations to the point of near-death. The direct failure can be felt by you or by others, but the end result is the same: death or near-death of someone or something.

You identify a mission critical system through your risk analysis and it is relatively simple to do. You simply imagine the situations where the system does not exist, where the system exists and suddenly does not exist, and where the system malfunctions and generates the wrong information or does the wrong thing. What is the impact or recovery from an inoperable system or a malfunctioning system? If the net result is not too bad, it is not a mission critical system. If there is minor pain, it is not a mission critical system; your mission is for all intents and purposes intact.

Once you realize that you have a mission critical system, you need to alter your normal processes. You need to do your analysis differently and spend more time on it; you will need to view your vendors differently, and consider your acquisition/deployment strategy in a different light. Every step, every process of your acquisition, deployment, and support will be different. One thing you will immediately figure out is that your budget and planning will be different and that a mission critical system will warrant extra investment and extra care!

There are several ways to view a mission critical system. It might be a STRATEGIC Information System, or it might be a STRATEGIC INFORMATION system. There is a difference. One deals with operations and execution, while the other deals more with decision making, business intelligence, and the information that is key to your business. This is where I would start. I would try to figure out what the mission critical system is. What is the mission, and why is it critical? What is its value equation? Until you start to scope or classify the situation, you are dealing with a big buzz phrase that can be misunderstood and mishandled. It is possible that a mission critical system can be a Class I –III system; one that can be purchased off the shelf, assembled from components, and that is relatively low risk with respect to planning, acquisition, deployment, and operation. It is also possible that a mission critical system can be a Class IV–VI system which will be more unique, low volume, and be risky. A mission critical system can also be inexpensive or costly.

A mission critical system might support your day-to-day operations, or support the tactical aspect of the venture, or contribute to the strategic governance or planning for the venture. An effective backup system is an example of the first. In most enterprises or activities, you need to have backups and a way to recover the data. If the whole disk array is contaminated in a business relying on customer files, design specifications, etc., what will the impact be? If your business is the accepting and shipping of customer items, your operational system is critical because it is how you make your money: tracking and controlling the movement of items. If the operational system provides functionality that helps differentiate your company, product, or service from your competitors, it has an escalated importance as well; a failure of the operational system can impact your brand and your whole business.

A mission critical system warrants extra thinking and planning at the start of the project. If you get it wrong, it will likely be hard to fix or change, and the whole venture might fail. For example, if you bet on certain emerging technology without a Plan B, you might be in deep trouble. If you lock in with a specific software vendor and the vendor goes bankrupt and the software is no longer supported, what will happen to you? The magnitude of the impact should give you a feeling of how much extra effort you need to put in. Will the impact be felt by thousands of customers? Will it affect your ability to make the payroll? Will customers go elsewhere or impose crippling penalties on you? Every situation will be different, but you should develop a short document that clearly identifies all of your current and planned mission critical systems and ensure that you are dealing with them appropriately. They need special care and feeding! They are not like word processing systems for creating inter-office documents, or expense tracking spreadsheets.

In perhaps the best scenario, you are in a Class I –III situation and there are several mature, long-standing vendors or third-parties to acquire a turn-key system from. You might host it yourself, or use the vendor to provide complete service, or use the vendor to provide a business continuity option. In these cases, you do the usual due diligence on supplier stability, reputation, and so forth. You pick the one you like and you make sure that you can get your data out and into another system easily and quickly if you have to. If there are existing systems that do the basic functions you need, it is not wise to build your version from scratch, or assemble a Tower of Babel solution yourself. Use a vendor or third-party integrator to provide the functionality and operational environment. Have them design your solution for you and use their expertise. You do not want to experiment or learn how to create a mission critical system if you have a choice.

When you get into a Class IV–VI situation, mission critical systems get tricky. Usually you cannot find a system off-the-shelf and you have to build up from the base technology or create the magic by assembling some existing and new parts. If you do not have prior experience designing or building a mission critical system, you should not consider doing it by yourself. There are many considerations and issues that will arise and if you are learning as you go, the path will be dangerous and painful for all concerned. You can do it yourself, work with others in a collaborative fashion, or outsource the complete project to an experienced organization. Three options. Each option has implications, risks, and costs. Even if you have done prior mission critical systems, a low volume, unique system that is implied by Class IV–VI, the path will be hard. You should not assume that you can pull it off — you will need Plan B or Plan C as insurance.

For extremely demanding mission critical systems, I do have a number of heuristics or concepts I try to use.

- I think and plan a great deal before making a build versus buy decision. In one system this took almost a half year. You need to be very careful when you decide to build yourself, not assemble, not buy. You need upper management to appreciate and understand the issues, risks, and benefits from each strategy.
- I try to clearly separate infrastructure from application layers with clear interfaces having loose coupling. Like the earthquake model.
- Choose a single vendor for as much as possible of the base infrastructure as reasonable. If possible, include development tools and operational tools for monitoring and manipulating the infrastructure. There is less finger pointing and a greater chance that everything will continue to work together as new versions are rolled out.
- If your organization does not have a track record, or lacks the skill and knowledge to build the system, then you work with an outside party to either build or acquire what you need. The infrastructure and application layers should be able to be dealt with as separate entities without tight dependencies on each other. That is, you can use an outside party to help with the application and/or the infrastructure aspects. A major vendor might have partners they rely on and you need to check credentials and their track record as well.
- I try to create a balanced system that will behave in a consistent fashion and that will work together in a seamless fashion. This often implies that you do NOT cobble a system together based on a myriad of best-in-breed tools from multiple sources. I have found it better to compromise on this aspect. I would rather have a robust system composed of similar work horses than have a set of software solutions where each acts like a fragile or self-willed thoroughbred race horse. I do not like mission critical systems where different key components are on different cycles, driven by different philosophies, owned by different companies. I have been there. It is not fun for anyone.
- If you can assemble suitable, compatible subsystems, that is better than building your own. You might have to compromise on certain features and functions and these need to be discussed at the business level. There are costs and risks when you brew your own and the company must be willing to invest and understand the extra costs and risks versus the functional value you might sacrifice.
- I try to avoid open source solutions for key components unless I am working with a very

large entity with a very large, very skilled, and very talented set of developers. A small firm with a handful of developers are playing a high risk form of Russian Roulette if they are betting their future on a open source base. They will have to take the open source base into their organization if they are going to rely on the functions, quality, robustness, and performance for their business; and for a tiny application, this might be OK. It might in fact be a business model: adopt and adapt an open source base as the product or as the key business software. This is a very risky way to do business. Some small firms do succeed, but it is my own observation that most fail.

- For the base infrastructure, I try to avoid the first release or version of anything. I like software that has aged a bit. I will take risks on some software but it is an acknowledged risk and I will try to have alternative plans and options if the newly released software fails. I will not bet my client's future on an untested base without a Plan B. I might use and test some new infrastructure components in the base development environment, but I will control the final production version and I will be conservative.

- I will not usually work with a vendor or third party who is going to use me or my client as the learning ground. Certainly not for a mission critical application. I will be a beta user or experiment on less critical applications, but not on the money maker. If I can create a parallel world and contain the possible impact, then I will do experiments and go into the land of mystery and risk.

- Assuming that I have a team that can build the mission critical system, I also have to consider who will maintain and support the system in the long run. This will also affect what I choose and what I do. I will try to avoid libraries and functions that will rapidly change and I will try to avoid exotic and complicated functions that might not be understood by the supporting team. A team of high-end geeks can develop a system using many innovative techniques and tools that cannot be maintained or understood by mere mortals. If you are going to use extreme coding techniques, it is then important to test out the theory that others can indeed understand the code and can support it. If others cannot, then you have to back off the highly-skilled functions and compromise the coding. Lower the geek-factor down. I have done this several times in my career; build something and then test to see if others can support it. In some cases, the code stays the same. In others, the code has to be reduced in sophistication and complexity.

- You have to run mental or paper simulations on how the system can fail and what can happen. How can it be prevented, detected, and how does recovery happen? For a big system, this can take weeks or months and is not something that can be done in minutes or hours.

- You have to plan out day-to-day operations, failure recovery, version control, and go through a complete life cycle thought process. What is likely to take place over the next decade? Will your vendor still be around? Will the software infrastructure still be supported in some compatible or close enough fashion? What might change? How does your solution take this into account?

- I will think through the system and think about what happens if any major subsystem needs to be replaced? Perhaps with a new version. Perhaps with new technology. Perhaps from a different vendor.

The more critical the system is, the more extreme I will be in my options and decisions. I will push more for single vendor support. One vendor for the software base, one vendor for the hardware. As much as possible. I will push for loose decoupling and encapsulation between and around subsystems so that I can pop out a system and replace it without causing the whole world to collapse. For any component that is strategic to the business and is meant to help brand or differentiate the business, I will try to do this in-house if possible. This is not always possible, but when it is, I will try it. You want to control your intellectual property and you need control over the key functions and your destiny. However, you need to understand the risks and the costs. If you have never built or used a certain type of complex software before, I would not advise doing it from scratch in your mission critical system. I would also be aware and careful about the second system effect where people swing from the one extreme of not knowing what to do and how to do it, to trying to do everything and over design and over build the system. If you are a novice in the domain, go find a real expert to guide you.

I will try to work with vendors and possibly third parties to understand the best and most appropriate way to use the infrastructure components. They are the experts and it is important to get their insights and guidance. They are invaluable when considering performance, availability, and future migration paths. They can help with strategies for how to layer and loosely couple the technology.

If I finally decide to actually build a system, I will control concurrent risks, types of risks and try to have backup plans for key elements. I will try to follow as much of this book as possible with respect to understanding the problem and designing a robust and flexible system. Not everything in this book will help you on a mission critical build, but most will. Building is my last option. There are times I will and there are more times I will not. It depends. No hard or fast rules.

Chapter 40

Final Thoughts

This is the last chapter. I have tried to hide nothing and to share all. The good, bad, and the ugly. I have tried to ask as many questions and to raise as many challenges as I can. If I had to sum up the main points from the book, they would be:

o It is not about you, it is about the value you give the users while addressing their comfort, experience, and evolution.

o Respect the user. Respect their time, their knowledge, and their desires. This does not mean that you do everything they want or suggest, but that you do not dismiss it out of hand.

o You might be good, very good. But drop any attitude. There is always someone better and you are not perfect.

o Do not confuse being quick-witted with being smart or knowledgeable. It takes time to develop the depth and breadth to be a good designer or architect.

o Do not let one success or one good experience get into your head. You have to be able to repeat successes — time and time again. Don't be a one hit wonder.

o Become a student of history: computing history, software design history, and management history. You can learn a great deal from the past.

o Always reflect and assume that you have room to improve on what you deliver and how you delivered it.

o Break things down to smaller, simpler components. Assume that constants aren't and variables won't.

o Go live with the users and understand them. Then build the system. Not the other way around.

o Try to engage and be with the user during the development. Really really be with them. Build tools and systems that allow you to do this in a quick and engaging way.

o Do not let methods and artifacts turn into religious icons and blind you from common sense. Understand the intent and concepts; analyze them in the context of your situation.

o Do not be pedantic, close-minded, literal-minded, or try to be a perfectionist when doing high end development in an Agile/Extreme way.

o Do not blindly follow what you read, what you hear. Do not let others do your thinking for you. Do not make decisions that will impact your client based on what you think your friends will think.

o Do not expect code or development in the early phases of a Class V or VI situation to look like polished code in a Class I –III development. It will not be pretty, nor polished. It will be efficient to bloat, copy, and be somewhat sloppy in a strategic way.

o Do not assume that claims of best practice makes it so. Do not assume that because it has not been done in the past it should not be done in the future.

o Do not assume that your assumptions are valid. Be aware of your assumptions.

o Do not assume that your situation is traditional or that it is non-traditional. Understand the problem and go with what it is, not what you wish it to be.

References

Abrahamsson, P., Salo, O., Ronkainen, J. and J. Warsta. (2002). *Agile software development methods — review and analysis*. Publication of VTT Technical Research Centre of Finland.

Ackoff, R.L. (1978). *The Art of Problem Solving*, New York: Wiley.

Adler, M.J. (1940). *How to Read a Book*. New York: Simon & Schuster.

Argyris, C., Putnam, R., and McLain Smith, D. (1985). *Action Science*. San Francisco: Jossey-Bass Publishers.

Aoyama, M. (1998). Web-based agile software development, *IEEE Software*. Nov/Dec.

Ashby, W.R. (1964). *An Introduction to Cybernetics*. London: University Paperbacks.

Aytug, H., Lawley, M.A., McKay, K.N., Mohan, S., and R. Uzsoy. (2005). Executing Production Schedules in the Face of Uncertainties: A Review and Some Future Directions. *European Journal of Operations Research*, 165, 1, 86-110.

Babbage, C. (1832). *Economies of Manufactures*. 2nd edition. London: Knight.

Barnes, J.G. (2001). *Secrets of Customer Relationship Management*. New York: McGraw-Hill.

Beck, K. (1999a). Embracing change with extreme programming, *IEEE Computer*, 32(10).

Beck, K. (1999b). Extreme programming explained: embrace change. Boston: Addison-Wesley.

Beck, K., Beedle, M. Van Bennekum, A., Cockburn, A., Cunningham, W., Fowler, M., Grenning, J., Highsmith, J., Hunt, A., Jeffries, R., Kern, J., Marick, B., Martin, R., Mellor, S., Schwaber, K., Sutherland, J. and D. Thomas. (2001). *Manifesto for agile software development*. Agile Alliance.

Beizer, R. (1988). *The Frozen Keyboard: Living With Bad Software*. Blue Ridge: Tab.

Boehm, B. (2005). Get ready for agile methods, with care, *IEEE Computer*, 35(1), 64-69.

Boehm, B. and R. Turner. (2004). *Balancing agility and discipline: a guide for the perplexed*. Boston: Addison-Wesley.

Bowker, G.C. and Star, S.L. (2000). *Sorting Things out — Classification and Its Consequences*. Cambridge: MIT Press.

Brooks, F.P. (1995). *The Mythical Man-Month* - Anniversary Edition, Reading: Addison Wesley.

Browne, M.N. and Keeley, S.M. (2004). *Asking the Right Questions, A Guide to Critical Thinking*. 7th Ed., New Jersey: Pearson.

Camerer, C.F. and Johnson, E.J. (1991). The process-performance paradox in expert judgment — how can experts know so much and predict so badly. In *Toward a General Theory of Expertise — Prospects and Limits*, K.A. Ericsson and J. Smith (eds.). Cambridge: Cambridge University Press, 195-217.

Chi, M.T.H. Glaser, R., and M.J. Farr (eds.). (1988). *The Nature of Expertise*. Hillsdale: Lawrence Erlbaum.

Cockburn, A. (2002). *Agile software development*. Boston: Addison-Wesley.

Colley, A.M. and Beech, J.R. (eds.). (1989). *Acquisition and Performance of Cognitive Skills*. New York: Wiley.

Couger, J.D. (1995). *Creative Problem Solving and Opportunity Finding*. Danvers: Boyd & Fraser.

Cusumano, M. and D. Yoffie. (1999). Software development in internet time, *IEEE Computer*, Oct.

Delaney, C. (2004). *Investigating Culture, An Experiential Introduction to Anthropology*. Malden MA: Blackwell.

Dewey, J. (1938). *Logic — The Theory of Inquiry*. New York: Holt.

Diebold, J. (1952). *Automation: The Advent of the Automatic Factory*, New York: Van Nostrand.

Easterby-Smith, M., Thorpe, R., and Lowe, A. (1991). *Management Research — An Introduction*. London: Sage Publications.

Ericsson, K.A. and J. Smith. (1991). *Toward a General Theory of Expertise — Prospects and Limits*. Cambridge: Cambridge University Press.

Ericsson, K.A., Charness, N., Feltovich, P.J. and R.R. Hoffman eds. (2006). *The Cambridge Handbook of Expertise and Expert Performance*. Cambridge: Cambridge University Press.

Emerson, H. (1911). *The Twelve Principles of Efficiency*. New York: The Engineering Magazine.

Galotti, K.M. (2004). *Cognitive Psychology in and out of the Laboratory*. 3rd Ed., Belmont CA: Thomson.

Gause, D.C. and G.M. Weinberg. (1990). *Are your lights on?*. New York: Dorset House.

Glass, R.L. (1997). *Software Runaways — Monumental Software Disasters*. New York: Prentice-Hall.

Glass, R.L. (2003). *Facts and Fallacies of Software Engineering*. Boston: Addison-Wesley.

Goldratt, E.M. (1997). *Critical Chain*. North River Press.

Grills, S. (ed.). (1998). *Doing Ethnographic Research — Fieldwork Settings*. London: Sage Publications.

Gross, T.S. (2004). *Why Service Stinks... and Exactly What to Do About It!*. Chicago: Dearborn.

Gummesson, E. (1988). *Qualitative Methods In Management Research*. Sweden: Chartwell-Bratt.

Haungs, J. (2001). Pair programming on the C3 project. *Computer*, 34(2), 118-119.

Hunt, A. and D.Thomas. (2000). *The Pragmatic Programmer*. Boston: Addison-Wesley.

Jackson, M.A. (1975). *Principles of Program Design*. New York: Academic Press.

Jewer, J.L. (2003). *Managing Risk in Software Projects: Prioritizing the Requirements*. MASc Thesis, University of Waterloo.

Jorgensen, D.L. (1989). *Participant Observation: A Methodology for Human Studies*. London: Sage Publications.

Kozmetsky. G and P. Kircher. (1956). *Electronic Computers and Management Control*. New York: McGraw-Hill.

Lavrakas, P.J. (1993). *Telephone Survey Methods, Sampling, Selection, and Supervision*. 2nd Ed.,

Thousand Oaks CA: Sage.

Liu, L. (2004). Case Study: Analysing the Impact of Community Health Information Network (CHIN) in Health Care in Ontario, MEng Project. University of Waterloo.

McCarthy, J. (1995). *Dynamics of Software Development*. Redmond: Microsoft Press.

McConnell, S. (1996). *Rapid Development*. Redmond: Microsoft Press.

McKay, K.N. (2003). Framework for Analysing the Impact of Information Systems. Course Notes for MAN. SCI. 442, 2003-2006.

McKay, K.N. and G.W. Black. (2006). Aversion Dynamics — Adaptive Production Control Heuristics Incorporating Risk. *Journal of the Operations Research Society of Japan*, 49, 3, 152-173.

McKay, K.N. and G. W. Black. (2007). The evolution of a production planning system: A 10-year case study. In press, *Computers In Industry*, 58, 8-9.

McKay K.N. and T.E. Morton. (1998). Critical Chain - E. Goldratt, Book Review. *IIE Transactions*, August, vol. 30, 759-763.

McKay, K.N. and V.C.S. Wiers. (2003). Planners, Schedulers and Dispatchers: a description of cognitive tasks in production control. *International Journal on Cognition, Technology and Work*, 5, 2, 82-93.

McKay, K.N. and V.C.S. Wiers. (2003). Integrated Decision Support for Planning, Scheduling, and Dispatching Tasks in a Focused Factory. *Computers In Industry*, 50, 1, 5-14.

McKay, K.N. and V.C.S. Wiers. (2001). 'Decision Support for Production Scheduling Tasks in Shops with Much Uncertainty And Little Autonomous Flexibility. In *Human Performance in Planning and Scheduling*, B. MacCarthy and J. Wilson (eds.), Taylor and Francis, New York, 165-178.

McKay, K.N. and S. Ng. (2004). Kore Ga Nani Wo Imisurunoka — What does it mean?. Working Paper V.2, University of Waterloo.

McKay, T. (2000). *Reasons, Explanations, and Decisions — Guidelines for Critical Thinking*. Belmont CA: Wadsworth.

Morton, T.E. and D.W. Pentico. (1993). *Heuristic Scheduling Systems*. New York: John Wiley.

Negroponte, N. (1995). *Being digital*. New York: Vintage.

Norman, D.A. (1988). *The Design of Everyday Things*, Basic.

Olson, J.R., and Biolsi, K.J. (1991). Techniques for representing expert knowledge. In *Toward a General Theory of Expertise — Prospects and Limits*, K.A. Ericsson and J. Smith (eds.). Cambridge: Cambridge University Press.

Orlicky, J. (1969). *The Successful Computer System*. New York: McGraw-Hill.

Nachmias, D. and Nachmias, C. (1987). *Research Methods in the Social Sciences*, 3rd Ed., New York: St. Martin's Press.

Nelson, B. (2001). *Please Don't Just Do What I Tell You! Do What Needs to Be Done*. New York: Hyperion.

Rodenburg, P. (2002). *The Actor Speaks — Voice and the Performer*. MacMillan.

Schank, R. (1991). *Tell Me a Story*: *A New Look at Real and Artificial Memory*. Athenum.

Schensul, S.L., Schensul, J.J. and M.D. LeCompte. (1999). *Essential Ethnographic Methods*. New York: Altamira Press.

Siegel, S., and Castellan, N.J. (1988). *Nonparametric Statistics for the Behavioral Sciences*, 2nd Ed., New York: McGraw-Hill.

Spradley, J.P. (1979). *The Ethnographic Interview*. New York: Holt, Rinehart and Winston.

Stewart, C.J. and Cash, W.B. (2003). *Interviewing principles and practices*, 10th Ed., New York: McGraw-Hill.

Sutton, R.I. (2007). The *No Asshole Rule*: *Building a Civilized Workplace and Surving One That Isn't*. New York: Warner Business Books.

Sweet, J.E. (1885). The Unexpected Which Often Happens. *Trans. American Society of Mechanical Engineers*, 7, 152-163.

Tse, D.C. (2006). *A Conceptual Model for Assessing the Value of Information Technology*. MASc Thesis, University of Waterloo.

Turk, D. and B. Rumpe, B. (2002). Limitations of agile software processes, *Proceedings of the Third International Conference on eXtreme programming and agile processes in software engineering*. May 2002, Sardinia, Italy, 43-46.

Voss, J.F. and Post, T.A. (1988). On the Solving of Ill-structured Problems. In *The Nature of Expertise*. M.T.H. Chi, R. Glaser and M.J. Farr (eds.). Hillsdale: Lawrence Erlbaum, 261-265.

Whisler, T. (1970). *The Impact of Computers on Organizations*. London: Praeger.

Wiers V.C.S. and K.N. McKay. (1996). 'Task Allocation: Human Computer Interaction in Intelligent Scheduling,' *Proceedings of the 15th Workshop of the UK Planning & Scheduling Special Interest Group*, November 1996, Liverpool (UK), 333-344.

Xin, H. (2006). A Preliminary Model of the Elderly in Relation to the Use of Consumer Electronics. Undergraduate Research Paper, University of Waterloo.

Yin, R.K. (1989). *Case Study Research*: *Design and Methods*. London: Sage Publications.

Zemke, R. and C. Bell. (2003). *Service Magic*: *The Art of Amazing Your Customers*. Chicago: Dearborn.